Agro-Animal Resources of Higher Himalayas

Agro-Animal Resources of Higher Himalayas

Editors

Deepa H. Dwivedi, Sanjai K. Dwivedi
and
Sandhya Gupta

2010

SSPH

SATISH SERIAL PUBLISHING HOUSE

403, Express Tower, Commercial Complex, Azadpur, Delhi-110033 (India)
Phone : 011-27672852, Fax : 91-11-27672046
E-mail : info@satishserial.com, hkjain1975@yahoo.com
Website : www.satishserial.com

Published by :

SATISH SERIAL PUBLISHING HOUSE
403, Express Tower, Commercial Complex, Azadpur, Delhi-110033 (INDIA)
Phone : 011-27672852 Fax : 91-11-27672046
E-mail : info@satishserial.com, hkjain1975@yahoo.com

ISBN 81-89304-79-8

Typeset at: **Harminder** *for* **Laxmi Art Creations** # 98 11 48 23 28
Printed at: Salasar Imaging Systems, Delhi-110035

BABASAHEB BHIMRAO AMBEDKAR UNIVERSITY
• LUCKNOW •
प्रज्ञा शील करुणा
ESTABLISHED 1996

बाबासाहेब भीमराव अम्बेडकर
विश्वविद्यालय
(केन्द्रीय विश्वविद्यालय)
विद्या विहार, रायबरेली रोड
लखनऊ – 226 025

प्रो. बी. हनुमय्या
एम.एससी.पीएच.डी. (मास्को),
कुलपति

Babasaheb Bhimrao Ambedkar
University
(A Central University)
Vidya Vihar, Raebareli Road
Lucknow-226 025

Prof. B. Hanumaiah
M.Sc., Ph.D. (Moscow)
Vice Chancellor

FOREWORD

The Trans-Himalayan geographic zone is located across the divide of the great Himalayas. In India, this zone comprises almost all of Leh and Kargil districts (Jammu and Kashmir), parts of Lahaul, Spiti and Kinnaur districts (Himachal Pradesh) and parts of Uttarakhand state. Outside the Indian region, the Trans-Himalayas is very extensive, covering parts of China, Tibetan plateau, Nepal, Pakistan, Afghanistan, Mongolia etc. A large percentage of area in the Himalayas is high altitude cold desert. The entire zone falls between 3,000 to above 6,000 meters above mean sea level.

Higher Himalayas is a region marked by poverty and developmental challenges, especially in its rural highland areas where residents are primarily engaged in subsistence-level agriculture and animal husbandry made difficult by a short growing season at high altitude. Few people are aware and sensitized to the harsh conditions viz. extremely low temperatures (reaching -40° C), hypoxic conditions, low precipitation, very high wind velocities etc., prevailing in the region which make the survival of the agro-animal resources very difficult. The climatic conditions existing in the region are specific and thus, recommendations for cultivation practices and livestock management from any other part of the country can be adopted in the region only after intensive experimentation specific to the region. Realizing the indispensability of encompassing the diversified agro-ecological zones of the region, the present publication **"Agro-Animal Resources of Higher Himalayas"** compiled to highlight the agro-animal resources, status, research on various related aspects and future strategies for their sustainable management.

I am delighted that an exclusive book on **"Agro-Animal Resources of Higher Himalayas"**, which encompasses historical perspectives and latest information, has been brought out. I complement the editors for their efforts to bring out such a comprehensive publication. I hope that the book will be of tremendous use to those who are associated with the Himalayas.

03 July, 2009 **(B. Hanumaiah)**

PREFACE

About 20 per cent of the world's total population is residing in the mountains regions at altitudes ranging from 1000 to 4000 meters. This proportion is like to increase in the present century as environmental degradation and pollution is driving the human population for settlement in the clean and serene environment of the mountains. However, major emphasis has been given on middle and lower mountainous regions world over. India is one of the few countries in the world where extensive research and development work is in progress in the Higher Himalayan Region. Research & Development is a challenge in such adverse areas but the diversity and the high quality agro-animal produce can be exploited scientifically at the national and international levels. However, the research activities in progress in the region have not been documented so far and a vacuum exists for reference for further projects. In view of the above it is felt that there is an urgent need for publication of research findings on agro-animal resources of Himalaya which will not only highlight the latest developments/contributions in the field but also provide principles and guidelines for researchers, teachers, policy makers, etc.

An attempt has been made to encapsulate latest and comprehensive scientific information pertinent to agro-animal resources of Higher Himalayas. The primary aim of this book has been to review latest information and viewpoints of eminent investigators, who have excellence in their own areas of research. In the book, 25 chapters from 12 institutions involving 36 eminent researchers of their field have been to compiled and critically reviewed. We hope that the present compilation would be extremely useful to the scientists, academicians, industrialists, environmentalists and people associated with ethnobotany, traditional medical system and natural products. The views and opinions expressed by the authors, who belong to different organizations, are entirely theirs and neither editors nor the publisher takes any responsibilities for the same.

We are grateful to all the contributors, who have contributed the chapters with valuable updated information on the topic. The editors would also like to convey their sincere thanks to BBA University, Lucknow, Defence Research and Development Organisation (DRDO) and SKUAST-Srinagar, for supporting the project. The editors are also grateful to Prof. B. Hanumaiya, VC, Prof. R.B. Ram, Dean, BBA University, Lucknow and Dr. A.K. Singh, Director, CEPTAM (DRDO), Delhi for their guidance, blessings and encouragement. The support rendered by Publisher and his team for their sincere efforts in bringing out the book thankfully acknowledged. We also solicit the co-operation of erudite readers to derive full benefit from the views of the learned contributors and also welcome suggestions from them.

15, August, 2009 **Editors**

CONTENTS

1

Introduction to Higher Himalayas

BRAHMA SINGH
Former Director Life Sciences, DRDO HQ, New Delhi
Former Director, DIHAR, DRDO, Leh
Former OSD to Rastrapati Bhawan, New Delhi

Himalayas [Sanskrit,=abode of snow], great Asian mountain system, extending c.1,500 mi (2,410 km) East from the Indus River in Pakistan through India, the Tibet region of China, Nepal, E India, and Bhutan to the southern bend of the Brahmaputra River in South East Tibet. For most of its length, the Himalayas comprise two nearly parallel ranges separated by a wide valley in which the Indus and Sutlej rivers flow westward and the Brahmaputra flows eastward. The northern range is called the Trans-Himalayas. The southern range has three parallel zones: the Great Himalayas, the perpetually snow-covered main range in which the highest peaks of average elevation (20,000 ft/6,100 m) are found; the Lesser Himalayas with 7,000 to 15,000 ft (2,130-4,570 m) elevations; and the southernmost Outer Himalayas, 2,000 to 5,000 ft (610-1,520 m) high, (http:// www.encyclopedia.com).

The Trans-Himalaya geographic zone is located across the divide of the Great Himalaya. Trans means the other side of, or beyond. Trans-Himalaya, means beyond the Himalaya. The region is generally termed as cold desert or cold arid due to scarcity of water/very low precipitation. Such areas in India are located 9000 ft above mean sea level, thus also called High altitude. In India the zone comprises almost all of Ladakh and Kargil districts (Jammu and Kashmir), parts of Lahaul, Spiti and Kinnaur districts (Himachal Pradesh) and parts of Uttarakhand State covering an estimated land area of 186,200 sq. km. A large percentage of area in the Trans-Himalaya is high altitude cold desert and is characterized by exceptionally low temperatures and precipitation. Outside the Indian region, the Trans-Himalaya is very extensive, covering a total of nearly 2.6 million sq. km. comprising the Tibetan plateau besides covering sizable areas in countries like China, Pakistan, Afghanistan, Mongolia, and Nepal. The entire zone is a high-altitude cold desert, altitudes varying between 4,000 to above 6,600 metres above mean sea level (msl). It is a sparsely populated domain, the Indian population being approximately 2.5 lakhs.

Though the area accounts for just over five per cent of the India's total land area, its value is tremendous as a drainage and feeder region for some of the greatest Indian river systems such as the Indus, Brahmaputra and Sutlej, as well as the Yangtse in China. Three mountain ranges, the Zanskar, Ladakh and the Karakoram, dominate the Trans-Himalaya. All these ranges lie in the rain-shadow region to the north. Their generally east-west running ridges, interspersed with some of the world's highest mountains and immense glaciers with river valleys cutting through them, create a sparsely vegetated terrain of great biological grandeur. The cold arctic desert of which very little is known, has climatic conditions like howling winds, cold night temperatures and frigid winters with temperatures dropping to -50° C. Hardly ideal conditions for life, one might imagine, but even here nature has been able to create a diversity of plant and animal life whose ability to economize and optimize resources is itself a strategy for survival.

About 20% of the world's total populations are residing in the mountains regions at altitudes ranging from 1000 to 4500 meters above mean sea level. This proportion is likely to increase in the twenty first century as environmental degradation and pollution is driving the human population for settlement in the clean and serene environment of the mountains. The entire area of Higher Himalayan region in India is approximately 74,809 sq. km of which a major part is confined to Ladakh in the state of Jammu and Kashmir (68,321 sq.km) followed by Lahaul and Spiti in Himachal Pradesh (6,488sq.km) and only a very minimal part being found in Uttarakhand. These regions in India are located in the high altitude zone which is described as a treeless world and an environment of low atmospheric pressure. All other environmental conditions are only chain effects of this reduced atmospheric pressure. Significant physiological, structural and ecological modifications are observed in plants and animals, all of which can be distinctly correlated with altitude. The regions are mainly characterized by barren mountains and uneven terrain with sandy soils. The climate is extremely harsh with sub zero temperature prevailing during the winter months and the region remains frozen from December to February. The summer days are extremely hot where temperature reaches up to 40° C but falls during night time. The atmospheric humidity in the region is low throughout the year (45%) and annual precipitation is very low (80-300 mm) which is primarily in the form of snow in the winter months, thus also called cold arid zones. The soils in this region are coarse and sandy with poor organic matter and thus, the vegetation in this region is sparse. The characteristics of cold arid region are given below:

Features of Cold arid Zones

- Extreme temperatures (-40°C in winter and +40°C in summer)
- Very low annual precipitation (80-300 mm) mostly in the form of snow
- Low relative humidity (20-40%)
- High wind velocity (8-10 km/hr)
- Low atmospheric pressure (493 mm Hg)
- Very low partial pressure of oxygen
- Thin atmosphere results in heavy influx of ultra violet and infra red radiations
- Immature, coarse textured and highly permeable soil
- Sparse plant density
- Short growing period (120-150 days)
- Long frozen winters (5-7 months)
- Maximum sunny days (300-310 days per annum).

Environmental Features and Physiological Effects

"Altitude" is the vertical height above sea level. The term "High Altitude" (HA) has no precise definition. Considering various aspects of mountain ascent, the army authority has taken an elevation of 3700 m (9000 ft) and above MSL as HA. They selected this height because above this point majority of the people develop signs and symptoms associated with low oxygen pressure.

The most important feature of HA is low barometric pressure, resulting in hypoxia (Oxygen deficiency) in the body. Other environmental hazards at HA are; (i) severe cold (ii) high velocity winds (iii) low humidity (iv) high solar radiation (v) increased ultraviolet radiation (vi) difficult terrains, etc. Oxygen concentration in air is 20.93 per cent, which remains same whether at sea level or on top of the mountains. As one ascent, the barometric pressure decreases with increase in altitude and the atmosphere becomes rarefied. This results in reduced number of oxygen molecules entering into the lungs with each breath and creates oxygen deficiency in the body. Reduced oxygen in the blood and the effect of lowered oxygen pressure on body tissues can lead to Acute Mountain Sickness (AMS) and high Altitude Pulmonary Oedema *etc.* in a few susceptible individuals. Traditional livelihood strategies in high mountains are no longer sustainable because of mounting demographic pressure and rapid climate changes which in turn may lead to food and nutrition insecurity. Due to strategic reasons, large number of troops are deployed in snow bound areas of high altitude. On induction to HA most of the people from the plains suffer from mild to severe forms of HA related ailments. Most of these generally occur due to lack of basic knowledge about HA and its environment. Before going to HA, one must therefore, be familiar with the environmental factors responsible for HA ailments, and take appropriate preventive measures against the ill effects of HA for a safe and happy living there.

Terrain

Ladakh covering 70 per cent geographical area of Jammu and Kashmir State forms the northern most part of the country in the outer Himalaya. The region is an elevated territory presenting a system of alternating valleys and mountain range except in the eastern part which is a plateau. The important ranges area, the Zanskar range, the Saltoro range and the Southern slopes of the Karakoram range. These ranges provide barren rocky surface largely devoid of soil and vegetation cover as well as human settlements. Nearly 68 per cent of its total land lies between 5,000 meters and 5,900 meters above sea level and is virtually unfit for human life and vegetation. About 5.8

per cent of total land lies between 4,500 and 5,000 meters, which exhibits some pastorals. Human settlement and agriculture is confined between 2,500 meters and 4,500 meters.

The area generally remains landlocked between November and June every year as the Srinagar-Leh and Manali-Leh highways which connect Ladakh with other parts of the country remain closed during this period. The major river is the Indus which is joined by its tributaries, the Zanskar, the Shingo and the Shyok. Ladakh is the land of lakes also. Important lakes in the region are PangongTso (150 km long and 4 km wide at a height of 14,000ft), Tso Moriri, Tso Kar, Yoye Tso Kium Tso and Amtila. Ladakh has two districts namely, Leh and Kargil which stand in contrast with each other in terms of geography and climate. Kargil is mountainous whereas Leh is valley.

Altitude

Altitude of a place indicates its height with respect to the mean sea level. Altitude is a topographical parameter which influences several climatic factors such as temperature, air composition, day length, solar radiation amount and quality and cropping season etc. All the factors have direct bearing on flora and fauna of a place. Due to altitude variation, the length of cropping season, type of crops and cropping pattern vary in cold arid region. The altitudinal variation is of evolutionary significance and has been a major factor, responsible for biodiversity in flora and fauna. These agro-zones differ not only in agricultural crops being cultivated but also in manners and timing of field operations that are needed for crop management. Hence it is essential that crop improvement programmes must be broad base embracing the climatic constraints being faced in each zone. A complete topographical study of each zone would be quite useful in formulating an effective integrated sustainable crop improvement programme. This demands mobilization of more resources both monetary and human if the cold arid zone is to be made self-reliant in terms of agricultural production.

Soils

The Physical and chemical composition of soils differ from one valley to another depending on the altitude and rocks. In Ladakh, soils range from gravelly and sandy loams on the alluvial fans to sandy and silty clay loams on the flood plains of Indus. By and large soil is sandy having varying quantity of pebbles. Sand particles are coarse.

The soils are characterized by low organic matter content and poor water retention capacity. The pH of soil ranges from 7.4 to 9.5. There is a potential problem of salinization in the Indus valley, especially on the flood plains of Indus. Number of saline plains are seen in valleys of Indus, Suru, Nubra and Shyok rivers. Soils of cold arid zones are immature, highly porous with a high proportion of sand. High permeability, poor nutrient status, poor biological properties and high salt concentration, wind/water erosion in certain patches are some basic soil properties. pH was found to range from 5.94-10.12, electrical conductivity was found to vary from 0.05-1.55 mmhos /sec and organic carbon ranged from 0.07-3.58%. A study conducted by DIHAR (DRDO), Leh, reports that available nitrogen is low in category because of the slow mineralization process. Available phosphorus is generally low in the soils of Ladakh (2.35 -137.4 Kg/ha), available potassium was high in the soils of Leh and Kargil districts of Ladakh (11.2-496.15 kg/ha), while the extractible micronutrients like Zinc, Copper, Iron and Manganese are generally deficient in this part of the cold arid region. Soil depth is problem especially for the plantation of fruit / forest trees. Presence of stone/ rocks under the soil surface poses serious problem in the plantation of fruit and forest plants.

Climate

The great Himalayan mountain, lying to the south, forms a barrier to monsoon rains in this area. Due to this reason Ladakh is an isolated cold desert region. Altitude in the region varies from place to place by several hundred meters and is the main factor affecting local climate. An extremely harsh climate prevails in and the winter temperature touches as low as -32° C (Leh and Kargil) and -50° C (Drass). Subzero temperature prevails from December to February, whereas, zero degree temperature is experienced during rest of winter months. This results in freezing of all conceivable water resources. During summer when mean maximum temperature is 23° C and 26° C (July & August), the sun can be scorching hot in its intensity. The nights remain cool throughout the year. The temperature influences almost all the growth parameters of plants by providing appropriate thermal environment. Air temperature primarily affects leaf temperature, photosynthesis, transpiration, respiration and translocation of metabolites. Temperature has complex relationship with yield contributing characters namely inflorescence size, rate of grain filling, duration of grain filling and grain/fruit ripening. Annual precipitation varies from 90 mm in Leh to 239 mm in Kargil. Also higher reaches near the glaciers often receive precipitation more than the distant area with the result that natural flora intensity and diversity exceed in such pockets compared to plains. Cold arid environments receive large part of summer rainfall during July and August with sharp decline in the beginning and end of the season. Rainfall is always for below the crop demand and in combination with temperature, radiation and relative humidity can create a drought syndrome at any time during crop growth and development. Winter precipitation is mostly in the form of snow. As such it has little agricultural value as there is no blade of grass in winter on account of extreme xeric conditions. Wild flora survive through short life cycle and closed canopy with few small leaves, flowers and seeds while plantation crops become devoid of leaves and undergo in dormancy.

The relative humidity is low and ranges from 31 to 64 per cent. Wind velocity in the after noon and night is of high order resulting in heavy soil erosion, dust storm and snow blizzards which make life very difficult. During summer months days are longer having sunlight for more than 14 hrs which is most suitable for crop production and higher productivity. Due to extreme long winter, the agriculture season is short and extends from April to September depending upon the altitude. The climate varies in different valleys / places which influences period of agriculture, type of farming nature and variety of crop, method of cultivation and farming system. Several different microclimate pockets make it difficult to generalize farming techniques and systems for the entire Ladakh. The atmosphere is comprised of reduced oxygen levels (thin air) and low concentration of carbon - dioxide. There is a heavy influx of infrared and ultraviolet radiations.

Wind Velocity

Wind velocity influences crop growth in a number of ways. Wind is also considered to be a clean, free an inexhaustible source of energy. It has been estimated that roughly 2x10 13 watt of energy is continuously available in the earth's winds. The utilization of this energy through various mechanical conversion devices could play a significant role in economic development of the countries where winds are strong and steady. In India as a whole, winds are neither strong nor steady. National average wind velocity is 9.4 km/hr. The economics and potentialities of wind energy depend on the cost of harnessing it, which in turn depends on the energy that is extractable at a particular site and the cost of the wind power plant. Hilly places are having high wind velocity all the year round. But the technological development in the use of wind energy is very poor as compared to solar energy. Use of wind power can be helpful for people of cold desert regions of India. Wind energy conversion system has been developed in the country and being used. Average wind speed of Ladakh is 8 m/s (afternoon/evening) however it is high in summer as compared to winter. It is more in afternoon. In some of pockets of Ladakh it is quite high and wind turbine can be run efficiently which can be used for generating power and other mechanical use. It has been observed that wind velocity varies from 4 to 12 m/s in the region thus wind equipment will run effectively from 1200 hrs to 2400 hrs. This it can be harnessed in this region for battery charging, lifting water, power generation etc.

Solar Radiation

Higher Himalayan region is rich source of solar energy. Bright sunshine hours in the region have been recorded and number of bright sunny days are 300-310 per year which is perhaps highest in the country. Bright sun shines hours per day ranges from 6 to 9 hrs. Since the altitude is very high and the air is thin consequently larger ultra violet radiation are received. It hasjbeen recorded that average direct radiation per day at Ladakh region is 6 kwh/(m2. From April to September it is more than 8 kwh/m2. This indicates that the region is ideal place to utilize solar energy. The average daily solar energy flux in India on a clear day is more than 5 kilo watt/hr/sqm. Solar radiation and its tapping as heat energy are now days most well known and economically viable sources of energy. The collections of solar radiation as usable heat energy by means of various devices are called solar thermal system. Solar thermal energy is being used for many purposes such as cooking, drying, heating, cooling & electric generation. Another aspect of solar energy which can be used to a great extent efficiently is the solar photovoltaic system. This can be used to produce the electricity directly from solar energy. There are several kind of solar cells under development. They can be used for lighting, lifting water etc. This energy can be utilized for keeping rooms / houses warm and also conserving energy by using green houses, animal/ poultry house for agro-animal production besides other uses.Sunlight is the major contributor to the biological energy in the world both directly and indirectly. Solar energy affects the crop productivity mainly by providing requisite thermal environments for physiological processes and it executes light environment so essential for the process of photosynthesis wherein solar energy in the form of electromagnetic radiation is converted into a useful form of chemical energy as plant organic matter. Solar radiation enhances water loss in arid climates by raising temperature and evaporation. Light not only drives photosynthetic supply of sucrose but also controls translocation and partitioning of metabolites from chloroplast. High light intensity coupled with low temperature during grain development stages increases grain weight in cereals.

Water Resources

The main rivers of Ladakh is Indus, which flows in a north-westernly direction between Ladakh and Zanskar ranges. It is joined by several major rivers like Zangskar, Suru and Shy ok before it reaches Pakistan. The main rivers of Ladakh is Indus river with annual water flow of about 1.56 billion m^3. Shay ok river, Suru river & Zanskar river are also there with a total flow of 4.2 billion rrd and average of 130 m^2/s discharge. Thus the potential will be 1500 MW. These rivers are In full flow from May to November and during rest of the period remains *frozen* or with meagre water. The underground water resources such as springs and seepage in Ladakh are highly viable owing largely to geology, climate and topography. In Ladakh number of natural lakes like Pangong tso (15000 ft) Tsomoriri (14000 ft) Tsokar (14000 ft) lake are there. Besides power generation by water resources, the direct utilization in the region is also encouraged by water mill. It can also be utilized for water lifting though hydraulic ram. Cultivation is normally found in valleys on alluvial fans and flood plains with sufficient water for irrigation. Nearly all irrigation systems are gravity reliant. Most of the irrigation is being done by canals/khuls. Skillfully constructed water channels carrying water from its source to the cultivated fields are the backbone of agriculture in Ladakh. In several places melted snow water is collected during night in reservoirs built by the community for irrigation during day time. Also melted snow water is channelised directly for irrigation. Melting of snow is obviously in the afternoon, resulting in flow of large quantity of water for harnessing and irrigation during night. There is a positive correlation between maximum day time temperature and rate of flow of water. The snow melted water from snow clad peaks is brought by the community to the villages and fields by constructing canals. These canals were made with local skill through rough land, sand and rocks. From the main canal a network of sub channels touches each and every field. No cement has been used in their construction. This irrigation system is working very well for centuries in Ladakh. Of late modern irrigation technology and material is being used in construction of big canals for irrigation. Most of the recently built canals are made of cement and stone masonry work. Khurbotang, Stakana, Igoo-Phey and Hunder are important canals. Natural water springs are also found in Ladakh and taped for irrigation. Nearly all the water used in irrigation is from melting snow banks and glaciers and is, therefore, most abundant in the spring and early summer. Depending on the source, amount of precipitation, soil type and crop grown, it may need to be rationed from about mid June. In some years shortage of water affects crop yield adversely. Besides other limitations, water is the major factor limiting the area that can be cultivated. Without irrigation vegetable farming is out of question in Ladakh. Indus River, its main and secondary tributaries have vast potential to provide gravitational irrigation. Scope of lift irrigation is limited. At certain places tube wells are being used successfully for irrigation. A systematic survey for exploitable under ground water in valleys of Ladakh is a pressing demand from farmers. Water samples from natural spring (FRL), glacier water collected from Leh town and Indus water collected at Ranbirpur were analysed. The important chemical constitutes in above samples are given below.

Table- Chemical constituents of water.

Sample	$Ecx10^6$	CO_3	HCO_3	Cl	Ca^{++}	Mg^{++}
Natural Spring	504	Nil	3.6	0.8	1.3	4.1
Glacier water	216	Nil	1.7	0.3	0.6	1.6
Indus water	360	Nil	2.5	0.6	0.6	2.0

Gravitational is the common traditionally adopted method for conveyance of irrigation water but the efficiency could be improved with advance scientific means to check the transit and operational losses. Mechanical methods are another effective techniques for water harvest. Hydropower, solar power and exploration of ground water for agriculture purpose are other important areas, which have great potential in Leh district. For better economic utility of the precious irrigation water the latest techniques of drip and various types of sprinkler irrigation systems are worth trying especially for horticulture crops including vegetables crops and pasturelands.

Conservation of water resources is the most important area to be dealt with and need immediate attention. All the glacier and snow deposits are reported to be shrinking, water discharges in river and rivulets are minimizing, recharging rate of natural springs are lowering, snowfall is getting less year by year. This could be attributed to the ecological imbalances, increase in human population, pollution, more vehicular moments and mainly due to the global warming. The snowfall received in the south facing hill slopes get blown away by the high speed wind and get deposited in the north facing sides, resulting generally scarcity of water in the south facing valleys during the crop season. Appropriate technologies need to be developed to arrest such as shifting snows on the desired sites.

Since all the agriculture activities are fully dependent on availability of irrigation water under Leh conditions, this area need to be tackled on priority. Appropriate techniques for conservation of snow, natural glaciers, creation of artificial glacier, diversion of waste going seasonal streams, water harvesting and management system and economic utilization of irrigation water are very important areas for research and development. Sound research based technology development in this area is urgently required, thus need to be taken on priority. It is for the scientists and Research Institutions to come forward to take the challenges and give technologies/solutions, to overcome the problems.

Land Size

The average land holding in Ladkh is about 1.56 hectare with only 23 per cent farms exceeding 2.0 hectare. Majority of the holdings (36.35%) ranges between 0.02 and 0.5 hectare. The average size of a field is about 0.05 ha and is mostly terraced. Small land holdings can provide higher returns by adopting commercial cultivation of vegetables, herbs, flowers, fruits.

Agricultural Labourer

Such areas are facing acute shortage of Agriculture labourer as the potential work force has joined military and paramilitary forces and J&K police. The younger generation is either engaged in educational institutions or has joined services. This has resulted in dependency on outside seasonal labourers. The number of outside labourers registered at Leh office is of the order of 7,000 to 8,000 each year, which is a very costly affair. Obviously, cost of production of all agricultural products is high. Agriculture has become a non-preferred venture. Hence, introduction of efficient agriculture machineries like tractor, power tiller, reaper/harvester, thresher and other light machineries, tools and implements to improve a working efficiency, to make Agriculture an easy activity and to cut down the high labour cost and to get more economic returns has become imperative. This area calls for proper research work for manufacture and introduction of efficient machineries, tools and implements suitable for varying topographic and field conditions of the district.

Agriculture and Animal Resources

Animal resources and animal production are important parts of agriculture. Cold arid region of Ladakh has unique position both in animal resources and production. Double hump camel is unique to the area and playing its role as an important animal resource. Yak is another animal. It used to be life line in these areas 50 years back as a source of draught animal, source of milk and fat, leather for cloth, meat as food, hair as wool and bones for tools and ornament. Ponies/ mules particularly, zhanskar pony, are quit useful and multipurpose horses. For travel and draught, ponies are being extensively used even in military and paramilitary forces in this area. Their importance was realized during KargilWar. DRDO has contributed in finding suitability of zhanskar ponies for defence forces. Ass having enormous diversity is useful poor man's ship in this part of the country. Sheep and goat are playing important role in the economy and livelihood in Ladakh. Besides food they are yielding fiber-wool of world class particularly Pashmina.

Cows specific to the area are found through out the area. They are good source of manures rather than milk and are there in large number. The area can sustain improved breeds of cows such as Jersy, Holstein-Friesien and their crosses with Shahiwal, etc. This aspect requires extensive research including fodder and feed production. Scarce feed and fodder in the area, their huge transportation cost from plain areas in J&K, HP, Punjab *etc.* are limiting factors in animal husbandry in the area. Green fodder during winter months is not available.

Poultry is uncommon in the area but has potential if feed is made available locally with local produce. Major constituent of poultry feed, maize, can be grown profitably in the region. Farmers need to be convinced and made to get remunerative price for the produce to have local poultry industry. Defence services are there as huge market. Quail, rabbit, ducks, fish, guinea fowl *etc.* can also be potential animal production ventures in the area.

Challenges and Opportunities

Cold arid regions of the country like Ladakh would always be having predominant challenges because of their remoteness, climate, sparse population, difficult terrain, altitudes etc. On the face of it the areas are not good human habitats. They are not ideal to sustain life. Environmental conditions mentioned in the chapter do not convince of many biological activities in these areas. But all this is not correct. Nature is great to create/provide local solution or opportunity to deal with if one wants to enjoy by defeating challenge. The opportunities in these areas for ago-animal production are long days in summer providing more than 12 hours sun facilitating photosynthesis for faster growth and larger produce. Since temperature during summer is moderate it provides opportunity of cultivation of all short of crops (temperate and tropical) by developing appropriate technologies. This has been amply demonstrated by DIHAR and other research agencies in greening of Leh , Nubra, Lahaul and Spiti valleys. Almost all sorts of important vegetable crops one can profitably grow in Leh valley. The same holds good for other areas. Fodder crop of the region, the perennial alfalfa, is nature's gift to be exploited using modern molecular tools. Wheat and barley besides having their large scale production have enormous genetic diversity awaiting gainful exploitation. Short duration local dwarf mustard is store of good genes. Crops like potato can be grown in large area with improved varieties to register highest yield in the country if not in the world. Prevailing agro-environment is an excellent condition to breed varieties and have true potato seed production. Among fruit crops apricot, seabuckthorn (natural growth) and strawberry can be the best produce of the world if cared for. The rare herbs, so many, rhodiola, mayapple etc are awaiting recognition and use. A wild species of gram, *Cicer microphyllum* having 4-5 grains pods profusely, presently very good natural fodder, is awaiting agricultural scientists to make fruitful use of it.

Forest tree and shrub species are many in the area. Major ones are willow, popular and seabuckthorn. All flora though become dormant during winter but compensate their growth and development during long days summers. Perennial flora in these areas is full of useful traits such as medicinal molecules, unique aroma, fuel, fence, food, fiber, fodder value *etc.*

Abundance of snowmelt water and under ground water provide potentiality of agro-animal production by harnessing it scientifically.

Likewise, though there are numerous challenges and opportunities in cold arid agriculture the major ones requiring attention are as follows:

Challenges

Road accessibility during winters- all season road(s) to the region can solve the problem. Extreme winter prohibits agriculture and animal husbandry activities mainly because of lack of electric power- nuclear power plants is the solution along with exploitation of solar and wind power. Power availability will take care of challenge of winter migration. Harness plentiful solar power for solar green houses during winter which has been demonstrated.

Opportunities

Mass scale vegetables and flower and their seed production which would provide quality produce to fetch better price. Massive forestation using seabuckthorn, popular, willows etc. This would make area much more conducive for habitation. Use fodder species of willow and other trees. Preserve organic farming prevalent in these areas. Help in certification of the produce. Systematic development of herbal gardens and farms for selected rare herbs for the industry and research as well as in situ conservation. Have plant gene banks under natural freezing conditions and conserve local flora and fauna in situ. Apricot crop needs attention in production, processing and marketing to make the area an important apricot region of the country if not the world. Promote solar green houses and protected cultivation including solar heating of animal houses. Fodder, feed and fuel crop production would promote animal production in the area. Pashmina wool production potentials in these areas are enormous by conserving and developing pastures. So is the case with cold fish farming. Agro-tourism needs to be popularized. Agro climate of these regions are suitable for advancing generation in certain crop improvement programmes. Because of lesser diseases and pests, low humidity and long days these areas are ideal for seed production of certain important crops.

□□□

2

Production Technology of Field Crops

DORJAY NAMGYAL
Regional Agriculture Research Station, S.K. University of Agricultural Sciences & Technology-Kashmir, Leh (Ladakh)-194101 (J&K)

In Ladakh region of Jammu and Kashmir, various crops like cereals, pulses, oil seeds and vegetables are grown from March to October. Wheat and barley are the major cereal crops in this region. However, area under barley is higher than wheat. Mustard is major oil seed crops which is grown on different regions of Ladakh like Thiksay, Say, Nimoo, Khaltsi area of Leh and some areas of Kargil district, and preferably grown in those areas which come under double cropping. Among the pulse crops, pea and lentil are important, however, area under pea is more than lentil. Farmers use the pea crop as vegetable not as pulse, but people in Zanskar valley leave the pod for maturity in the field. In some of the areas, farmers have started to grow lentil. After cereals, vegetables come at second rank, having sizable area under cultivation. Some of the vegetables which farmers grow are cabbage, cauliflower, tomato, potato, capsicum and onion. Area under different vegetable crops varies between two districts, *i.e.*, Kargil and Leh. In Kargil district, farmers grow like vegetables, turnip, potato, onion, cabbage while in Leh district onion, tomato, potato, carrot, cabbage, cauliflower and pea are dominant.

In cold arid region of high altitude, the crop production practices are different because of different agro-climatic conditions, soil type and availability of irrigation water. Besides these, small and patchy cultivated land areas which are very common in Ladakh districts. Farmers of both the districts still use the old production practices with their local varieties and obtain meagre yield and are unable to meet the annual requirement. Therefore, scientific cultivation must be adopted to fulfill the demand of wheat, pulses and vegetable for their own consumption. Production technology of main crops is given below:

1. Wheat

Wheat is the world-widely cultivated food grain crops. In India, it is the second important staple food crop after rice. Wheat is eaten in the form of chapatti where wheat is staple food in states like Uttar Pradesh, Bihar, Punjab, Rajasthan and Madhya Pradesh. In Ladakh, wheat is the most important crops which is grown on wider area compared to the other crops. Besides the high nutrient, wheat has relatively more protein, high in niacin and thiamine compared to other cereals grain. It has a characteristic substance called glutien which is important for bakeries. It provides structure for the spongy texture of bread and its products. Wheat cultivation in Ladakh is very old (Cunningham, 1983) and grown on wide area like Khaltsi, Nimoo, Saspol, Leh, Thiksay, Say, Chaoglamsar, Gayamirue, Kahalsar, Somur, Nubra, blocks of Leh district and Tsuru, Drass, Chiktan, blocks of Kargil district with different time of sowing and maturity.

1.1 Climatic Requirement

Wheat crop can survive in wide rang of climatic condition and can grow in tropical and subtropical areas. However, the optimum temperature for wheat ranges from 20 to 25°C. In Ladakh, wheat crop is grown from April to September in various unfavorable conditions like scanty rainfall, high temperature fluctuation, low humidity, sandy soil having very low water holding capacity. Fluctuation in high and low temperature during flowering stage may also be harmful. In general, at the time of sowing seeds suffer from low temperature (which occurs during seed germination) while temperature remain high during vegetative growth stage and again comparatively low during grain maturity stage. This fluctuation of temperature alters the sowing and harvesting time of crop which results in delayed germination of seeds, leaf burning and grain shriveling and ultimately leads to the low production.

1.2 Soil

The ideal soil for wheat cultivation is clay loam or loam texture with adequate water holding capacity. However, most of the cultivated land of Ladakh is sandy loam and some of the area is clay loam which reduces the production to great extent. Bali (1994) reported 16.5 qha^{-1} differences in yield between the production on sandy loam and sandy clay loam. The high soil pH (7.5 to 8.5) with low organic matter ultimately reduces the yield.

1.3 Soil Preparation

Generally farmers of this region, in single cropped areas leave the field as such after harvesting of crop and in the next season after irrigation they start preparation of sowing. This provides the good shelter for soil insect to pass winter as hibernation stage. Because of this, after harvesting of previous crop ploughing is must. Before sowing one deep ploughing followed by a light ploughing is sufficient for sowing of field. With the onset of crop season heavy irrigation is desirable before pre-sowing deep ploughing. Sufficient moisture is essential for germination of seeds.

1.4 Sowing Time

In Ladakh condition, it is very difficult to say a particular time or month for sowing of wheat. Early sowing can cause poor germination of seeds while late sowing crop has not sufficient time for grain maturity because of the short cropping season. However, in this region sowing time varies according to altitude. For example, in Leh district, sowing takes place from 15th May to 15th June while in Kargil district it takes place in the month of March. On the basis of altitude of different area, sowing time is as following:

- Lower agricultural zone: April
- Central agricultural zone: May
- Upper agricultural zone: April

Generally, sowing time depends on availability of snow melt irrigation water and weather condition (when temperature goes up towards normalization).

1.5 Varieties

In most of the areas farmers used to grow local variety of wheat which gives low yield compared to improved varieties. Several new high yielding varieties have been identified for Ladakh. Some of them are given below:

Table 1: Important Wheat Varieties of Ladakh

S.No.	Variety	Maturity (days)	Height*	Grain	Yield (qha^{-1})	Response to disease
1	Leh local	100-110	Dwarf	Medium bold	18	Susceptible to yellow rust, smut, bunt and lodging
2	Sonalika	110	Medium	Bold, amber, shiny colour	30-32	Susceptible to yellow rust and ear cockle disease
3	Kailash	125	Medium	Bold and amber colour	40	Resistance to rust
4	Mansarovar	115	Medium	Bold	35	Moderate resistance to rust
5	Singchen	115	Medium	Bold	40	Resistance to rust and lodging

* Dwarf: below 90 cm, Medium: 90 to 120 cm, Tall: above 120 cm

1.6 Manure and Fertilizer

Wheat consumes heavy quantity of nutrients from the soil. Farmers add fertilizer, Farm Yard Manure (FYM) and night soil to accomplish the requirement of nutrients. Addition of night soil and FYM is very common practice in this region to improve the quality of soil. Besides, it also improves the physical and chemical property of soil, makes micronutrients available to plant, reduces the crust formation and maintains favourable temperature of soil. The soil of this region is sandy or sandy loam. Therefore, addition of high dosage of FYM improves the soil quality. However, addition of 20-25 tones well decomposed FYM before the sowing is sufficient in addition to the fertilizer. As the soil is poor in nutrients and farmer apply nitrogen, phosphorus and potash in higher dosage. The application of 120 kg nitrogen, 50 kg phosphorus and 40 kg potash per hectare is effective.

1.7 Methods of Application

FYM should be applied during preparation of field for sowing. The fertilizer may be applied as below:

- Full amount of phosphorus, potash and half-dose of nitrogen should be applied at the time of field preparation;
- Remaining half-dosage of nitrogen should be applied in two parts, one-fourth as top dressing after 20-30 days sowing, *i.e.*, crown root initiation stage;
- Remaining one fourth dosage of nitrogen should be applied at boot stage of crops, *i.e.*, after 50-55 days after sowing as top dressing.

1.8 Methods of Sowing

In most of the areas of this region, farmers sow the crop by broadcasting method to reduce the labour expenses and time. Farmers simply broadcast the seeds over the soil followed by harrowing. In small area, which is less than one *kanal*, this practice is very common. Second method of sowing is from behind the local plough, which is common in almost all areas of this region. In this method, fertilizer mixed seeds are placed in furrow by following the plough. This method is more common than broadcasting method. Third method is by using seed cum ferity-drill which is very effective method for placing seeds and fertilizer uniformly and this method requires less amount of seed compared to broadcasting method. However, this method is not common and mainly limited to Government farms and to those farmers who are availing this facility by Government.

1.9 Seed Rate

Seed rate depends on germination percentage, tillering ability, time and method of sowing. In Ladakh, most of the farmers used to sow the seed by broadcasting method. Mostly a very high rate of 250-300 kg seeds are required for one hectare (20 *kanal*) area. In this area, proper tillering does not take place due to short cropping season. In the experiments carried out at SKUAST, Leh, it was found that sandy loam or loam soil having sufficient availability of nutrients and irrigation water increased the tillering of wheat, thereby the seeds rate was reduced up to 25-50 kg per ha. But in that area where the aforesaid facility is not available higher seed rate is required. Some time farmers willing to get higher amount of biomass, decide the seed rate and use higher dosage of seeds so that they can harvest good amount of straw, which is very costly in this region.

1.10 Spacing

Spacing depends on variety and time of sowing which provides favourable microclimate for crop growth, ventilation and reduces the disease and insect-pest infestation. For normal sown crop, spacing of 20 cm row to row is recommended.

Seed Treatment

Diseases like rust, bunt and smut are very common and prevalent in this region. Therefore, seed treatment is needed. Covered and loose smut are very common in warmer belt like Nubra valley, Btalik, Khaltsi and Kargil. Carbendazim or Vitavax @ 2 g per kg seed for seed treatment is found effective to minimize the disease severity.

1.11 Irrigation

Irrigation is an important aspect for successful cultivation of any crop in this region. In some of the areas of Ladakh, irrigation water decides the sowing time of crop because availability of irrigation water is dependent on snow melt. In this region, sandy soil having very less water holding capacity and high temperature with high wind velocity cause water loss from field rapidly, thus 8-10 irrigations are required to harvest the wheat crops. However, farmers of this region are unaware about the requirement of irrigation to a crop. Farmers used to irrigate their field by flooding method whenever snow-melts water is available. The flood irrigation causes loss of nutrient. This running water takes away the nutrients and leaves them somewhere else thereby causing uneven distribution of nutrients in the same field which ultimately influences the yield. It also causes gully erosion of soil.

1.12 Intercultural Practices

(i) Weeds

Weeds are major problem which cause heavy losses by competing with crop for nutrient, water, sunlight and other resources. Wild oat (*Avena fatua*), *Rumex* spp., *Chenopodium album* (bathua), *Malva* sp. and wild sarson are very common weeds in wheat crop beside those which remain unidentified. Hand weeding is very effective method to manage the weed. These weeds can be utilized as feed for cattle or be stored for stall feeding during winter. Eradication of weed is necessary because cultivated land is of small size. Hoeing is also useful to manage the weeds. Some chemicals like selective herbicides have been found to be effective to kill some weeds. 2, 4-D is very effective herbicides and application of 625 g/ha in 600 liter water reduces the *Chenopodium* and wild oat, which are the major problem in this region. Though, chemical control is very effective method but is not very much practiced due to use of weeds as fodder for animal, which is high in demand in Ladakh.

(ii) Disease

Some of diseases like bunt and smut are very common in this region. However, following diseases has been recorded in this area:

Table 2: Important Diseases & Their Control

S.No.	Name of disease	Causal organism	Symptoms	Control measures
1.	Yellow rust	*Puccinia* spp.	Crop gives yellow look and forced result in maturity	Growing of resistant varieties like Kailash, Mansarover and Singchen; Three spray of Dithane Z-78 (0.25%) at 15 days intervals is found effective.
2.	Loose smut	*Ustilago nuda tritici*	Ear with black powder, in place of wheat grain, which is spore of smut fungus	Growing of resistant variety is effective Seed treatment with carbendazim or Vitavax @ 2 g. per kg seeds reduces the infestation Rouging of infected ear reduces the further spread of disease in next season.
3.	Ear cockle	*Anguina tritici*	Relatively small ear, twisted leaves and grains area replaced by dark brown galls of nematode in ear	Rouging and burning of infected ear before harvesting; Removing of infective propagules from seed by dipping in salt solution reduces the disease severity; Sowing of resistant variety like Kailash and Singchen.
4.	Hill bunt	*Tilletia* spp.	Grains are filled with powdery mass of black spores of *Tilletia* spp.	Growing of resistant varieties is effective; Seed treatment by Bavistin or Carbendazim (2 g per kg seed) reduces the disease.

(iii) Insect-pests

Aphid is the major insect pest of wheat in this area which sucks the sap from developing ear. Spraying of Endosulfan, Chlorpyriphos or Phosphamidon @ 0.05 per cent reduces the infestation of aphid.

1.13 Harvesting

Most of the high yielding varieties become mature in 120-130 days. During maturity wheat requires high temperature. Harvesting should start when ear becomes brown and harvested crop should be left leave in field for few days for drying. After threshing, through animal or thresher, grain and straw can be separated.

1.14 Production

Yield may vary according to variety, availability of fertilizer and other resources. Local variety produces 5-10 qha^{-1} while high yielding varieties produce 30-40 qha^{-1} grain and 80-90 qha^{-1} straw.

2. Barley

Barley is one of the important cereal of the world which is grown on wide area of different countries like China, France, Canada, USA, Spain *etc*. It is staple fodder of people of different countriesy like Tibet, Nepal and Bhutan. In India, most of the states like U.P., Rajasthan, Punjab and Bihar produce high quantity of barley. However, in cold arid region of Ladakh, barley is staple food and covers wide area. It is considered as traditional crop of this region that people of all ranks consume (Cunningham, 1953). It is largely consumed in the form of *Sattu*. *Sattu* is used in a variety of food preparation with tea, coffee, milk or water (Mir and Mir, 2000). It is also used in preparation of local beverage called as Chhang. Barley can be cultivated form 8,500 to 14,000 feet ASL and can grow in all types of soil. This crop is best suited for this region due to short cropping season (which is sufficient for crop maturity), type of land having small amount of fertility, and less water requirement compare to wheat.

2.1 Climatic Requirement

Like wheat, barley needs cool weather condition during early growth stage and during maturity it requires hot and dry weather condition. However, in Ladakh weather conditions are not same as barley crop requires. During the time of sowing low temperature is prevalent in most of the cropping area like Drass, Zanskar and Leh which delays the sowing to some extent. Similarly, at the time of harvesting rarely high and dry temperature coincides with crop harvesting. However, barley is the most suitable crop for this area and can complete all crop stage successfully without any effect of cold and high temperature in comparison to wheat.

2.2 Soil

For barley, cultivation requires well drained, fertile loamy soil. However, in Ladakh soil is alkaline and sandy loam having very little organic matter in which barely survives well compared to wheat. Salt which is high in soil is tolerable to the crop. Therefore, it can be said barley can grow almost all soil types occur in Ladakh.

2.3 Soil Preparation

Onset of cropping season, a pre-sowing irrigation is required, where ice over the field is not sufficient or not coincides with snow melt and sowing time. After irrigation, one deep plough is must, followed by one or two light ploughing because proper tilth soil provides uniform distribution of seeds and germination.

2.4 Fertilizer and Manures

The soil of this region is sandy in nature having very low organic matter and water holding capacity, therefore, addition of organic manure is must to harvest a good crop. Higher dosage of organic manure is not harmful to the crop and it releases nutrients slowly. Addition of 10-20 tones of well decomposed FYM at the field preparation is effective. It is very common that people use un-decomposed FYM which invites various pests which cause harm to the crop. Besides the FYM, inorganic manure is also required. The inorganic fertilizer should be applied on the basis of nutrients available in soil. However, 100 kg nitrogen, 40 kg phosphorus, and 30 kg potash is sufficient for one hectare area. Half-dose of nitrogen and full-dose of phosphorus and potash should be applied before the sowing of crop. Rest of the nitrogen dose should be applied in two split doses. One-fourth dose of remaining nitrogen should be applied as top dressing after 20-25 days of irrigation. Rest of the nitrogen should be applied at booting stage, *i.e.*, 45-55 days after showing.

2.5 Improved Varieties

Cold arid region is completely different from the rest of the India having different types of soil, climatic condition, and food habit, and single crop in a year. Therefore, these conditions compel farmers to grow a good crop to accomplish their requirement. For this, farmers should grow high yielding improved barley varieties which are as follows:

Table: Important varieties of barley for cultivation in Ladakh

S. No.	Variety	Plant Height	Yield (grain & straw) (qha^{-1})	Response to disease
1.	Tuckchu min	80 cm	25 and 40	Susceptible to rust and lodges
2.	Sermotok tok (necked barley)	80 cm	25 and 52	-do-
3.	Sindhu (necked barley)	100 cm	35 and 60	Less susceptible to rust
4.	Nurboo (SBL-4)	100 cm	40 and 70	Moderate resistance to rust
5.	SBL-1 (husked barley)	90 cm	40 and 55	Resistant to rust but susceptible to leaf blight and moderate resistance to lodging
6.	SBL-5 (husked barley)	100 cm	35 and 60	Resistant to yellow rust and some extent to leaf blight

2.6 Sowing Time

In Ladakh, sowing time is decided by weather condition and availability of irrigation water. Optimum temperature for sowing time varies one year to next, therefore, sowing time also varies in each year. This influence on the sowing time ultimately influences the production. However, sowing of barley in month of April to June is common depending on weather condition, availability of irrigation water, and altitude of crop growing area.

2.7 Sowing Methods

Like wheat, sowing of barley can be done by line sowing, broadcasting or through seed cum fertilizer drill method. Broadcasting method is common method of sowing in this area. In this method, high seed rate is required and that results in improper placing of seeds in the soil. This can be avoided to some extent by adopting line sowing. In line sowing, sowing behind the plough is common in this area. Farmers use this method for sowing as well as for fertilizer placing by mixing with seeds. In some areas, where seed drill is available, farmers use it for sowing, but limitation with this is the high cost of seed drill, and small land size, which creates operational problem.

2.8 Seed Rate

Farmers of this region use very high quantity of seeds due to low tillering percentage. This is because of short cropping season which hampers the formation of tiller. Seeds for sowing may go up to 400 kg per hectare. However, 250 to 300 kg per hectare seeds is sufficient in this region. Spacing between line to line must be maintained 22 cm as this provides proper ventilation and microclimate for the growth of crop. Line sowing reduces the seed rate which comes down up to 200 to 250 kg per hectare.

2.9 Seed Treatment

Barley crop is also infected by several diseases. Most of them are seed born diseases for which seed treatment is needed to minimize the infection. For the seed treatment, 2 g Carbendazim fungicides should be used to treat one kg of seeds. This will be effective for the control of leaf blight, loose smut, and covered smut disease.

2.10 Sowing Time

Sowing time varies according to weather condition and altitude. However, in lower agricultural zone sowing takes place during second half of April while in central agricultural zone sowing is carried out from 25th May to 15th June. Early sown crop gives more straw and better grain quality in all barley growing areas.

2.11 Irrigation

Due to high altitude with high sun shine and sandy soil having very low water holding capacity, water loss is very high in this region. However, barley requires less number of irrigation compared to wheat. Generally, 7-10 irrigations are required from sowing to harvesting of barley crop in this region. Irrigation is must at stages like tillering, booting and milking. However, during crop growth period frequent irrigation is required.

2.12 Intercultural Operations

(i) Weeds

Weed is a major problem in this region which limits the production of barley. It is major problem in those fields having sufficient organic matter and moisture. Some of the major weeds are wild oat (*Avena fatula*), bathua (*Chenopodium album*), wild sarson (*Brassica* spp.), Tikttimo (*Convolvulus arvensis*), *Lepidium latifolium*, Sounchal (*Malva* spp.), besides some minor unidentified weeds. These weeds compete with crop for nutrient, water, light and space thereby reduce the yield. Management of weed is important to avoid the loss. Sowing of fresh and clean seeds avoid the infestation of weeds. It can be controlled by hand weeding. Use of chemical herbicides like spraying of 2,4-D (650 g in 600 l water for one hectare) after 30-35 days after sowing is found effective to manage broad leaves weeds. Similarly, Isoproturon is found effective against the narrow leaves weed. However, in this region manual weeding is found suitable because it is environmentally safe and weed can be used as fodder.

(ii) Disease

Diseases like rust, smut and blight are major problem in this region. In Rust, yellow rust is very common which is caused by *Puccina gumarum*. Infection of this disease results in shriveling of grain. Leaf blight disease caused by *Helminthosporum graminerum,* which produces symptoms as small chlorotic and irregularly scattered lesion on the leaves shows blight appearance, can be observed from tillering stage onward. Infection hampered spike emergence and less number of grains and poor quality. Loose smut disease caused by fungus, *Ustilago nuda,* which is common in Leh and Kargil district of Ladakh region. Symptom of this disease appears in late stage of crop. The fungus grows within growing crop and symptoms produce at the time of ear formation. The fungus produces huge amount of spores which are covered by silvery membrane in spikelets. This cover ruptures before emergence of ear releasing the spores which cause further infection.

Covered smut disease causes severe loss to barley crop in Leh and Kargil district of Ladakh region. It is caused by *Ustilago hordei.* Like loose smut, the spores of the fungus are covered by a membrane and being persistent till the harvesting of crop. It ruptures at the time of harvesting, spores disseminate and adhere to healthy grains which causes infection after sowing of these infected seeds. Ear cockle disease caused by a nematode, *Anguina tritici* and commonly occurs in almost all barley and wheat growing areas of this region. However, higher damage of this disease was recoded in Leh district compared to Kargil. Symptoms of this disease are: twisted head and leaves of barley, and grains are replaced by nematode galls.

Various Diseases can be Managed by Following Treatments

- Seed treatment with Carbendazim 50 WP 2 per kg is found effective to manage the disease
- In the case of loose smut of barley, rouging of effected ear heads before rupture of the membrane and burning them is effective to reduce the disease spreading
- Hot water treatment is found effective to reduce the infestation of loose smut disease
- In the case of covered smut disease, rouging out of infected ear which shows black appearance should be done before the harvesting of crop
- For the management of ear cockle disease, sowing of clean seeds, free from the nematode galls, is found effective to reduce the infestation or seed floatation in (5 per cent) brine to remove the nematode galls is effective method to reduce the infection of nematode.

(iii) Insect-pests

Cutworm and aphid are the major pests of barley in cold arid region. Cutworm attacks on foliage of young seedlings at evening while aphid causes infestation by sucking the sap from ear. Clean field and flood irrigation is found effective to minimize the cutworm infestation. Spraying of Monocrotophos 36 WP or Endosulfan 20 EC @ 0.05 per cent is found effective to manage the pest problem.

2.13 Harvesting and Production

Barley crop should be harvested after maturity, which takes place in August-September, depending on the weather conditions. Harvesting should start at morning hours because after maturity, ear become fragile and start braking through short disturbance. Therefore, timely harvesting is must to avoid the grain loss. Using high yielding variety, 30-40 qha^{-1} grain and 45-55 qha^{-1} straw can be obtained.

3. Mustard

Mustard is an important crop of the region and accomplishes the oil requirement for daily consumption. The oil is used in frying of vegetable and pickle. This crop is reported to be grown in farmer's field in this region since long (Cunningham, 1853; Stewart, 1917). The area under this crop is very less compared to wheat and barley.

This crop is grown as second crop in those areas where two crops are taken in a cropping season. However, production is less due to low temperature during maturity which inhibits the proper ripening of grain. The low area under the cultivation of mustard is due to production of less biomass (straw) from mustard and higher quantity and value of wheat and barley straw which farmers use as fodder. Also due to unawareness about other oil corps, farmers do not grow other oil seed crops. Mustard is suitable for almost all cultivated area of this region.

3.1 Climatic Requirement

Mustard requires almost similar weather conditions as required for wheat. It prefers moderate temperature around 22-25°C for growth. However, in cold arid region this temperature does not remain constant. During sowing time crop suffers from lower temperature while at maturity it gets appropriate temperature in single cropped areas. Thus, it can be said that this crop has adaptability to grow in different altitude of Ladakh. But in double cropped areas crop suffers from low temperature at the time of seed maturity which results in incomplete grain filling.

3.2 Soil

Mustard has wide variety of soil adaptability and can grow in sandy loam to heavy clay soil. However, rapeseed gives better performance in loamy soil while mustard is hard to grow in such soil. Well drained sandy loam soil of this region is well suited for growing of mustard.

3.3 Soil Preparation

A pre-sowing irrigation is required for preparation of soil. At proper moisture a deep ploughing followed by one shallow ploughing is sufficient for mustard sowing because it does not require fine tilth as required by wheat and barley.

3.4 Sowing Time

Due to less cultivated land and mustard growing as second crop having less time for crop growth, farmers do not express their keen interest for mustard cultivation. Because of this, they do not apply organic and inorganic fertilizer in appropriate quantity. However, crop utilizes sufficient amount of nutrient from soil (Aulakh *et al.*, 1985;

Jain *et al.*, 1984 and Maiti *et al.*, 1988). Therefore, sufficient quantity of fertilizer should be applied to obtain excellent yield. Regarding the manure, 10-12 qha^{-1} well decomposed FYM is required at the time of field preparation that improves the soil quality and provides the micronutrient. About organic fertilizer, 60 kg nitrogen, 40 kg phosphorus and 20 kg potash per hectare is sufficient to fulfill the requirement of nutrient of mustard. The half quantity of nitrogen and whole amount of phosphorus and potash should be used at the time of field preparation as basal dressing. Rest of the nitrogen should be applied as top dressing in two times, *i.e.*, first half of nitrogen should be applied after 20-25 days after sowing and remaining dosage should be applied at flowering stage (after 55-65 days of sowing) as top dressing. Sulfur increases the oil content in grain, therefore, application of 20 kg zinc sulphate in one hectare is found effective.

3.5 Varieties

Farmers mostly grow local varieties of mustard with traditional cultures methods. Following varieties, which are area specific, should be used to increase the yield:

a. *Local mustard:* It is under cultivation throughout mustard growing area. As main crops it matures in 100-120 days while in 80-90 days as second crop. The grain yield of main crop is 8-10 qha^{-1} while 5-8 qha^{-1} of second crop.

b. *SML-3:* It is a dwarf variety which matures in 90-110 days and suitable for late sowing. Seed size is medium and oil content is up to 38 per cent with grain yield of 15 qha^{-1}.

c. *SML-6:* It is tall variety of which the plant height ranges from 120 to 130 cm. It is suitable for timely sowing. The grain yield potential is up to 14 qha^{-1}.

d. *RML-514:* It is high yielding tall variety with yellow flower. It is suitable for timely sowing with high fertility condition. It matures in 120 to 125 days. It is tolerant to aphid infestation. The Grain yield potential is up to 25 qha^{-1} under Ladakh condition.

e. *SSL-10:* It is brown sarson variety which matures in 100 to 115 days. Seed size is medium with 15 qha^{-1} yield.

f. *TL-59:* It is dwarf brown variety of toria which matures in 85-90 days. It is suitable for late sown condition in central agricultural zone as second crop in double cropped area of cold arid region. The grain yield potential is average yield of 9 qha^{-1}.

3.6 Sowing Time

Cropping season in this region ranges between March to September and temperature varies according to altitude. In Kargil district sowing time is earlier as compared to Leh district. Great variation in weather condition is also recorded in the same district. In double cropped areas, sowing should start form March-April while in single cropped areas sowing should start from May-June depending on weather condition. Late sowing may go up to 10th June in single cropped area while in double cropped area it is up to April. However, late sowing caused lower grain filling in siliqua thereby reduction in yield was recorded.

3.7 Method of Sowing

Broadcasting method is very common for sowing of mustard. This method is quick, time and labour saving. In this method, seed and fertilizers are broadcasted on soil surface followed by light harrowing. The drawback of this method is the irregular seed germination, variable plant population and low nutrient use efficiency. Second method of sowing is line sowing. This method is scientific and recommended for this region. In this method, a person drops seeds in opening furrow by following plough. This method provides uniform placement of seed, proper spacing, increases water and nutrient use efficiency and possible comfortable intercultural practices.

3.8 Spacing

Optimum spacing provides better microclimates for crop growth and avoids pest infestation. For mustard crop in cold arid region spacing between row to row - 30-35 cm and between plant to plant -4.0-5.0 cm is recommended sowing depth of seed should be 5-7 cm depending on availability of moisture.

3.9 Seed Rate

Seed rate depends on germination percentage, time of sowing and method of sowing. Normally, 12-15 kg (600–750 g for one kanal) seeds of mustard or brown saison is sufficient for sowing of one hectare area while 10 kg (500 g in one kanal) seeds of toria is required for one hectare.

3.10 Irrigation

Water loss in soils is very fast in this region, which increases the water demand for the crop. Generally, 9-15 irrigation must be given to mustard crop to harvest excellent yield. However, number of irrigation varies according to availability of moisture in soil in different area. Critical stages of mustard crop for irrigation are flowering and grain filling stage. At this stage adequate moisture must be maintained.

3.11 Intercultural Practices

(i) Weed

Common weeds of mustard crop are *Chenopodium, Convolvulus, Sisymbrium, Rumex* etc. It is important to suppress the weed at initial stage of crop otherwise they will cause economic loss. Manual weeding or hoeing is beneficial and economic in this region. Herbicides like 2, 4-D (625 g per 20 kanal or hectare) or Pendimethalin (1.25 kg per hectare) is found effective to control the weeds. Application of Isoproturan (0.75 kg per hectare) as post emergence herbicides is found effective.

(ii) Insect-pest

Aphid and cabbage butterfly are major insect-pests of mustard crop which cause economic losses in cold arid region. It starts appearing in last week of May. Aphid causes damage by sucking the sap from plant while cabbage butterfly causes defoliation. Application of Endosulfan or Dimethoate or Phosphamidon at the rate of 0.05 per cent is found effect for the control of these pests.

3.12 Harvesting and Threshing

The appropriate time of harvesting is when pods on plant turn yellow in colour. However, late harvesting and threshing may cause shattering of seed from siliqua. If there is moisture in harvested crop at the time of harvesting then it should be left in field for 5-6 days to dry. Threshing may be with the help of stick beating or through the thresher.

3.13 Production

The high yielding recommended varieties of rapeseed and mustard can produce 10-12 qha^{-1} depending on moisture content in the grain by following the scientific cultural practices.

References

Aulakh, M.S.; Sidhu, B.S.; Arora, B.R. and Singh, B. (1985). Content and uptake of nutrients by pulse and oilseed crops. Indian J. Ecol. 12(2): 238-242.

Bali, S.V. (1994). Soil and water conservation practices in wheat under rainfed conditions. In: Wheat Production Technology. State level training organized by Plant breeding Section, SKUAST, RARS, R.S. Pura, Jammu.

Cunningham, A. (1953). Ladakh: Physical, Statistical and Historical. Gulshan Publishers, Srinagar (Kashmir) 495p.

Jain, G.I.; Sahu, M.P. and Somani, L.I. (1984). Secondary nutrient research in Rajasthan, Proc. FAI (NRC) Seminar, Jaipur, pp. 147-174.

Maiti, S.; Hegde, M.R. and Chattopadhyay, S.B. (1988). Hand Book of Animal Oilseed Crops, Oxford & IBH Publishing Co. Pvt. Ltd. New Delhi, pp. 68-108.

Mir, M.S. and Mir, A.A. (2000). Ethnic foods of Ladakh. In: Dynamics of Cold Arid Agriculture (Sharma, J.P. and Mir, A.A. Eds.) Kalyani Publishers, Ludhiana, India, pp. 297-306.

Stewart, R.R. (1917). The flora of Ladakh-Western Tibet. Bull. Torr. Bot. Club. 43: 571-590.

□□□

3

Vegetable Production Technology

ANAND KUMAR SINGH, B. K. SINGH AND V. K. YADAV*
Department of Horticulture, Faculty of Agriculture, Banaras Hindu University, Varanasi (UP), *GB Pant University of Agriculture and Technology, Hill Campus, Ranichauri, Tehri Grahwal (Uttarakhand)

Ladakh is a high altitude cold desert and one of the most remote and rugged regions of the world lying between 2,800 m and 4,500 m above mean sea level on the western edge of the Tibetan plateau, and covers about two third of the state of Jammu and Kashmir. Its area is 96,701 sq. km, mainly mountainous, having a population of over two lakh. The winter is very cold, sometimes below –30 to –40°C while, days can be very warm in summer, over 30°C, and mild in winter. In this extremely cold and dry climate, not a single plant can grow in open during winter. The short summer growing season, from April to September, is only suited for agricultural work and production of maximum edible biomass through vegetable cultivation.

Vegetables are an important component of the daily diet and the nutritional value of vegetables is a vital source of micronutrients. They are pro-vitamin A, C, B_6, E as well as folic acid, iron and magnesium (Singh *et al.*, 2001). They provide not only nutritional security but are also capable of producing 5 times more biomass of food per unit area as compared to cereal crops. Most of the cold arid regions are remote pockets and vulnerable to malnourishment problems. Thus, self-reliance for food and nourishment in these areas is only possible by introducing suitable high yielding varieties and adopting new production technologies (Yadav *et al.*, 2005). Vegetables are becoming increasingly important as cash crops with great potential for export and improving the nutritional status and health of the people.

With such a huge potential, production of vegetables in adequate quantity can help in reducing levels of malnutrition and under-nutrition in the cold arid region.

1. Historical Background

In Ladakh, few vegetables like turnip, radish, carrot, onion, potato, local pea and certain leafy vegetables were grown before 1962 but their cultivation was confined to home gardens. However, some direct sown vegetables were grown prior to 1950 but transplanted vegetables became popular only after 1965. Systematic experimentation and demonstrations on different vegetables, particularly on the production of cabbage, cauliflower, knoll khol and tomato started after the establishment of Field Research Laboratory (FRL) at Leh in 1962. FRL pioneered in protected vegetable cultivation through experimentation in glasshouse, polyhouse and soil trench etc., which helped in production of leafy vegetables during frozen winter. Till date, they have identified about 63 types of suitable vegetables and their improved varieties for cultivation and standardized their agro-techniques for the region (Singh *et al.*, 2000). By adopting these technologies, local farmers are now able to supply 40 per cent fresh vegetables required for troops locally and thereby have reduced the transportation cost by air @ Rs. 22.00 per kg. In the year 1989, a Regional Agricultural Research Station of SKUAST was established in both the districts of Leh and Kargil with the objectives of evaluating and identifying important varieties and developing production technology for different crops. Scientists of SKUAST identified some important varieties of tomato, pea, French bean, cabbage, cauliflower, kale, potato, cucurbits; and production technology for tomato, cauliflower, cabbage, potato, beet leaf and pea. Besides, popularization of trench vegetable production technology for year-round vegetable production (Singh, 2005a), mulching in vegetable crops (Singh and Mir, 2005a; Singh, 2006a) and double cropping in cabbage has been also standardized.

2. Present Status

In the past, vegetables and fruits were imported in Ladakh. Now the situation is different. Vegetable production in the region has gone up. The experimentation on protected structure (polyhouse, soil trench, mulching etc.) in agriculture sector have succeeded in raising a variety of vegetable crops. The days have gone when green vegetables were rarely available to the people of Ladakh. With Ladakh, witnessing a green revolution, the farmers of the region are now supplying fresh vegetables, to the army. Supplies worth Rs. 550 lakhs are being made through five cooperative societies of Leh district. Similarly, Kargil district is also a potential supplier of fruits and vegetables to the Defence Forces.

3. Production Technology

Plenty of sunshine is available in Ladakh with long days and the plants can trap and accumulate more light energy for vegetative and reproductive growth which results in 2 to 3 times more biomass production and yield as compared to plains. Now a days, more than 63 types of vegetables are grown in different parts of the Leh and Kargil districts. With some minor modifications in cultivation techniques all types of vegetables of temperate and tropical regions can be successfully grown in Ladakh from April to September and cultivation of 51 vegetables has been found quite successful either in open field or in protected structure (polyhouse, soil trench, mulching, Ladakhi polyhouse etc.).

3.1 Varieties

On the basis of recommendations made by FRL and RARS, SKUAST Leh/Kargil the vegetable varieties suitable for Ladakh region are given in (Table 1).

3.2 Cultivation Practices

The climate of Ladakh region is entirely different from the rest of the country. It is endowed with cold arid climate with short cropping period and hence, specific agro-technologies which are appropriate for the region should be employed. Through long experimentations on various vegetable crops, the different cultivation practices for Ladakh have been worked out by FRL, Leh and SKUAST, Leh/Kargil and are presented in (Table 2).

Table 1: Promising varieties of vegetables for cold arid zone of Ladakh

Group	Vegetable	Suitable varieties	References
Cole Crops	Cabbage	Golden Acre, Pusa Drum Head, Early Drum Head, ARU-Glory, Pride of India, Summer Express, Late Drum Head	Singh and Mir, 2005
	Cauliflower	Pusa Snowball, K-I, Pusa Snowball-I, Silver King, Early Snowball, Snowball-16	Singh *et al.*, 2000 Singh, 2004
	Knol-khol	White Vienna, Purple Vienna	Singh *et al.*, 2000
	Broccoli	Italian Green Sprouting	Singh *et al.*, 2000
	Brussel's sprouts	Hilds Ideal	Singh *et al.*, 2000
	Kale	Kashmiri Local, Kawadari, Khanyari	Singh *et al.*, 2000
Solanaceous crops	Tomato	Pusa Ruby, Pusa Early Dwarf, Punjab Chhuhara, Sel-7, Arka Vikash, Shalimar-I, Shalimar-II, Summer Market, DARL-Hybrid 304, FRL Sel-l, FRL Sel-2, FRL Sel-7, Sioux, Marglobe, H.T. 6, Rupali, Pusa Hybrid-I, Pusa Sheetal	Singh and Narayan, 2004; Singh, 2005
	Brinjal	Pusa Purple Round, Pusa Purple Long	Singh *et al.*, 2000
	Capsicum	California Wonder, HC-202,	Singh *et al.*, 2000
	Chilli	Jawala	
Leguminous crops	Pea	Bonnevillea, Arkel, Harbhajan, FRL Sel H-3578, DMR-11, VL-7, Azad P-3, Azad P-1, NDVP-4	Singh and Mir, 2005
	French bean	Ageta, Holland-84, Contender, HUR-37, Premieer, Pusa Parvati, Jwala,	Singh, 2006
	Broad bean	HUR-4, Contender, VL-I, FRL Sel Lincoln, FRL Sel, Local	
Root Crops	Radish	Pusa Himani, Scarlet Globe, Local, Japanese White	Singh *et al.*, 2000
	Turnip	Purple Top White Globe, Golden Ball	Singh *et al.*, 2000
	Carrot	Nantes, Local, Yamdagni	
	Beetroot	Detroit Dark Red	
	Swede	FRL Swede Sel-l	
Leafy Vegetable	Beet leaf	K. Raya, FRL Sel.	Singh *et al.*, 2000
	Spinach	Prickly Seeded	Singh *et al.*, 2000
	Palak	Mongol, All Green	Singh *et al.*, 2003
	Mustard	ARU-Black, Zem, TM-4	
	Fenugreek	Pusa Early Bunching, Kasuri	
	Amaranth	Methi	
	Karam sag	Green and Red	
	Chinese cabbage	Kashmiri	
	Orich (Ustakh)	Pusa Sag	
	Coriander	Kashmiri	
	Mint	PD-I, Pant Haritima Local	

Contd...

Bulb crops	Onion	Local White, Pusa Red, Nasik Red, Agrifound Dark Red	Singh *et al.*, 2000
	Garlic	Local	
	Leek	Local	
Salad crops	Celery	Early Zem	Singh *et al.*, 2000
	Lettuce	Great Lake, Dum (Local)	
	Parsley	Local	
Tuber crops	Potato	Kufri Chandarmukhi, Kufri Jyoti, MF-IJ, TPS-67, MF-TITPS-3, MF-III, MF-IVTPS-13, TPS-7I, TPS-13, TPS FRL-1 (UP) TPS FRL-2 (UP), TPS-C-3 (UP)	Singh *et al.*, 2003b
Cucurbits	Pumpkin	Arka Chandan	Singh *et al.*, 2000
	Summer squash	Australian Green	
	Bottle gourd	Pusa Summer, Prolific Long, Pusa Naveen, Arka Long, Pusa Sandesh	
	Ridge gourd	Pusa Nasdar	
	Spong gourd	Pusa Chikni	
	Long melon	Tar Deshi, Green Long	
	Cucumber	Poinsette, Pusa Sanyog	
	Musk melon	Hara Madhu, Punjab Sunehra	
	Water melon	Sugar Baby, Arka Marnik	
	Okra	Pusa Swani, Parbhani, Kranti, Harbhajan, Varsa	

Some viable technologies have been developed through long experimentations which has made possible the production of vegetables successfully in this region. For example, in beat palak, 150 kg/ha N coupled with three cuttings gave maximum green leaf yield of 103.33 q/ha while, seed yield of 11.29 q/ha was obtained with 100 kg/ha N with no cutting (Singh *et al.*, 2003a). Likewise, potato crop yielded 305.24 q/ha with big size seed tuber (50-75 g) and row to row spacing of 25 cm (Singh *et al.*, 2003b). In tomato, a spacing of 30 x 45 cm combined with 120 kg/ha N, 80 kg/ha P_2O_5 and 100 kg/ha K_2O increased the maximum fruit yield and number of fruits/plant (Singh, 2004a). Similarly, in cauliflower, an application of 140 kg N/ha and 80 kg P_2O_5/ha yielded highest marketable curd yield, highest net return and improved the cost: benefit ratio (1:6.8) under cold arid conditions of Ladakh (Singh, 2004b). Application of 100:60:80 kg/ha NPK in onion crop significantly increased the bulb depth, bulb diameter and bulb yield under cold arid conditions of Ladakh (Anonymous, 2003). Cabbage (Eclipse Drum Head) has successfully yielded 900 kg/ha of seed in Ladakh (Singh *et al.*, 1997).

Table 2: Sowing time, seed rate, spacing, manure, fertilizer, days to harvesting and yield of different vegetable for Ladakh

S. No.	Vegetables	Sowing time	Seed rate (kg/ha)	Spacing (cm) between		Manure (t/ha)	Fertilizers (kg/ha)			Days to harvesting	Av. yield (q/ha)
				rows	plant		N_2	P_2	K_2		
	Cole Crops										
1.	Cabbage*	March-June	0.5-0.6	60	45	30	100	50	50	90-100	350-600
2.	Cauliflower*	March-May	0.4-0.5	60	45	25	100	50	50	90-100	250-300
3.	Knol-khol*	March-June	1.0-1.5	30	30	25	80	50	50	80-90	200-250
4.	Chinese cabbage*	April-May	0.5-0.6	45	20	20	80	40	50	60-80	150-200
5.	Brussels sprouts*	April-May	0.5-0.6	60	45	20	80	60	50	100-120	125-150
6.	Kale*	April-May	0.6-0.8	45	30	20	80	50	50	90-110	100-150
7.	Broccoli*	April-May	0.5-0.6	60	45	25	100	50	50	100-110	200-250
	Solanaceous crops										
8.	Tomato*	April-May	0.6	60	45	30	100	60	60	90-110	250-300
9.	Brinjal*	April-May	0.5-0.6	60	45	25	80	60	50	100-110	80-100
10.	Chilli*	April-May	1.0-1.2	30	20	15	80	40	40	100-120.	40-60
11.	Capsicum*	April-May	0.7-0.8	60	45	20	100	40	50	100-120	80-100
12.	Potato	April-May	2500	50	20	35	30	60	80	110-130	350-550
	Pea and legume crops										
13.	Peas	April-June	75-100	30	2.3	15	80	60	50	90-100	120-150
14.	French bean	April-June	50-75	30	10	15	20	80	50	70-80	100-150
15.	Broad bean	May-June	80-100	30	10	15	30	80	50	100-120	150-200

Contd...

	Root crops										
16.	Radish	March-August	10-12	25	10	25	70	40	50	70-80	350-800
17.	Carrot	April-June	4-5	30	10	25	50	40	50	70-80	350-550
18.	Swede	April-June	4-5	45	15	25	80	50	60	80-110	350-450
19.	Beetroot	May-June	10.0	45	15	25	50	40	50	60-80	350-400
20.	Turnip	March-June	3-5	30	15	25	50	40	50	60-80	300-550
21.	Chinese radish	April-May	3-4	45	30	10	100	50	40	65-70	800-900
	Cucurbits crops										
22.	Bottle gourd	May-June	5	200	90	20	60	40	40	90-100	100-150
23.	Cucumber	May-June	2.5-3.0	150	100	15	80	40	40	70-80	150-200
24.	Gherkin	April-May	3.0-3.5	125	90	15	80	50	40	100-110	100-150
25.	Long melon	April-June	2.5-3.0	150	60	15	80	30	30	70-80	100-120
26.	Musk melon	April-May	3-5	160	80	15	120	50	40	90-110	100-125
27.	Pumpkin	May-June	7-8	250	150	20	60	30	30	80-90	200-350
28.	Long melon	April-May	2.0-2.5	150	75	15	50	40	40	3-100	100-125
29.	Ridge gourd	April-May	4-5	150	60	15		50	50	100-120	50-60
30.	Sponge gourd	April-May	4.5-5.5	150	60	15	75	50	60	95-120	60-70
31.	Summar squash	May-July	8-10	90	90	20	60	30	30	60-80	250-300
32.	Water melon	May-June	2.5-3.0	200	150	15	50	30	30	80-90	100-150
	Bulb crops										
33.	Onion*	April-May	8-10	20	10	25	60	50	60	100-110	200-250
34.	Garlic	May-June	600	25	15	20	80	60	40	100-110	40-60
35.	Leek	April-May	10-12	15	10	20	60	50	60	100-120	150-200
	Malvaceae crops										
36.	Bhindi	May-June	12-14	30	15	20	50	40	40	80-90	80-120

Contd...

	Leafy vegetables										
37.	Beet leaf	April-June	12.0	20	10	20	60	40	40	30-40	200-250
38.	Palak	March-August	10-15	20	10	20	60	30	30	30-40	100-150
39.	Spinach	March-August	30-40	20	5	20	60	30	30	40-50	60-80
40.	Karam sag	March-June	0.7-0.8	30	30	20	60	50	50	80-90	200-250
41.	Veg. Mustard	March-August	5-6	30	20	20	50	40	50	45-60	150-200
42.	Dum (Local Lettuce)	April-June	1.2-15	20	10	15	30	40	40	50-60	150-200
43.	Coriander	March-August	20-25	20	10	15	60	60	60	50-60	120-150
44.	Fenugreek	April-June	10-12	30	5	15	80	40	40	60-80	100-150
45.	Amaranth	April-June	3-4	30	5	10	50	50	50	30-40	80-100
	Others										
46.	Artichoke	May-June	1000	60	45	25	60	60	60	100-120	150-200
47.	Lettuce	March-July	0.7-0.8	30	20	15	80	50	50	40-45	75-100
48.	Orich (Ustakh)	April-June	5-6	20	10	15	75	50	40	60-80	150-200
49.	Celery	March-May	0.3-0.400	45	20	15	50	40	80	150-200	60-80
50.	Parsley	March-August	0.5	30	10	15	60	30	30	40-60	80-100
51.	Mint	April-June	50-60	20	10	15	60	40	40	50-80	100-150

* Transplanting should be done after 6 to 7 weeks of sowing or when seedlings are at 4 leaf stages

- Irrigation is applied at an interval of 6-10 days in all vegetables depending upon soil and weather condition
- Weeding and interculture operations should be done 2-3 times during early stage of crop growth (20-25 and 40-45 days after sowing or transplanting) depending on weed population and duration of crops
- 25-35 t/ha well decomposed organic manure (FYM or compost or night soil) would reduce requirement of inorganic fertilizers
- Use fertilizers after getting soil analysis of the field
- Control insect-pests and diseases by using suitable plant protection measures

An extended cropping season of about 20 to 25 days can be attained through protected nursery raising as compared to the open field conditions (Singh *et al.*, 2000). The trench technology has responded significantly and a cropping intensity of 400% has been obtained with round the year vegetable production and its availability (Singh, 2005a).

Some of the important technologies are described below:

3.3 Nursery Raising

For vegetable production in Ladakh, nursery raising is one of the most important operation. With short growing season from April to September one month of valuable time is lost in nursery raising. In order to utilize the limited time of favourable growing season nursery should be raised under protected conditions. The FRL, Leh and SKUAST, Leh/Kargil has devised specialized nursery raising techniques in polyhouse, poly tunnel and soil trench by which seedlings are available to the farmers 20-25 days earlier and as soon as the season starts. This enables the farmers to harvest their crops more than one month in advance giving them high remunerative value for their produce. Vegetable nursery in Ladakh can be raised in the following way:

3.4 Nursery Raising in Protected Structure

Different structures for raising nursery under protected conditions are being used in Ladakh region. The growers are able to get vegetable seedlings 20-25 days in advance by using protected nursery raising technique. The different protected structures employed for this purpose are glasshouses, polyhouses, polytunnels, fiberglass tunnels and soil trenches, and are of different shape and size. The different shape and size of the structures found suitable for the cold arid region of Ladakh are given in (Table 3). Among these structures soil trench has been found to be more effective and economically viable than the other structures (Singh *et al.*, 2000). It also reduces the cost of vegetable production, i.e., Rs. 6.17/kg as compared to other methods (Singh *et al.*, 1998). The different structures are helpful in providing conducive environment for raising nursery as well as for cultivation of vegetables as they facilitate early germination, proper growth, protect the seedlings from severe cold and provide better environment for vegetative and reproductive growth. The nursery is raised either on flat beds or in pots/trays of size 5 feet length, 2 feet width and 2 feet deep inside the structures. The sowing time of nursery in protected structures depends on the type of vegetable crops and cultivation practices to be employed either under protected or open field cultivation (Table 4).

Table 3: Shape and size of protected structure

Structure	Shape	Suitable size
Glasshouse	Span roof type	24 feet length, 12 feet width and 7 feet height
Polyhouse	Arch type/ quonset Leanto-type	(I) Polyhouse: 30-50 feet length, 15 feet width and 7 feet height (ii) Ladakhi polyhouse : 40 feet length, 20 feet width and 7 feet height of rear wall and 1.5 feet height of front wall
Polytunnels	Semicircle	6 feet length, 3 feet width and 2.5 feet height
Fiber glass Tunnels	Semicircle	10 feet length, 3 feet width and 2.5 feet height
Soil trench	Rectangular	12 feet length, 6 feet width and 2.0 feet deep
Hot bed	Rectangular	10 feet length, 4 feet width and, ½ feet deep (surrounded with sloppy wooden frame)

Source: Singh *et al.*, 2000

Table 4: Sowing time of nursery

Vegetable	Sowing time	
	Protected	Open
Cole crops (Cabbage, Cauliflower, Knol-khol)	Mid February	Last week of March to 2nd week of May
Onion	Last week of October (for green) 1st week of March (for bulb)	1st week of April
Tomato, Capsicum, Chilli	Mid March	Last week of April
Brinjal	Last week of March	Last week of April
Cucurbits	Mid March to first week of April	Last week of April to 1st week of May
True Potato Seed (TPS)	……..	1st week of April

* Brinjal and cucumber do not bear good fruits in open field conditions

3.5 Surface Nursery Raising under Polythene Covers

In surface nursery system nursery beds of 2.5 m x 1.5 m are made on the ground surface. The beds are manured with well rotten farm yard manure at the rate of 15 kg per bed and a mixture of fine soil and FYM, 1 cm thick, is spread over the bed. The nursery sowing operation commences around so that the seedlings become available in first week of May. Immediately after sowing light watering is done and is repeated only after germination or in case of a dry spell. After watering, the beds should be first covered with a transparent polysheet erected on a frame to avoid the sheets touching the ground and then by a black polysheet. Soon after germination the black cover is removed to allow sunlight for the seedling growth. However, seedlings raised by this method are useful for only those farmers who have irrigation facilities but the majority of farmers are dependent on snow melted water which is available only by the month of May (Rao, 1984).

3.6 Nursery Raising in Open Field

The vegetable nursery under open field conditions of Ladakh region is raised in the same manner as it is raised elsewhere in the country using standard practices. Sowing of the seed is done in the last fortnight of April. However, these seedlings grow at a faster rate but they can not compete with seedlings raised under protected conditions in respect of earliness. Nursery sown during April does not require protection against low temperature. The sowing time for late transplanted vegetables like cabbage, cauliflower and knol-khol is during May.

4. Mulching

Mulching is an agricultural and silivi-cultural technique aimed to create a barrier for the moisture loss, alter temperature by a few degrees, keep weeds away and create favourable micro climatic conditions for plants particularly near the root zone. For mulching, locally available materials like organic materials (plant straw, dry leaves, saw dust, animal manures etc.) and synthetic materials (polythene, paper, wax coated paper and aluminum and steel foils etc.) are used for covering the soil surface. In Ladakh, summer season vegetable crops like cucurbits, tomato, brinjal, okra, cabbage and capsicum etc. are known to respond better to mulching and give early and increased yield (Anonymous, 1996). The cultivation of cucurbits is only possible by the use of mulch. Black polyethylene (150 gauge thick) has been observed to be very beneficial for the crops like cucurbits, brinjal, okra, capsicum, tomato, cabbage and chilli (Anonymous, 1996). Besides these, it has also been evaluated with cabbage and cauliflower at FRL experimentation stations. It was found that in mulched crop of cucumber, irrigation is needed at 10 to 15 days interval against 7 days interval in un-mulched crop. On the basis of the experiments, it concluded that the black polyethylene mulch is relatively more beneficial in several vegetable crops like tomato, cabbage, cauliflower, capsicum and cucurbits. It significantly increased yield up to 41 to 47%, suppressed the weed growth by 97.56%, increased ground temperature by 2-3° C and conserved greater soil moisture in tomato and cabbage (Singh and Mir, 2005; Singh, 2006).

5. Winter Cultivation

Growing of vegetables under controlled and modified environment is called protected vegetable cultivation. For this purpose different type of protected structures, *viz.*, glasshouse, polyhouse with metal frame, *Ladakhi* polyhouse, soil trench, plastic tunnels and polyethylene crop cover suitable for one or other aspects are being used for vegetable cultivation in cold arid region of Ladakh. However, soil trench or *Ladakhi* polyhouse is most effective, viable and economical as compared to other structures for leafy vegetable production during winter (October-April) (Singh *et al.*, 2000; Singh and Yadav, 2001; Singh, 2005a; Singh, 2008). These protected structures are of different shapes and sizes and are covered by transparent or translucent glass or plastic film (usually UV stabilized 200 gauge LDPE film). The plastic film allows solar radiation to pass through, and traps the thermal radiation emitted by the objects within and does not allow it to escape. The carbon dioxide released by the plants at night is also trapped inside increasing its level and making it available to the plants during the day time thereby enhancing photosynthesis. During the winter months (October- March) due to extremely low temperature in Ladakh there is absolutely no greenery in the open fields. However, vegetables like beet leaf, palak, fenugreek, vegetable mustard, dum (local lettuce), orich, coriander, mint, celery, lettuce, parsley etc. can be successfully grown under protected conditions during the winter months. The structures can also be used to grow vegetables like cucurbits (bottle gourd, cucumber, long melon, ridge gourd, summer squash, musk melon, water melon etc.), early tomato, brinjal, okra, capsicum and chilli during the summer months (April-September). Due to cooler night and low relative humidity these vegetables show sub-optimal growth in open fields even during the main cropping season as they show restricted growth and fruit development. Almost all type of vegetables become ready for marketing in advance by 30-45 days under protected cultivation as compared to open field conditions.

It is a challenging task to grow crops when environmental temperature dips to -25°C to -30°C and not blade of grass can be seen outside during peak winter months (December–January) affecting the crop growth adversely under greenhouse also (Dwivedi *et al.*, 1999; Singh *et al.*, 2000). A decade of experimentation has led to certain recommendations on leafy vegetables for production at subzero temperature under protected structures. The recommended varieties, sowing time and planting distance of vegetables suitable for winter months are presented in (Table 5), (Singh, 1995), in which sowing time is the most important factor. The crop should be sown before the onset of winter enabling it to resist low temperature as delay of even one day in sowing causes significant reduction in yield. Certain production tips required for winter vegetable cultivation are listed below:

Table 5: Vegetables, varieties, sowing time and planting distance for winter cultivation

Crop	Variety	Sowing time	Planting distance (cm)
Beet leaf	All Green, Mongol	1st fortnight of Sep.	20
Beetroot (leaf)	Detroit Dark Red	-do-	20
Celery	-	-do-	25
Bathua		1st fortnight of Sep.	10
Coriander	RIL-195	End of Sept.	15
Dums	Local	End of August	20
Fenugreek	Kasuri	2nd fortnight of Sep.	15
Karam sag	Kashmiri	2nd fortnight of August	20
Lettuce	Great Lake	End of Sept.	25
Mint	-	-do-	20
Parsely	Curled Leaf	-do-	25
Radish (leaf)	Pusa Himanil, Local	-do-	20
Veg. Mustard	ARU Black	End of Sept.	20

Source: Singh, 1995

- Incorporate FYM at 1.5 to 2.0 kg/m^2 30 days before sowing and avoid the use of chemical fertilizer. However, DAP 10-12 g/m^2 and urea 5-7 g/m^2 before sowing enhances the growth of plants
- The soil under cover is very precious and so all ridges and channels should be utilized for growing crops
- Seed soaking in water for 36 to 48 h before sowing gives faster and higher germination percentage; moreover instead of seed sowing if seedlings are transplanted at that time it will be more beneficial
- Light and frequent irrigation enhances the crop growth and improves its tolerance against low temperature
- During the peak winter months when greenhouse temperature falls below 0°C, cold protection measures are essential. An additional polyethylene cover reaching 2 feet above the plants should be used during the night (December and January) as it improves the micro-climate inside and enhances crop growth. This covering can be removed during day time depending upon environmental conditions.

6. Vegetable Storage During Winter

The severe winter months lasting about 7 to 8 months with temperature dipping to as low as -40°C (NAEDB, 1992) virtually makes growth of any kind of vegetation impossible. The precipitation during these months is mostly in the form of snow which makes most of the mountainous peaks inaccessible and surface transportation becomes very difficult. Hence, storage of vegetables is another important aspect for the cold arid region. From October onwards when the night temperature goes to subzero (-4 to -10°C) and day temperature is around 10°C, fresh vegetables undergo freezing during the night and thawing during the day resulting in its spoilage. At that time they are protected against desiccation by storing under covered mudhouses and not allowing the night temperature to go below 4°C. Natural freezing conditions prevail during December and January and so vegetables like cabbage, knoll khol, and cauliflower can be kept in a frozen state. Vegetables such as potato, carrot, radish, turnip, beet root and Swede are stored underground for supply during February to June. For preserving these vegetables they are kept in underground pits or cellars during the month of November with 2 feet of sand piled above them. They are taken out during the month of February and stored in mudhouses at temperature below 10° C and above freezing. Like this vegetables can be stored to meet the local demand during the cut off period of winter. However, the storage aspect requires more intensive research for sustained vegetable production. The softer varieties of vegetables such as cauliflower, tomato, pea etc. which were hitherto being sundried, are now also being tinned fresh.

7. Opportunities

Until recently, *Ladakhis* grew and consumed their own grains, cereals, vegetables, manure, seeds and other agricultural inputs, reared their own animals and prepared their own farms in a well integrated, coordinated and balanced form of farming system evolved in response to the agro-climatic conditions. However, as Ladakh opens up to the world, its ancient and reliable agricultural system faces disruption and agriculture takes a back seat as it is difficult at high altitude and arid climate to the more easier and fast growing sectors like tourism and related service. The cold desert area of Ladakh is now a thing of the past. The area has undergone transformation successfully through what we may call a "green revolution". Today, the area produces 63 vegetable crops in summer in open fields including some exotic crops. The vegetables grown during winter in the rest of the country can easily be grown during the summer months in the Ladakh region due to low summer temperature. Potato is grown in summer in Ladakh region, particularly during May to September and meets 75% of the needs of the troops. Standardized technology has been demonstrated and transferred to farmers and new vegetable crops like parsley, celery, sarda melon, Japanese bunching onion, sweet turnip, karam sag, chenopodium and garlic have been introduced.

8. Limitations

In the cold arid region of Ladakh, vegetable cultivation is getting popularity among the progressive farmers due to adoption of new techniques such as use of high yielding cultivars, nutrient management and crop protection measures. But the cultivation is limited to only those areas where assured irrigation is available. Weed management during summer is also a major problem in vegetable cultivation as it enhances the production cost on one hand and competes with the crops for nutrients and water on the other, consequently reducing the yield. Some of the major problems in vegetable cultivation are listed below:

- Transfer of technology is a major bottleneck as majority of the growers are unaware of the production technology, nutrient management practices and plant protection measures
- Lack of availability of high quality seed/planting material
- Lack of assured irrigation facilities giving farmers a very short period (last week of May to September) for vegetable cultivation
- Inadequate supply of fertilizers, insecticides, pesticides and fungicides
- In cucurbits, low relative humidity restricts the movement of insects (pollinating agent) resulting in low fruit set and forcing hand pollination which is tedious and costly
- The protected structures and mulching materials are quite expensive and beyond the reach of common farmers
- Lack of proper storage technologies and storage facilities for winter storage of vegetables.

9. Future Strategy for Research

Though vegetable production in Ladakh has great potential, research on many production aspects is lacking due to mobilization of technical and financial resources. The vegetable produced in the region could fetch better price in northern plains during off season as there is scarcity of these vegetables at that time. Therefore, better communication and transport systems need to be strengthened preferably by engaging refrigerated vans and containers for smooth and uninterrupted supply. The crops in great demand in the market of plains are cabbage, carrot, pea, beetroot, cauliflower and potato.

Research priorities for enhancing vegetable production in cold arid region are:

- To develop suitable varieties/hybrids suited to different agricultural zones of cold arid region of Ladakh
- To develop suitable production technologies of vegetable cultivation under protected conditions for the cold arid region

- Inexpensive and easily accessible infrastructure modeling of green houses, trenches, tunnels etc.
- Standardized integrated nutrient management system should be developed for sustainable vegetable production
- To develop seed production techniques for different vegetable crops
- Use of appropriate plant protection measures against prevalent diseases and insect- pests should be worked out
- The transfer of technology aspects need to be given more emphasis and strengthened
- Use of suitable cropping pattern and mulching material for soil and moisture conservation
- To develop technologies/facilities for post-harvest management and processing of vegetables
- To develop technology and create infrastructure for winter storage of vegetables.

References

Anonymous (1996). Research report. SKUAST, RARS, Leh. 45 p.

Anonymous (2003). Annual report of SKUAST-K, Shalimar, Srinagar. 34 p.

Dwivedi, S.K; Sareen, S.K. and Paljor, E. (1999). Trench for winter vegetable production in Ladakh (In Hindi). Phal-Phool 21(4): 23-24.

NAEDB (1992). Action plan on cold deserts—An integrated approach for sustainable development. National Afforestation and Ecological Development Board, Dr. Y.S. Parmar, University of Horticulture and Forestry, Nauni, Solan (H.P.) India.

Rao, S.K. (1984). An oasis in the cold desert-Field Research Laboratory, Leh. Def. Sci. J. 34(2): 213-220.

Singh, A. K. (2004a). Effect of spacing, phosphorus and potassium fertilization on yield of tomato under Cold Arid Condition of Ladakh. Vegetable Sci. 31(1): 75-77.

Singh, A.K. (2004b). Effect of nitrogen and phosphorus on growth and curd yield of cauliflower var. Snowball-16 under Cold Arid Region of Ladakh. Haryana J. Hort. Sci. 33 (1&2): 127-129.

Singh, A.K. (2005a). Standardization of suitable vegetable cropping sequence for trench cultivation in cold arid region of Ladakh. Prog. Hort. 37(1): 188-193.

Singh, A.K. 2005. Genetic variability, correlation and path coefficient studies in tomato (*Lycopersicon esculentum* Mill.) under cold arid region of Ladakh. Prog. Hort. 37(2): 437-443.

Singh, A.K. (2006). Genetic divergence in French bean (*Phaseolus vulgaris* L.). Vegetable Sci. 33(1): 103-105.

Singh, A.K. (2006a). Response of different types and colour of mulch on growth and yield of tomato (*Lycopersicon esculentum* Miller) cv. Punjab Chhuhara under Cold Arid Conditions of Ladakh. Hort. J. 19(2): 93-96.

Singh, A.K. (2008). Trench mein sabji utpadan-unche sushak Himalayee kheshetro ke paripechhya mein (Hindi). Pahari Kheti Bari 13(1&2): 73-75.

Singh, A.K. and Mir, M.S. (2005). Genetic variability, heritability and genetic advance studies in pea (*Pisum sativum* L.) under cold arid region of Ladakh. Environ. Ecol. 23(3): 445-449.

Singh, A.K. and Mir, M.S. (2005a). Effect of different mulches on growth and yield attributing traits of cabbage (*Brassica oleracea* var. *capitata*) cv. Pride of India under cold arid region of Ladakh. Environment and Ecology 23(2): 398-402.

Singh, A.K. and Narayan, R. (2004). Variability studies in tomato under cold arid condition of Ladakh. Hort. J. 17(1): 67-72.

Singh, A.K. and Yadav, V.K. (2001). Ladakh ke lia vardan "trench mai sabji utpadan". Phal Phool 24(4): 4-6.

Singh, A.K. and Yadav, V.K. (2005). Character association and path coefficient analysis studies for certain metric traits in pea (*Pisum sativum* L.) under cold arid region of Ladakh. Hort. J. 18(1): 35-38.

Singh, A.K.; Mir, M.S.; Yadav, V.K. and Khan, Z.H. (2003a). Standardization of nitrogen fertilization and cutting management in beet palak (*Beta vulgaris* L.) under cold arid condition of Ladakh. Environ. Ecol. 21(3): 628-631.

Singh, A.K.; Yadav, V.K. and Mir, M.S. (2000). Trench vegetable production technology for throughout year vegetable production in Ladakh. Indian Farmer Digest 33(11): 26-28.

Singh, A.K.; Yadav, V.K.; Mir, M.S. and Khan, Z.H. (2003b). Standardization of seed size and spacing for improving yield of potato (*Solanum tuberosum* L.) under cold arid condition of Ladakh. Environ. Ecol. 21(3): 639-641.

Singh, B. (1995). Greenhouse development in Leh. Paper presented in National workshop on commercialization of tissue culture and greenhouse technology, Banglore, India 5-6 May 1995.

Singh, B.; Singh, N. and Singh, D.V. (2000). Perspective of vegetable production in Ladakh. In: Dyanamics of Cold Arid Agriculture, Edited by J.P. Sharma and A.A. Mir. Kalyani Publication, Ludhiana. p. 91.

Singh, J.; Kalloo, G. and Singh, K.P. (2001). Vegetable Crops: Nutritional Security. Tech. Bull. No. 6, Published by Indian Institute of Vegetable Research. p. 1.

Singh, N.; Singh, B. and Sareen, S.K. (1997). High altitude cold desert beet and seed production. Indian Horticulture. Jan-Mar: 57-58.

Singh, N.; Dwivedi, S.K. and Singh, B. (1998). Protected Vegetable Cultivation in Ladakh. Book in Hindi-Field Research Laboratory-Leh C/O 56 APO.

Yadav, V.K.; Singh, A.K.; Singh, H. V. and Khan, Z. H. (2005). Crop diversification for increasing crop production in cold region of Jammu and Kashmir. Indian Farmer Digest 38(9): 12-14.

□□□

Singh, I., Kalloo, G. and Singh, K.P. (2001). Vegetable Crops: Nutritional Security. Tech. Bull. No.6. Published by Indian Institute of Vegetable Research. p. 1.

Singh, N., Singh, B. and Sareen, S.K. (1997). High altitude cold desert beet and seed production. Indian Horticulture. Jan.-Mar. 57-58.

Singh, N., Dwivedi, S.K. and Singh, B. (1998). Protected Vegetable Cultivation in Ladakh. Book in Hindi. Field Research Laboratory-Leh C/O 56 APO.

Yadav, V.K., Singh, A.K., Singh, H.V. and Khan, Z.H. (2005). Crop diversification for increasing crop production in cold region of Jammu and Kashmir. Indian Farmer Digest 38(9): 12-14.

4

Vegetable Seed Production

ANKUR AGARWAL
Defence Institute of High Altitude Research (DIHAR), (DRDO), C/o 56 APO
(Presently at Defence Institute of Bio Energy Research, (DIBER), Pithoragarh, Uttarakhand)

Quality seed is the foremost and the most cost-effective input for crop production. Therefore, production, certification and flow of improved seed to the farmer is always a high priority in agricultural growth and development. After Independence, significant work has been done on the development and standardization of agro-techniques to achieve self-sufficiency in the seed production of vegetables but the goal is quite far. With the achievement of Green Revolution, India has made significant advances in agriculture where the role of seed sector is substantial. In comparison to the annual global seed trade (US $21 billion), Indian market share is around US $900 million (about Rs 4000 crores) (Sarkar and Prince, 2006). According to Verma *et al.* (2002), about 100 tonnes of breeder seed is being produced annually against the average requirement of about 36 tonnes in the country, which forms the basis of multiplication of superior quality foundation and certified seed. The largest beneficiaries of breeders seed, *i.e.,* NSC is presently producing only 650 tonnes of certified seed whereas private seed industry contributes about 2,000 tonnes and contribution of other certified state seed agencies amount to 500-1000 tonnes. The demand of vegetable seed is presented in (Table 1).

With the introduction of new seed policy in 1988 and liberalized seed import under open license with the objective of providing best planting material to the farmers, growing of F_1 hybrid became popular in India. Though the hybrids have started getting a substantial hold due to their uniformity, better adaptability, tolerance to biotic and abiotic stress and market acceptability but in national perspective, seed production of open pollinated varieties has got special emphasis. According to Verma and Singh (2001) in all the public sector endeavours for vegetable seed production, the major share is of open pollinated varieties as presented in (Table 2).

Temperate vegetable constitute a formidable share in our daily vegetable consumption but seed production of these is restricted to some pockets of India and demand is mostly met by imports. Katrain (Kullu valley, Himachal Pradesh) is the most favourable region for seed production of temperate vegetables. Besides Kullu and Kashmir valley, other areas fulfilling the climatic requirement of seed production of temperate vegetables and would widen the scope of this industry are presented in (Table 3). Ladakh, comprising of Kargil and Leh district, is spread over a geographical area of 96,701 km^2 that accounts for 43% area of the state and 75% of the cold arid region of the country. The net sown area in the region is 29,200 ha of which 2,555 ha is sown more than once (Mir, 2000). The double cropped area which is mild in climate and with fertile soil offers the potential area for vegetable seed production. Seed industry on commercial scale will be profitable in this area due to cheap labour and also generate employment opportunities.

In vegetable crops, seed production has been divided into two groups, *viz.*, (a) European or biennial or temperate vegetables requiring temperate climate especially during a specific stage of their growth to meet the vernalization (chilling) requirement and (b) Asiatic or annual or tropical vegetables which produce seed satisfactorily in plains. European or biennial vegetables include cabbage, knol-khol, Brussel's sprouts, sprouting broccoli, kale, collard, snowball group cauliflower, European carrot, radish, beet and turnip. Rest of the vegetables come in Asiatic group.

Table 1: Demand of vegetable seeds in India

Vegetables	Area (ha)	Requirement (ton)
Tomato	499449.70	249.72
Brinjal	492879.00	369.60
Chilli	420948.30	420.98
Cabbage	312861.70	156.43
Cauliflower	312861.70	156.43
Cucumber	22256.30	55.63
Gourds	420984.30	2104.90
Leafy vegetable	142207.00	5688.29
Melons	174197.20	696.78
Okra	382302.95	3058.42
Onion	399054.10	3990.54
Peas	233663.90	23366.39
Radish	306558.70	767.14
Beans	244402.70	24444.27
Others	1529938.00	3824.80
Total	5894530.46	69350.32

Table 2: National scenario of vegetables hybrids and open pollinated varieties

Crop	Total Area (ha)	Hybrids		Open pollinated	
		Area (ha)	Share (%)	Area (ha)	Share (%)
Tomato	485520	153000	31.51	332520	68.49
Brinjal	479095	85300	17.80	393795	82.19
Chilli	573529	14000	2.44	559529	97.56
Cabbage	242140	76000	31.39	166140	68.61
Cauliflower	304156	10000	3.29	294156	96.71
Gourds	409270	10000	2.44	399270	97.56
Melons	169350	6800	4.02	162550	95.98
Okra	371665	20000	5.38	351665	94.62

Table 3: Temperate vegetable seed production regions and there climate in India

Region	Climate
Srinagar valley (J&K)	Transitional climate, cold winter, warm summer, 50-75 cm rainfall distributed over about 60 days
Upper Kullu valley (HP)	Cold winter, warm summer, 100-125 cm rainfall distributed throughout year
Lahaul valley (HP)	Severe cold winter, cold summer, 10-25 cm rainfall
Kalpa valley, Kinnaur (HP)	Severe cold winter, cold summer, 10-25 cm rainfall
Saproon valley, Solan (HP)	Cold winter, mild summer, 75-100 cm rainfall
Kumaon hills (Uttaranchal)	Cold winter, mild summer, 100-125 cm rainfall
Alimpong-Darjeeling Hills (WB)	Cold winter, warm summer, 200-250 cm rainfall
Nilgiris (Tamil Nadu)	Cold winter, warm summer, 110 cm rainfall

Time of transplanting for seed production of all temperate vegetables in Ladakh is invariably during March-April as per the altitude of the area after over-wintering of vegetative parts in underground cellar house.

History of Seed Production in India

The history of seed production dates back to start of civilization. Systemic efforts in India have started with the colonial times when Imperial Govt. encouraged the seed production of European types of vegetables at Quetta in Baluchistan (now in Pakistan) during 1942-43. At about the same time Department of Agriculture, Punjab opened a small farm at Katrain (Kullu valley) in HP (part of Punjab before Independence). Till Independence, the seed production was mainly carried out by the private seed growers but with the partition in 1947, Quetta centre went to Pakistan and seed supplies to India were cut off mainly for temperate vegetables. Thus, it became necessary to strengthen the temperate seed program at some other suitable location in India. In 1949, Ministry of Food and Agriculture, Govt. of India started a research station. Central Vegetable Breeding Sub Station, at Katrain (HP). This station was transferred to Indian Agriculture Research Institute (IARI), New Delhi during 1955 with the sole objective of intensifying research work on the temperate vegetable and that later named as IARI Regional Station.

Climate

Climate is utmost important for successful production programme in any agriculture activity. Main features of climate for satisfactory seed production of temperate vegetables are as under:

a) *Temperature*: Low temperature ranging from 4.4 to 10°C for 60 days during winter is necessary for seed production of European types of vegetables. This provides necessary requirement for the breaking the dormancy of biennial vegetables and to induce the flowering. Plant growth stage, prevalence of temperature and time of exposure have interacting effect of bolting. Summer should be mild for high seed set because high temperature during summer results in pollen sterility and mortality. Mild temperature below 30°C favours efficient pollination, fertilization and seed setting.

b) *Rainfall*: In general well distributed rainfall of 50-80 cm over the year is good for seed crop. Maturity and harvesting stage of seed should not coincide with rainy season rather dry weather with low humidity is required for good quality seed.

Soil

All types of soils except problematic soils, which allow satisfactory vegetable production, are generally good for seed production also. Soil should be fertile and have sufficient organic matter. It is advisable to analyse the soil for its nutrient status so that deficiency, if any, can be corrected timely. Clay soils are difficult to prepare and normally not considered good for root crops. Heavy soils delay the maturity of seed crop considerably whereas light soils hasten the maturity and ripening. Poor soils require considerable quantity of plant nutrient to be added externally as organic manure to make them fit for seed production.

Soil Preparation

The land should be worked properly and thoroughly to make it loose and friable. Field should be ploughed 3-4 times, harrowed and then properly levelled to facilitate irrigation and arrangements must be made for drainage. All the stubbles and weeds should be removed to make the field clear. Organic matter should be incorporated thoroughly to the soil during field preparation about 30 days in advance of planting. Physical properties of soil can be improved by adding soil amendments particularly organic amendments which improve soil structure and fertility. Plant growth is directly related to tilth of soil and under poor tilth it tend to become hard and crusty after drying and sticky upon irrigation. Soil should not be worked to such a fine tilth that crusting occurs after rain. Field should not be ploughed when wet otherwise it results in hard clods and undulation in field. The deeper the soil is prepared, the greater is its capacity for holding air and moisture that in turn provides congenial microclimate for growth of roots and microorganism.

Manures and Fertilizers

Vegetables being a short duration crop need optimum and timely fertilization for proper development. However, seed production demands heavy doses of fertilizer to support satisfactory vegetative and reproductive phase. Nutrition should be distributed judiciously during both the phases and one phase should not be over-manured at the expense of other. Though the flowering will be there even in the absence of fertilization but plant will not bear the sufficient load of flowers and seed. Furthermore, the quality and the vigour of the seed will be poor.

Green and organic manure are the two good sources of soil fertility restoration. Cowpea can be grown successfully as green manure crop in hilly regions (Singh *et al.*, 1959a). Although incorporation of green manure crop in intensive crop rotation looks difficult due to less lean period between two consecutive crops but feasibility of this must be checked as and when possible. Selection of green manure crop depends upon: a) climate, b) time of planting for the crop, c) ease in plant material incorporation in soil, d) duration of green manure crop to reach suitable stage to turn over, e) quantity of vegetative matter produced, f) root growth pattern and g) stage at which crop to be turned under, i.e., green or dead. As far as organic manure (FYM) is concerned care has to taken to apply well rotten manure to the field otherwise crop suffers from nitrogen starvation at early stage of development due to deficiency of nitrates owing to their utilization by microorganism for organic matter decomposition and high C:N ratio. C:N ratio of more than 15:1 usually results in competition between microorganism and crop plants for nitrates already present in soil and secondly burning effect on the plants due to rapid decomposition of urine in the manure. The main advantages of organic manure are:

- Improvement in soil texture and therefore soil is more friable and subjected to less backing and crusting
- Water holding capacity of soil is enhanced due to increase in surface area of colloidal particles and thus allow retention of more water and nutrient in light soils

- Organic matter, cement the clay particle together in to aggregate that makes the soil more permeable to water and improves aeration of the soil
- It supports the microorganism population and activity. Soil inhabiting microorganism like bacteria and fungi changes the plant nutrients from unavailable to available form while other fix them in soil for further use
- Hastens mineralization process through CO_2 emission during organic matter decomposition. This CO_2 forms mild acid in the soil which acts as an solvent to make more minerals available to the plants. This also increases cat-ion exchange capacity of soil to retain more nutrients.

Judicious and balanced use of organic and inorganic fertilizers is always recommended. Manure is generally deficient in phosphorus (present only in protein). Though it is possible to apply entire phosphorus through manure, it will result in wastage of surplus nitrogen and potash. Therefore, use of inorganic fertilizers is very important to compensate the essential mineral nutrients in plant growth and development. Many physiological disorders results from deficiency of essential elements particularly the minor elements because most of these elements are the constituents of enzymatic systems. Though fertilizer application should be based on the fertility of the soil, irrigation facility, cropping pattern and crop itself but the major nutrients (N, P and K) used by the plants in large quantities need to be applied regularly.

Method and time of fertilizer application is very important and usually nitrogenous fertilizers are applied in split doses mainly to avoid leaching and volatilization losses and increasing availability to plants whereas phosphatic and potasic fertilizers are applied in single doses at the time of planting. Phosphorus does not move in soil water and therefore must be placed in the root zone of plant rather than on soil surface. Whereas potash leach is quickly in light soil and deficiency is less likely to occur in heavy soil. Fertilizer recommendations for some vegetable crops are presented in (Table 4).

Spraying of boric acid (30 ppm) at flowering stage was found suitable to increase seed yield in radish cv. Japanese White (Roy and Seth, 1971). Spraying of boric acid (30 ppm) in turnip var. Pusa Chandrima and cabbage at pre-flowering stage was found to increase the seed yield as this enhances the pollen germination (Bhagchandani *et al.*, 1982). Foliar spray of urea (1-2%) is highly suitable to counter the poor growth because of its high solubility, easy and quick absorption by plant tissues besides containing more nitrogen. This is beneficial in replanted crop where the crop makes slow growth after replanting due to cold climate of hills in the crops for better seed production. In cauliflower, foliar spray of borax (0.25-0.50%) or soil application of borax (15-20 kg/ha) is beneficial in acid soil to counter the boron deficiency. Similarly, soils with pH less than 5.5 show molybdenum deficiency as whiptail leaves and to correct this foliar spray of 200-300 g ammonium molybdate at the concentration of 0.01-0.1% or soil application of ammonium molybdate (1-2 kg/ha) is beneficial.

Table 4: Fertilizer recommendations for various vegetable crops

Crop	Recommendation	Reference
Cabbage	250:125:100 kg NPK/ha in var. Golden Acre	Thakur and Gill, 1978
	60 kg N/ha and Azotobacter culture for increasing seed yield in addition to recommended dose in var. GA	Verma *et al.*, 1997
Cauliflower	375:250 kg NP/ha in var. Snow Ball-16 for best yield	Gill *et al.*, 1975
Knol-khol	100:80:85 kg NPK/ha for better seed yield	Verma and Sharma, 2000
Brussels sprout	200:80:100 kg NPK/ha with N in 4 equal splits	Verma and Sharma, 2000
Sprouting broccoli	200:100:100 kg NPK/ha with N in 4 equal splits	Verma and Sharma, 2000
Kales and Collards	200:80:100 kg NPK/ha with N in 4 equal splits	Verma and Sharma, 2000
Carrot	100:50:60 kg NPK/ha for *in situ* method and 50:50: 60 kg NPK/ha for root to seed method in addition to recommended dose required for root crop	Verma and Sharma, 2000
Radish	70:40:50 kg NPK/ha for roots production and additional 50 kg/ha for further seed production	Roy and Seth, 1971
Turnip	120:50:50 kg NPK/ha for *in situ* seed production and 120:100:100 kg NPK/ha for root to seed method	Verma and Sharma, 2000
Beet root	100:50:100 kg NPK/ha for root production and additional 50 kg N/ha for *in situ* seed production. For replanted seed crop 100:50:50 kg NPK/ha required additionally	Verma and Sharma, 2000
Lettuce	100:60:60 kg NPK/ha with manure	Thompson and Kelly, 1957
Celery	150-200 kg N/ha for higher seed yield	Saini *et al.*, 1989
Swiss chard	80:40:40 kg NPK/ha for leaf purpose	Singh, 1997
Spinach	100:50:40 kg NPK/ha for leaf purpose	Singh, 1997

Method of Sowing and Transplanting

Two major methods of seed multiplication of vegetable crops are *in situ* and replanting method. The basic principle of each method is same for each crop with some adjustment. In cole crops, it is referred as seed to seed (*in situ*) and head to seed (replanted) method whereas in root crops it is called as is seed to seed (*in situ*) and root to seed (replanted) method. Generally, both the methods of seed multiplication are followed with varying production efficiency in theses crops. Seed to seed method is commonly followed for certified and foundation seed production at farmers' field using high quality breeder's seed. Whereas replanted method is used for nucleus seed production.

In crops like cauliflower, Brussel's sprout, sprouting broccoli and kale and collard, only seed to seed (*in situ*) method of multiplication is followed whereas in cabbage and knol khol both the methods are followed. Depending upon the plant part left for seed production in cabbage, *in* situ method can be classified into three; a) head intact method, b) stump method and c) modified stump methods. Head intact method is the most common method in commercial seed production of cabbage. At the place of original transplanting as seedling, plants are allowed to over-winter with the mature head intact. Head development is complete before the onset of winter and with the rise in the temperature during March-April, the developing flowering stalk exerts an internal pressure and the head starts bursting. To facilitate the flower stalk emergence, one cross cut with sharp knife is given to the head vertically as it takes dome shape appearance without injuring the central growing point. Earthing up is required at this time to support the weight of head and flowering stalk. This method is used with similar efficacy for late cauliflower (Snowball group) and broccoli. Since head to seed method is not successful in these crops because there is practically no rest period between vegetative and reproductive phase. Dormant period before the development of seed stalk makes the establishment after transplanting easy. Large quantity of seed for Snowball group is produced in Kullu valley (HP) where time of planting is adjusted in such a way that curd formation is complete before onset of frost or now. Transplanting of nursery is done during first fortnight of October. If transplanting is done before this, plants make necessary growth for curd development before onset of winter and these curd starts growing as they get optimum temperature (10-16°C) during November-December and switch to generative phase leading to damage by frost or snow. In delayed transplanting, plants do not make sufficient vegetative growth to support curd formation before onset of winter and therefore lead to poor yield (Singh *et al.*, 1960). El-Hafez and Said (1985) also reported that curd weight is closely related to leaf weight, as former is dependent on photosynthesis occurring in the leaves. Boswell (1929) found that the size of the plant in cabbage at exposure to low temperature is of considerable importance.

Sprouting broccoli (biennial type) is easier to grow for seed production as it is hardy and exhibit better resistance to frost than cauliflower as well as takes lesser duration for seed maturation. Brussels' sprout and kales are similar to cabbage in seed production method but transplanting method is not preferred in these crops. The thumb rule it that heads should be fully mature before the onset of conducive temperature for bolting because head size decides the seed yield. Verma *et al.* (1996) reported that larger heads resulted in higher seed yield and bold seeds in cabbage cv. Golden Acre grown *in situ*.

In knol-khol, seed production method resembles as in cabbage. In replanted method, all leaves around the knob are cut carefully leaving the expanded basal part of petiole (leaf scar) intact with knob to avoid injury to knob and invasion by microbes leading to rotting. Crown portion should not be damaged as this allows development of flowering stalk.

In stump method, fully grown heads are decapitated just below the base leaving the outer whorl of leaves intact with the stump. The stump thus left allows development of flowering stalk from the axillary buds. This method allows extra income from the sale of heads though require extra labour compared to the top head intact method. Flowering and seed maturity is advanced by 14 days and seed yield is more than the head intact method. But this is suitable for regions with less frost or snow fall because cut end of stumps are liable to rotting. In modified stump method, heads are chopped off on all sides in such a way that the central core is not damaged. The flowering shoots arise from the terminal and axillary buds. This method requires extra labour for cutting operation with recovery of unmarketable produce but higher seed recovery and early maturity than stump method.

In replanted (head to seed) method, true to type compact heads of cabbage or cauliflower or knol-khol are selected, uprooted and replanted in a separate field. Before replanting, outer leaves are removed and whole stump is buried in the soil allowing head to rest over the soil surface. Plants are firmly pressed to prevent tilting of plants after irrigation due to weight. Among the various factors, head size also affects the seed yield and heads weighing 1401 to 1700 g were found suitable for seed production in cabbage cv. Pride of India (Pathania and Negi, 1993). Similarly, *in situ* head intact method was found superior over others including transplanting method in cabbage *cv.* Golden Acre (Verma, 1998).

Verma and Sharma (2000) reported a modified method in areas like Kinnaur where land remains covered with snow foe long time due to heavy snow fall. In such areas planting is adjusted in such a way that the head development is complete before onset of winter (December). The true-to-type heads are selected, uprooted and stored in trenches for over-wintering after removing the outer whorl of leaves. Too early and too late planting of seedlings result in heads for poor storage capacity either due to splitting in the field due to over maturity or undersized grown heads. Trenches of 90 cm wide, 75 cm deep and 300 cm long are convenient for storage. Heads are stored in single layer in slanting position and roots are buried 5-7 cm deep in the soil. Trenches are covered with wooden planks and 15 cm soil layer over it leaving holes for ventilation. As the winter is over, heads are taken out and replanted in well prepared field during March-April. Flowering takes place during June-July and seeds mature during August-September. Similar method is followed in Ladakh traditionally to multiply the seed of various vegetable of cole group. Fully grown heads of cole crops are stored in underground cellar houses during winters, which protect the heads from frozen winters. These cellar houses are made up of stones, mud and wooden roof plastered with mud with provision of storing vegetables when surplus and also for safe establishment the roots of cole crops for over-wintering.

Similar to the cole crops, seed in root crops is also multiplied using *in situ* (seed to seed) and transplanting (root to seed) method. In both the methods, roots are allowed to attain full edible size during first season. In seed to seed method, the roots are allowed to remain and bolt at the same place where the seed was sown initially. As the roots reach to desirable maturity, the plants are thinned to have recommended spacing between the plants for better seed production. This practice makes the method more labour intensive and time consuming. Earthing up is also required to support the plants as roots are pushed over the soil surface while development. The seed yield is higher in this method compared to the root to seed method but the main disadvantage of this method is poor seed quality due to difficulty in rouging and selection for true-to-type root traits. The other disadvantage is that the land remains occupied for the full life cycle of crop and cannot be utilized for other crops. This method is mainly suitable for certified seed production, which is used by the farmers for raising commercial root crop. Even though highest quality seed should be used to multiply by this method. In root crops, if the sowing is delayed, there is less tendency of the crop developing typical root characteristics of the variety which make selection difficult whereas in early sowing, pithiness is the main problem. The drawback of replanting pithy roots is that when buried in soil, the hollow space inside the root is not suberized and enhances the chance if disease infection (root rot) by soil microbes.

Root to seed method is the back-bone of the quality seed production and is the preferred method for nucleus seed production. This method involves uprooting of fully matured roots and then selecting true-to-type roots by discarding pithy, hairy, cracked, fingered, discoloured and misshapen roots. The selected roots are trimmed without injuring the crown portion by giving 1/4 to 1/3 root tip cut and about 2/3 shoot tip cut in carrot whereas cutting 1/2 to 3/4 root portion and 2/3 of leaves in radish leaving 2-15 cm long roots in extra long radish varieties. Similarly in turnip, roots and shoots are trimmed adequately leaving 10-12 top leaves and trimming 1/3 of roots. In beet root, whole root planting is considered best because of the lower portion, from which secondary roots arise. Upper portion of hypocotyl throws no or very less secondary roots. Since secondary roots develop from tap root, larger the tap root is removed longer the time will be required for the root to establish which will further affect the growth and seed yield. Loomis (1925) reported that the plants that are safely and readily transplanted are the ones whose damaged and destroyed roots are speedily replaced by new growth, which in turn depends on the ability of the roots to absorb water freely.

In carrot, cut end is inspected for self colour and thin core generally called 'coreless' roots. Whereas in radish, turnip and beet traits examined are stumpyness, colour and other varietal traits. Incidence of winter injury is minimized by complete burial of roots in the soil. Some disadvantages encountered in this method are low seed yield, utilization of more labour while uprooting, preparation of stecklings and replanting. Root rot causing microorganism may enter the roots through cut ends if not treated with fungicide. But, this method ensures considerable reduction in planting area for other vegetables due to sorting of undesirable roots which otherwise occupy the area in seed to seed method. In heavy snowfall areas like Kinnaur and Ladakh, where roots cannot be left in open (as in *in situ* method) due to damage caused to the delicate growing point, this method is used for seed production.

In crops like spinach, swiss chard, lettuce, celery and packchoy, seed multiplication is done mainly by transplanting method so that plant to plant spacing can be maintained easily. In cold regions like Ladakh, nursery of these crops is grown in underground trenches covered with the polythene sheets during winter. For this, seed is required to sown during September-October to facilitate seed germination which otherwise is not possible if sown late in the winter season due to extreme cold condition temperature dipping to sub-zero. After germination plants grow very slowly under extreme low temperature. As the temperature increases during March-April, as per the altitude of the area, these plants are uprooted from the trenches and transplanted in well prepared fields in open fields at the recommended spacing.

Seed Rate and Sowing / Planting

Seed rate is important to ensure the proper crop stand and to avoid the need of resowing the nursery for transplanted crop. The quantity of seed to be required depends upon its viability, purity, planting geometry, soil condition and possible ravages of diseases and insects. If longer time is required for germination of any kind of seed, the seed rate should be higher to obtain a proper stand because it is directly related to damage to germinating seed. Time of sowing and planting decides the success of any crop to a considerable extent. The planting time should be determined according to soil and weather condition of the locality, duration of the crop and stage of harvesting. When crop is grown for seed production, ripening and harvesting stage should coincide with dry spell. Along with these, plant to plant spacing is also important to achieve a good crop stand with healthy and sturdy plants with maximum production potential. If they are too crowded, interplant competition increases leading to decreased yield and quality. And if plants are planted with too much spacing then plant population becomes lesser then optimum again leading to low yield per unit area. Recommended seed rate, spacing and time of planting/transplanting for different vegetable crops are given in the (Table 5).

Table 5: Seed rate, spacing, time of transplanting and isolation distance for some vegetable crops

Crop	Seed rate/ha	Spacing (cm × cm)	Time of transplanting*	Isolation distance (m)
Cabbage	600-750 g	45-60 × 45	Last wk Aug. to 1st wk Sep. (Early cv) Mid July to 1st wk Aug (Late cv.)	1600 (F) 1000 (C)
Cauli-flower	500-600 g (Snowball)	45 × 45	1st F/N of Oct. (Kullu & Saproon valley)	1600 (F) 1000 (C)
Knol-khol	400-500 g	45-60 × 45	1st F/N of Sep. (Kullu & Saproon valley)	1600 (F) 1000 (C)
Brussel's sprout	500-600 g	90 × 90 (bold var) 60 × 45 (small var)	Mid Aug. to Last wk of Aug.	1600 (F) 1000 (C)
Sprouting broccoli	600-750 g (Early var) 500-600 g (late var)	45 × 45 (early) 45-60 × 45 (late)	1st to 3rd wk of Oct.	1600 (F) 1000 (C)
Kale & collards	500-750 g (dwarf) 400-500 g (med) 350-400 g (tall)	60 × 45-60 (dwarf) 75 × 45-60 (medium) 90 × 60 (tall)	Aug.-Sep.	1600 (F) 1000 (C)
Carrot	5-6 kg	45 × 30-45	July-Aug. (sowing) March (replanting)	1000 (F) 800 (C)
Radish	10-12 kg	45 × 45	1st F/N Aug. to 1st wk Sep. (seed to seed) Mid Nov. to Last wk Nov (root to seed)	1600 (F) 1000 (C)
Turnip	2 kg	45 × 45	Last wk Aug. to 1st wk Sep. (sowing) Mid Nov. to 1st wk Dec. (replanting)	1600 (F) 1000 (C)
Beet root	5-6 kg (*in situ*) 1.5-2.0 kg (root to seed)	45-60 × 45	Mid June (sowing in high snowfall area) 2nd F/N July (Kullu & Srinagr) Nov. (replanting)	3000 (F) 1000 (C)
Celery	125 g	70-75 × 20	Aug. to Sep. (Plains) Dec.-Feb. (Hills)	50
Lettuce	400-500 g	45 × 30	Aug. to Sep. (Plains) Dec.-Feb. (Hills)	50
Spinach	25-30 kg	20-22×5-7	Aug. to Sep. (Plains) Dec.-Feb. (Hills)	1600 (F) 1000 (C)

Isolation

Isolation means separation of the seed crop from various sources of contamination to maintain its genetic as well as physical purity. Genetic contamination depends upon the breeding behaviour of the crop (cross or often cross pollinated) and results from natural crossing with other varieties or kinds of the same crop or species growing in the adjacent field and off-types present with in the seed crop. Mechanical contamination can occur while sowing, harvesting, threshing, processing and handling of seed in any crop. Self-pollinated crops are prone to physical/mechanical contamination. In self-pollinated crops, undesirable off-types and not true-to-types plants, when unnoticed, grow and multiply its population in subsequent generations. Since some degree of cross pollination always occur, it is therefore advisable to discard border rows of seed plot. Problem of contamination is acute in cross pollinated crops and therefore isolation distance has to be maintained from all possible contaminants to maintain the genetic purity of seed crop. Isolation distances for various crops are presented in (Table 5).

Maturity and Harvesting

Among the various factors which decides the longevity of seeds, proper maturity in terms of both stage and standards is of utmost importance. Maturity standards used to vary with the species of plant. Austin (1972) studied the effect of environment before harvesting on the seed viability. The ripening process is interrupted if the seeds are harvested too early and seed quality is affected adversely. Whereas in vegetable like *Brassica* spp. and lettuce delay in harvesting can lead to seed loss due to shattering.

Generally, later the harvest date, higher the seed yield, but losses increase as the harvesting is delayed once the optimum percentage of seeds reach maturity. Thus, the ideal time for harvesting is immediately before the threshold level where the loss of mature seeds exceeds the amount of seeds yet to reach maturity. Seed crops which are prone to shattering should preferably be harvested at the time of comparatively high relative humidity. In arid areas this can be achieved early in the day. To prevent shattering of seed, use of poly vinyl acetate (PVA) has been found beneficial (Williams, 1977). This compound when sprayed on seed crop acts as an adhesive after drying and prevent shattering when subsequent seeds develop.

At harvesting stage, maturity of the seed is sufficiently advanced and generally not affected by subsequent drying. Biddle and King (1977) reported that pea seed quality is reduced if seeds are harvested when their moisture content is above 30-35%. Globerson *et al.* (1981) studied the nature of flowering and seed maturation in relation to mechanical harvesting in onion and observed that the best time for mechanical harvesting was when the seeds have dry matter content of 60-70%. Seeds which were dried in capsules still attached to the stalks germinated better than those dried in their capsules after separation from the umbels. It is generally accepted that seed should be allowed to develop fully in the fruits before extraction particularly in fleshy fruits. Maturity standards for seed crop of some vegetables are presented in (Table 6).

Table 6: Maturity standards for seed crop of some vegetables

Crop	Maturity stage	Remark	Reference
Cole group	Plants starts drying out and become orange-brown on ripening. Oldest pod become brown first	Considerable loss from shattering and birds. Early morning is best for harvesting	Nieuwhof, 1969
Carrot	Second and third order umbel turn brown	Shattering occurs on delayed harvesting. Harvest primary umbel if rain is possible	Gray, 1979
Radish	Pods turn brown and parchment like when mature	No shattering problem so harvested in day time	Watts and George, 1957
Turnip	Plants haulms turn from green brown to parchment colour	Prone to shattering so harvested early morning	McNaughton, 1976
Beet root	Fruits (seed ball) at inflorescence base turn brown Cut the seed ball transversely if milky then unripe, mealy is ripe	Seed takes 1-2 week to dry Uneven ripening. Harvest in phases	Chroboczek, 1934
Onion	Seeds become black in silvery capsules	Prone to shattering Harvested during morning hours	Jones and Mann, 1963
Spinach	Plants start turning yellow at seed ripening	Shattering occurs on delayed harvesting	Parlevliet, 1968
Lettuce	Seed heads show bursting and presence of the pappus (hair like structure) on seed	Uneven ripening and highly prone to shattering. Cut at 50% seed head maturity	Soffer and Smith, 1974

Post Harvest Handling and Storage

Curing is an important operation in maintaining the seed quality and storability. Harvested seed lots are kept on threshing floor or on tarpaulin for 3-4 days then whole lot is turned upside down and allowed to cure for 2 days. This helps in uniform curing and maturity of seed. Curing is followed by threshing. Threshing is done by rubbing, flailing, beating or rolling the dried plant material. It is important to avoid unnecessary fragmentation of plant material which otherwise produce debris, that is difficult and costly, to separate from seed lot during processing. Any damage to seed should also be avoided while threshing. Threshing can be done with the motorized thresher. Speed of 1,100 rpm is generally optimum for most small seeded vegetable crops whereas 700 rpm is used for large seeded legumes. After this seed is cleaned and graded either manually or by cleaning and grading machines.

Before storing the seed, important step is drying of seed to the safe moisture content. By the time seed is separated from mother plant the seed moisture content is below 50 per cent and there after the moisture content is in equilibrium with the storage atmosphere. Sun drying is the most common method of seed drying. Seeds are spread on the threshing floor or tarpaulin for a few days in the sun till the safe moisture level reached in the seed. Rate at which a seed lot can be dried artificially depends upon the packing need and the initial moisture content of the seed. Lettuce is considered to be quick drier; carrot, beet and tomato are medium drier while brassicas, legumes and onion are slow driers. Generally at 43.5°C, about 0.3% seed moisture content is removed per hour by an air flow rate of 4 m^2 per cubic meter of seed. For open storage, moisture content of starchy seeds is reduced to 12% and in oily seeds to 9%. The moisture content for seeds to be stored in sealed containers is less than for unsealed storage of specific vegetables. Agarwal (1980) suggested that brinjal seed should be dried in partial shade to a moisture content of 8% or below.

In commerce, seeds in store represent a significant proportion of companies or farmers material asset and a vital link between the successive crop generations. Therefore, it is necessary to keep the seed in best possible conditions to ensure the maximum vigour, germination and other quality factors. Seed moisture content has direct bearing on the seed longevity but temperature and relative humidity plays an important role in seed storage. At 1 per cent decrease in moisture content the storage life of seed is doubled. High seed moisture content coupled with high temperature increases the biological activities in the seed and produces more heat and damages the seed. Therefore, dry and cool conditions in storage are essential. Seeds remain viable for long at 7-8% seed moisture content, 4-5°C temperature and 50% RH.

Seed Yield

Seed yield depends mainly on the variety, environmental condition and agronomical practices but effect of size of vegetative parts is also prominent for example head/curd size in cole crops, root size in rot crops, bulb size in onion etc. Therefore, to achieve the maximum yield, recommended cultural practices should be followed. Experiments carried out at Field Research Laboratory (FRL), Leh revealed seed production potential of capsicum, carrot and carrot var. Nantes (Table 7, Annon, 2002). According to a report in cabbage heads replanting was done during April and seed was harvested during August. About 600-650 silique/plant with 18-20 seeds/pod were harvested. Test weight of seeds was 3.875 g and yield per plant was 50.26 g giving a yield of 13.96 q/ha. Similarly in carrot, replanting was done in April and seed was harvested in August. About 45-50 umbels/plant were harvested with seed recovery of 75-80 g/plant. Test weight of seeds was 2.390 g and yield was 9.67 q/ha (Annon, 2003).

Table 7 : Seed production potential of capsicum at FRL, Leh

Variety	Seed yield (g/kg fruit)	Seed yield (g/m^2)	Seed yield (kg/ha)
CW	6.75	15.74	125.9
FRL Sel- 1	3.15	14.35	114.8

Source : Annon, 2002

According to Singh and Singh (2003b) seed yield of some root vegetables at experimental farm at FRL, Leh is as under

Crop	Variety	Seed yield (q/ha)
Carrot	Nantes	19.6
	Yamdagni	17.3
Radish	Local	18.2
	Pusa Himani	14.3
Turnip	PTWG	7.4
Beet root	Detroit Dark Red	24.2
Swede	FRL-Sel	5.7

Seed Production in Leafy Vegetable

Some important leafy vegetables which grow successfully in Ladakh conditions are spinach, swiss chard, lettuce, celery and karamsag. Palak this is beet root is more prolific seed producer than spinach (*Spinacea oleracea*). It yields more seed since it is bisexual and all plants are seed bearing compared to spinach where 50 per cent are male. Palak var. All Green produces seed stalk in 75 days and seed can be harvested in about 180 days. Since there is no problem of seed shattering, farmer can wait till the majority of fruit balls. Average seed yield is 5-6 q/ha. Spinach is comparatively early in bolting and seed can be harvested in about 150-160 days. Average seed yield is 5 q/ha. Seeds remain viable for 3-4 years under proper storage condition.

Lettuce is the most popular of the salad crops grown in India. It thrives best in a relatively cool climate with monthly mean temperature of 12.7° to 18.3°C and bolting occurs at 21.1° to 26.6°C. in hot and dry weather, leaves become tough and develop bitter taste. Based on seed production, lettuce is of two types: a) early flowering and b) late flowering type. Early flowering types are reported to have good seed yield. Lettuce is self pollinated but a small amount of out-crossing has been reported. Loose leaf types are transplanted at recommended spacing and they bolt easily. In case of headed type lettuce, it is desirable to give cross incision to facilitate the emergence of seed stalk. Rouging is done before onset of flowering in loose leaf types whereas in head types at completion of heading. Direct bolters, ill formed and off types are rogued out. The maturity of seed is irregular in lettuce. Because of unevenness in the ripening of seed even on the same shoot, it is rather difficult to assess the full maturity stage of plant. Therefore, harvesting should be started when substantial number of heads show bursting and the pappus (white hairs on the seed) and mostly have turned brown if not showing pappus. Whole plant or matured seed branches can be cut at this stage and staked for few days for drying up and later shaken for seed collection. In case of heading varieties like Great Lakes, it is desirable to collect the mature seed periodically by shaking the tops in a vessel because they do not flower so profusely.

Celery is the second most important salad crop but not commercially popular as vegetable crop in India. Leafy stalk or petioles are eaten as raw salads or used in soup and stews. In India, celery cultivation is limited to high hills region having relatively cool growing season. The seeds are small and about 2,470 seeds constitute one gram. Germination is slow and can extend to one month. Blanching is done in celery by earthing up the soil around the leaf stalk while growing or by wrapping the paper to make them devoid of chlorophyll to increase the market acceptability. But blanching is discouraged as the nutritive value is reduced by this process (Yawalkar and Ram, 2004). There are self blanched type varieties available now. Salad is ready for harvesting within 75-90 days with average yield of 40/ha.

Seed yield of some leafy vegetables at FRL, Leh experimental research farm has been 12.6 q/ha (Spinach), 5.6 q/ha (Parsley) and 6.8 q/ha (Celery) as reported by Singh and Singh (2003b).

True Potato Seed

Food security is an important national objective today and potato (*Solanum tuberosum* L.) produces more food per unit area and time than all other major cereal crops, here lies its importance. Not only the yield is more compared to wheat, rice or maize but production of dry matter and protein/kg/ha/day is also about 4.0 and 2.4 times more than these crops. India's share in world potato production is about 6%. Total potato production in India is 25 million tonnes from an area of 1.40 million ha with an average productivity of 18.0 t/ha (Pandey *et al.*, 2005). The productivity in India is quite low compared to the Netherlands (45.0 t/ha), the UK (41.3 t/ha) and Germany (31.6 t/ha) where it is grown as long duration crop (180 days) with long photoperiod (12-16 hrs) compared to India (90 days with 10.3 hrs photoperiod).

Traditionally, potato is grown through seed tubers. Several years of research and surveys indicate that potato seed alone accounts for 40-60% of total cost of cultivation in many countries if grown through tubers (Strunik and Wiersema, 1999). Therefore, ensuring genuine quality and quantity of disease-free potato seed tuber (required 25-30 t/ha) is major constraint and is of utmost importance in increasing the potato productivity. The high cost of seed tubers, problem of storage and transport and virus infiltration in tubers leading to degeneration of seed stock have been the main reason to utilize True Potato Seed (TPS) technology for crop production.

The concept of True Potato Seed was first realized in early 50's to raise the commercial crop in India by Dr. S. Ramanujam, the first Director of the Central Potato Research Institute at Patna (Bihar). The work on TPS has been conducted jointly by CPRI, the International Potato Centre and All India Coordinated Potato Improvement Project. This has led to the identification of few promising TPS families, *viz.*, TPS C3 and TPS C17 from Patna and HPS 1/13, HPS 2/67 and HPS

7/67 from Modipuram. The details on the TPS technology have been documented by Grewal (1990) and Gaur (1991).

The basic requirement to induce flowering in potato is high light intensity, moderate temperature with long day condition (Moorby, 1978). These conditions are naturally available in high hills during spring-summer season whereas in plains these conditions can be achieved by artificial illumination for 4-5 hrs after sunset during winter months (Dec-Jan) through 150 watt sodium lamp. One bulb for 100 m^2 surface area is sufficient to provide desired illumination (Ram, 1997). This technology has been found successful and economical in Ladakh also (Singh, 1995) and through several years of research, agro-techniques has been standardized for TPS production (Singh and Singh, 2003b). Along with the natural availability of temperature and climatic requirement, lack of rains and disease makes Ladakh region ideal for economical TPS production during summers.

Since producing TPS and growing commercial crop using TPS are two different practices, emphasis here is given on production of TPS. All types of soils suitable for potato production are also suitable for TPS. After ploughing and harrowing, soil is made friable and field is levelled. FYM is added @ 20 t/ha. Since Hybrid TPS have shown promise over open pollinated TPS, separate but adjacent blocks are made for parental lines. The ratio of female to male plants is kept between 4:1 to 6:1 depending upon the pollen production ability of male plants. Planting in male block is done about a week earlier than female. Singh and Singh (2003b) reveled that best time for potato tuber planting for TPS production in Ladakh is during March–April. Planting is done at 60 cm × 20 cm and earthing up is done. Female block ideally should have beds with three rows to facilitate crossing, to be done at later stages. Walking space of 80 cm is kept between the two adjacent beds for ease in agricultural operations. But for open pollinated TPS seed beds of convenient size can be made with proper arrangement for irrigation and drainage. Tubers of about 25-30 g are planted at a spacing of 15 cm in the rows. Other field operation and management practices are done as required for general crop. Stacking is required in female block to prevent logging of plants and berries touching the soil to minimize the pathogen attack on berries. Regular inspection and rouging is required to maintain the purity.

The important step is hybridization for making hybrid TPS. According to Singh and Singh (2003b), flowering in potato in Leh

(Ladakh) commences during first week of July. As the flower starts opening in female block, 5-6 healthy unopened buds are selected on each branch and rest are removed. No more flowers are allowed to appear on selected branches and removed as soon as they appear at any stage of plant development. Pollens are collected one day prior to hybridization from male bloc. For this purpose freshly opened flowers with pollen sacs and fully swollen buds likely to open next day are collected and kept overnight at room temperature to allow bursting of pollen sacs. Pollen grains are collected next morning either manually or mechanically in a small vial. Small painting brush is ideal for transferring pollen on the stigma of already selected flower buds. Surplus pollen grains can be preserved in a cool place (6-7°C) in a desicator having calcium chloride so that they can be used next day for hybridization. After hybridization, fertilized flower clusters should be covered with muslin cloth bag (8 × 12 cm) carefully.

The berries are ready for harvesting after about 45-50 days of fertilization. After harvesting, berries are kept at room temperature for 2-3 weeks till they reach to full ripe stage. These ripe berries are crushed manually or mechanically and allowed to ferment for 24-48 hrs. seeds are separated from pulp using 3.13 mm mesh sieve. Seeds are cleaned by stirring in to 10% HCl for 20 min and then thoroughly rinsing with tap water to wash out HCl. Seed is dried in partial shade for 3-4 days. Fully dried seed can be stored in moisture proof containers or polybags in a cool place. Freshly harvested seed has 5-6 months dormancy, which can be broken by dipping in GA_3 (2000 ppm) for 48 hrs (Ram, 1997). About 30-35 fruits/plant having 200-300 seeds per fruit can be harvested. One gram TPS may contain about 1,500-1,700 seeds. Experiments carried out at FRL, Leh revealed that TPS yield may vary 45-60 kg/ha (Singh and Singh, 2003b). According to seeds produced in Ladakh are not only healthy but also bold with more vigour. Leh valley is ideal for production of disease-free true potato seed.

Extensive research carried out in different countries showed that use of TPS for the production of seed or ware potatoes not only reduced the production cost, but also increased the net profit of the farmers (Upadhya, 1994). However, TPS technology is limited by high requirement for good seed germination, low vigour of seedling, slow initial plant development, small size of tuber produced, genetic variation within the crop and high labour requirements. These are the reasons as to why TPS has not been so successful in developing countries despite its promotion by the International Potato Centre (Pandey *et al.*, 2005).

This technology has gained momentum in Karnatka, Maharastra, Madhya Pradesh, Orissa and North-eastern hill region where good quality tubers are either not available or cannot be produced. But for wider application of this technology, research efforts are needed in the following areas:

- Development of superior parental lines with profuse flowering and ability to set seed both in hills and plains
- Development of highly fertile parents with abundant pollen
- Male sterile lines to be used as female parents
- More sturdy TPS population to bear transplanting shock with vigourous seedling
- TPS with high tuber yield potential and relatively homogeneous population (Chadha and Grewal, 1993).

Seed Production Onion

Onion (*Allium cepa* L.) is one of the most important bulb crop of India and is cultivated under an area of 0.40 million ha with the estimated production of 4.30 million tonnes. Its use in the culinary as well as in medicinal purposes has been documented since ancient times. Supply of fresh onion to the Ladakh sector of J&K is of paramount importance because onion has been the part of balanced daily diet. Dieticians have recommended about 80 g of fresh onion per day to a solider in the high altitude area like Ladakh. Supply of fresh onion is mostly maintained through air transport for most period of the year leading to heavy expenditure in the tune of about Rs 27 per kg as air-fright charges and Rs 5-6 per kg for road transport to Ladakh. This leads to the poor availability of onion bulbs to be used for the seed production and therefore leading to poor availability of seed. Considering the fact that seed to seed production method of onion reduces the production cost, saves the onion bulbs for consumption and reduces the expenditure on harvesting and replanting of onion bulbs which otherwise are involved in bulb to seed production method, has tried the possibilities of producing onion seed in Ladakh sector using seed to seed method and its standardization. Success of the experimental results has ensured the vast potential of producing onion seed in this sector to meet the local as well as the national requirement which is about 4,000 tonnes whereas the availability is quite low, *i.e.*, about 600 tonnes (Annon, 2005). The success of onion seed crop in this sector is also favoured by the virtual absence of purple blotch and stem-phyllium disease, which otherwise is a devastating disease to seed crop of onion in the other areas of the country (Tomar *et al.*, 2005 and Singh, 1995). The agronomic practices for successful seed production are as under.

All types of soil are suitable for onion seed production but fertile soils devoid of water logging are the most suitable because onion requires sufficient root aeration for its growth. Generally onion is planted during mid October in most parts of North India but in Ladakh, best time for transplanting is last week of March to April. For seed production using bulb to seed method, March is best for planting whereas for seed to seed method transplanting is done during October November. This allows establishment of seedlings before onset of winter. At the onset of summer, these established seedlings pick up growth quickly during March-April. Trench cultivation has been found more effective for seed production because the polythene cover over the trench provides protection to the dormant seedlings during the severe winter as well as regulates the temperature and moisture inside the trench. If seed production is done in open field, seedlings are required to be covered with the polythene low tunnel to protect them against frost and snow. Other cultural practices are same as done for bulb crop. In bulb to seed method, recommended spacing is 30 cm × 30 cm to 45 cm × 30 cm whereas in seed to seed method spacing is kept at 20 cm × 10 cm to 20 cm × 20 cm. Earthing up is required as the bolting starts to avoid the falling down of seed stalks due to winds at the time of seed development and maturation. Bolting starts in last week of May and seeds can be harvested during last week of July to August.

Experiments conducted at FRL, Leh revealed that bulb yield was higher at closer spacing and Brown Spanish was the most suitable cultivar with low percentage of twin bulbs as presented in (Table 8) (Anon., 2003).

Table 8: Performance of various onion cultivars at two spacing at FRL

Cultivar	Bulb yield (q/ha)	
	20 cm × 10 cm	20 cm × 20 cm
Brown Spanish	422	339
White Creole	240	158
Agrifound Dark red	311	290

According to Singh (1995) productivity of onion seed crop is low in Ladakh mainly due to the low population of insect pollinators which are required for cross pollination in onion. At experimental plots of FRL, seed yield of onion cv. FRL Sel was 2.14 and 2.50 q/ha during 1992 and 1993, respectively. Singh and Singh (2003a) reported that seed yield of onion through bulb to seed method varied from 6.2 to 9.4 q/ha at experimental field of FRL, Leh as presented in (Table 9).

Table 9 : Seed yield of various onion cultivars at FRL

Cultivar	Seed yield (q/ha)
Local White	6.2
Agrifound Dark Red	8.6
Agrifound Light Red	9.4
Brown Spanish	6.8
White Creole	6.6

In another experiment carried out at FRL, Leh, Singh and Singh (2003b) reported the significant effect of bulb size and spacing on seed yield of onion cv. Sindhu Sweta. The study revealed that 3.0 cm bulb size and 30×20 cm spacing combination gave the highest seed yield in Leh condition. The detailed results are presented in (Table 10).

Table 10 : Effect of bulb diameter and spacing on seed yield of onion

Treatment	No. of sprout per hill	Seeds/umbel	Seed yield (q/ha)
Bulb diameter			
1.5 cm	3.55	335.65	7.24
3.0 cm	4.28	585.50	10.50
4.5 cm	5.28	638.41	8.17
Spacing (cm)			
30×10	3.27	273.30	6.03
30×15	4.34	534.9	8.69
30×20	5.53	751.3	11.20

Diseases

Seed production pockets should be free from pest and diseases because cyclic production programme allow build up of pest and disease pressure to economic threshold level and may pose a grave challenge to the seed industry. According to Verma and Sharma (2000) some diseases those have caused great loss to seed growers are black rot and downy mildew of cole crops and sclerotinia rot of cauliflower in Saproon valley, Kullu valley, Kashmir valley and Uttaranchal Hills. Generally, Ladakh has naturally disease-free environment and therefore to avoid ecological pollution chemical control measures are not used. Some important diseases and insect-pest along with their control measures are given here for reference.

Early and Late Blight: These are the major diseases of potato. Control measures for both the diseases are almost same. Rouge out and destroy the diseased plants as they appear in the field. Spray of Mancozeb or Zineb (2.5 kg/ha) or Copper oxychloride (3.0 kg/ha) must be done in 800-1,000 l water/ha as the first symptom appears. 4-5 sprays must be done at an interval of 10-15 days. Ridomil MZ-72 (2.5 kg/ha) is also effective in controlling the late blight.

Leaf Curl: This is transmitted by insect vector-white fly, which suck the sap of the plant. Therefore, rough out and destroy the diseased plants and to reduce the incidence transplanting must be done after mid October in North India. Spray of Metasystox or Rogor (1.0 ml/l water) at 10-15 days interval help in reducing the infection. If the problem was seen in the previous years also then apply furadan granule (5 g/m^2) in the nursery as well as main field (35 kg/ha). And spray the Metasystox or Rogor right from nursery stage as prophylactic measure. Three sprays of growth hormone (2,4-D @ 5 mg/l water) after 30, 45 and 60 days of transplanting has been found beneficial in promoting the growth and yield if crop is infected.

Black Scurf: It is caused by the fungus and leads to development of brown-black layer on the tubers which do not separate up on washing of tuber. Treat the tuber with mercurial fungicide like Aretan-6 or Agalol or Tefsan (2.5 g/l water) before planting. Dip the tubers in the solution for 15-20 minutes. Dipping in Boric Acid solution (30 g/l water) is also found effective.

Alternaria Leaf Blight: This disease affects cole crops, radish and turnip. Seed treatment with Aretan-6 or Agalol or Tefsan fungicides (2.5 g/kg seed) for 10-15 minutes is effective in reducing the early occurrence of disease. Spray of Mancozeb (2.0 g/l) or Carbendazim (1.0 g/l) or Ridomil (2.5 g/l) can be done when disease appears in the field.

Black Spot: This disease affects most cole crops. Seed crop is affected more by this disease and to reduce the spread of disease, uprooting and destruction of diseased plant is necessary. Spray of Mancozeb or Zineb (2.5 kg in 1,000 l water/ha) can be done as the symptoms appear in the field.

Black Leg: This disease is transmitted through seeds, therefore, treatment of seed with warm water (50°C) for 30 minutes and then treatment with fungicide mixture of Thiram: Carbendazim (2:1) @ 3.0 g/kg seed is effective in reducing the incidence. Along with this use disease-free certified seed and crop rotation must be incorporated to reduce the further incidence.

Black Rot: All the measures as said for black leg must be followed and spray of Streptocycline (50 g in 1,000 l water/ha) is effective in reducing the disease.

Purple Blotch: The most devastating disease which is caused by the fungus *Alternaria porri*. Disease is characterized by the appearance of purple coloured spots on the infected leaves which later turn to black-brown colour. This black colour powder can be seen on the finger upon touching. The fungus survives on the plant debris in the soil. Early symptoms can be seen in the last week of January which get severe as the temperature rises. Onion seed crop is more prone to disease and in severe cases no seed set is reported to occur. To control the disease it is necessary to purchase the disease-free certified and treated seed from authentic source. Prophylactic spray of the mixture of insecticide (Metasystox or Rogor) @ 1.0 l/ha water and fungicide (Mancozeb or zineb) @ 2.5 kg/ha must be done as soon as attack of thrips is seen in the crop. Addition of sticker (sandovit or titron or washing soap) @ 1% in the solution is important for effective control. About 2-3 sprays for bulb crop and 5-6 sprays for seed crop are needed.

Insect and Pest

Aphids and Thrips: These are the small insects which suck the sap from the tender parts of the plants. Thrips attack the onion and garlic while aphids can be seen mostly as white mass at the terminal portion of the plants. These lead to stunting of plants and crinkling of leaves. To control the pest, spray of Metasystox or Rogor (1.0 ml/l water) must be done.

Cut Worm: Attack of this pest is also common during the nursery stage and early stages after transplanting as well as in potato. This pest is active during the night hours and cuts the emerging tender stems near the collar region leading to great loss. To control the pest, make the small heaps of leaves in the field to allow the pest to hide beneath it. During the morning hours pest can be trapped and destroyed. Application of Thimet granules (1.5 kg/ha) in soil is beneficial.

White Cabbage Butterfly : The caterpillar of this insect is yellow coloured with black spots. They feed on the leaves of cabbage and cauliflower. Infestation at early stage lead to poor development of head and at later stage head size used to get reduce. Female used to lay yellow colour eggs on lower surface of leaves in group which after hatching do gregarious feeding. The best method to check the spread of infestation is regular inspection and destruction of egg mass and hatched larvae by plucking the leaves. To control this insect spray Endosulfan 0.07% (2 ml/l) on the crop.

Cabbage Leaf Webber : This insect attacks cole crops, lahi and radish etc. and make great loss to crop. Green caterpillars feed on leaves and make holes in the leaves. Caterpillars feed on the tender growing point in cabbage and cauliflower and stick to lower surface of leaves. For effective management of the insect, lahi should be grown as trap crop in cabbage field and spray of Endosulfan (2 ml/l water) on lahi. Two lines of lahi after every 15 lines of cabbage are sufficient. One line should be sown 15 days prior to transplanting of cabbage so that initial attack of the insect is on lahi. Second line of lahi should be sown about 25 days after transplanting of cabbage so that lahi leaves are available for full life cycle of the insect and attack on cabbage crop can be minimized. As soon as the larvae is seen in the field, cabbage crop should be sprayed with neem based insecticide such as nimbicidine, Neemigon or Achuk @ 5 ml/l water. Surfactant like Sandovit or Teepol or soap solution (0.5 ml/l sol) should be added to insecticide for increasing the efficacy of insecticide.

Other Pest : Other pest which cause damage are cabbage semilooper, leaf webber, diamond back moth (DBM) and bihar hairy caterpillar in cole crops while saw fly in radish. The caterpillar of the butterfly are very voracious and cause much damage to cole crops. They start feeding from the margin of the leaves and proceed towards the centre. The caterpillar of the DBM at first tunnel inside the leaf tissue and subsequently attack the lower surface of the leaf and can be seen on upper surface in severe cases. Saw fly is small dark green or black colour inset which attack the crop at 2-3 leaf stage and may cause severe damage. To control these insects apply carbofuran in the nursery @ 4-5 g/m^2 soil and spray endosulfan (1.0 ml/l) or carbofuran (2.0 g/l) after germination. To control saw fly, apply malathion dust @ 15-20 kg/ha. Biocontrol agents like Thuricide, a *Bacillus thuringiensis* formulation, has been found effective against most of the Lepidopteran pests in the larval stage particularly when applied at the hatching time and/or first to third larval instar during intensive feeding stage. Since BT must go inside the stomach of the insects for killing, its spray must be thorough on the leaves. Use of egg parasitoids like *Trichogramma* spp. against these pests is also effective and for this Trichcards are available carrying the hatchable eggs of this control agents. These cards can be punched in the field to control the population of the target pest.

References

Agarwal, R.L. (1980). Seed Technology, Oxford and IBH Publishing Co., New Delhi. pp. 198-201.

Anonymous (2002). Annual Report, Field Research Laboratory (DRDO), Ministry of Defence. Field Research Laboratory (DRDO), C/o 56 APO. 23p.

Anonymous (2003). Annual Report, Field Research Laboratory (DRDO), Ministry of Defence. Field Research Laboratory (DRDO), C/o 56 APO. 39p.

Anonymous (2005). Onion seed production. Krishi Chayanika. 48.

Austin, R.B. (1972). Effect of environment before harvesting on viability. In: Viability of Seeds, Roberts, E.H. Chapman and Hall, London, pp. 114-149.

Bhagchandani, P.M.; Verma T.S. and Singh, N. (1982). Effect of boron on vegetable and seed yield of cabbage. Hort. J. 1 (1/2): 32-36.

Biddle, A.J. and King, J.M. (1977). Effect of harvesting on pea seed quality. Acta Hort. 83: 77-81.

Boswell, V.R. (1929). Studies of premature flower formation in wintered over cabbage. Bulletin Maryland Agricultural Experiment Station 313: 69-145.

Chadha, K.L. and Grewal, J.S. (1993). Potato Research in India - History, Infrastructure and Achievement. In: Chadha, K.L. and Grewal, J.S. *Advances In Horticulture Vol-7 Potato*, Malhotra Publishing House, New Delhi. pp. 1-20.

Chroboczek, E. (1934). A study of some ecological factors influencing seed stalk development in beet (*Beta vulgaris* L.). Cornell Univ. Agri. Expt. Sta. Mem. 154: 1-84.

El-Hafez, A.A. and Said, M.S. (1985). Leaf thinning effect and correlation study between curd weight and certain other characters in cauliflower. Acta Agronomica Academiae Scientiarum Hungaricae 34 (1/2): 147-153.

Gaur, P.C. (1991). Manual for TPS Production and Utilization. Extension Bulletin No. 30 (E). CPRI, Shimla. 11 p.

Gill, H.S.; Thakur, P.C. and Bhullar, B.S. (1975). Effect of nitrogen and phosphorus application on seed yield of late cauliflower. Indian J. Horticulture 32 (1/2): 94-97.

Globerson, D.; Sharir, A. and Eliasi, R. (1981). The nature of flowering and seed maturation of onions as a basis for mechanical harvesting of the seeds. Acta Hort. 111: 99-114.

Gray, D. (1979). The germination response to temperature of carrot seeds from different umbels and time of harvest of the seed crop. Seed Sci. Technol. 7: 169-178.

Grewal, J.S. (1990). True Potato Seed for Potato Production in India. Central Potato Research Institute, Shimla, India. 13 p.

Jones, H.A. and Mann, L.K. (1963). Onion and Their Allies. Leonard Hill, London. Inter Science Publishers, New York.

Loomis, W.E. (1925). Studies in transplanting of vegetable plants. Memoir Cornell Agriculture Experimental Station 87: 63.

McNaughton, I.H. (1976). Turnip and Relatives. In: Evolution of Crop Plants Simmonds, N.W., (ed.) Longman, London and New York.

Mir, M.S. (2000). Potential and problems of fruit crop production in Ladakh. In: Dynamics of Cold Arid Agriculture. Sharma, J.P. and Mir, A.A., Kalyani Publishers, New Delhi. pp 187-213.

Moorby, J. (1978). The Potato Crop. In: Harris, P.M., ed. Chapman and Hall, London. pp. 153-187.

Nieuwhof, M. (1969). Cole Crops. Leonard Hill, London.

Pandey, S.K.; Singh, S.V. and Sarkar, D. (2005). Potato (*Solanum tuberosum* L.) for sustaining food and nutritional security in developing world. Indian J. Agri. Sci. 75: 3-18.

Parlevliet, J.E. (1968). Breeding for earliness in spinach (*Spinacea oleracea*) as based on environmental and genetic factors. Euphytica 17:21-27.

Pathania, N.K. and Negi, C.P. (1993). Effect of head weight on seed yield in cabbage. Vegetable Sci. 20: 38-39.

Ram, Hari Har (1997). Vegetable Breeding: Principles and Practices. Kalyani Publishers, New Delhi. pp. 378-396.

Roy, R.N. and Seth, J. (1971). Nutrient uptake and quality of radish as influenced by levels of nitrogen, phosphorus and potassium and methods of their application. Indian J. Hort. 28 (2): 144-149.

Saini, S.S.; Cheema, S.S.; Randhawa, G.S. and Mahey, R.K. (1989) Indian J. Agron. 34: 341-5.

Sarkar, N.C. and Prince, S.J. (2006). Career in Seed Industry. Employment News, Vol XXXI (8): 1.

Singh, Brahma (1995). Vegetable Production in Ladakh. Technical Bulletin No. 4. FRL (DRDO) Ministry of Defence, Leh C/O 56 APO. 67p.

Singh, D.K. (1997). Important tips for cultivation of leafy vegetables in winter. Indian Farmers' Digest XXX (9): 17-22.

Singh, H.B.; Thakur, H.R. and Bhagchandani, P.M. (1959a). Vegetable seed production in Kullu valley-I. Indian J. Hort. 16: 92-101.

Singh, H.B.; Thakur, H.R. and Bhagchandani, P.M. (1959b). Vegetable seed production in Kullu valley-I. Indian J. Hort. 16: 153-161.

Singh, H.B.; Thakur, M.R. and Bhagchandani, P.M. (1960). Vegetable seed production in Kullu valley -III. Indian J. Hort. 17 (1): 38-47.

Singh, N. and Singh, Brahma (2003a), Effect of bulb spacing in onion (*A. cepa*) seed production in Leh (Ladakh). In: Approaches for Sustainable Development of Onion-Garlic, Singh, K.; Lawande, K.E.; Pandey, U.B.; Singh, L. and Bhonde, S.R. eds., NHRDF, Nasik. pp. 285-288.

Singh, N. and Singh, B. (2003b). *Ladakh Mein Sabji Utpadan* (Hindi). Field Research Laboratory (DRDO) Ministry of Defence, Leh C/O 56 APO. 139p.

Soffer, H. and Smith, O.E. (1974). Studies on lettuce nutritional quality V. Nutritional effects. J. Amer. Soc. Hort. Sci. 99: 459-463.

Struik, P.C. and Wiersema, S.G. (1999). Seed Potato Technology. Wageningen Press, Wageningen, The Netherlands, 383 p.

Thakur, P.C. and Gill, H.S. (1978). Effect of different levels of nitrogen, phosphorus and potash on seed yield of cabbage. Indian J. Hort. (33) 40: 262-265.

Thompson, H.C. and Kelly, C.W (1957). Vegetable Crops. Tata McGraw-Hill Co., Inc., New York.

Tomar, B S.; Singh, Balraj and Negi, H.C.S (2005). Onion seed production under drip irrigation. Indian Hort. 49 (4): 24-25.

Upadhya, M.D. (1994). True Potato Seed: Propagules for potato production in the 21st century. In: Potato: Present & Future Shekhawat, G.S.; Paul Khurana, S.M.; Pandey, S.K. and Chandla, V.K. eds., Indian Potato Association, Shimla, Himachal Pradesh, India. pp. 15-22.

Verma, A. and Singh, P.M. (2001). Yojana.

Verma, A.; Kallo, G.; Singh, K.P. and Banerjee, M.K. (2002). Vegetable Seed Production. In: Production, Productivity and Export of Vegetables. Technical Bulletin # 7. IIVR, Varanasi, India. pp. 10-11.

Verma, T.S. (1998). Epiportion of cabbage plants affecting the seed production *in situ* and under transplanted conditions. Seed Research 26 (1): 19-22.

Verma, T.S. and Sharma, S.C. (2000). Producing Seeds of Biennial Vegetables in Temperate Regions. ICAR, New Delhi, 181 p.

Verma, T.S.; Thakur, P.C. and Singh, A. (1997). Effect of biofertilizers on vegetable and seed yield of cabbage. Vegetable Sci. 24: 1-3.

Verma, T.S.; Joshi, S. and Thakur, P.C. (1996). Effect of *in situ* and replanting in cabbage with different head sizes on seed productivity and its quality. Recent Hort. 3: 147-150.

Watts, L.E. and George, R.A.T. (1957). Classification performance of varieties of radish. Eighth Annual Report of National Vegetable Research Station, pp. 3-15.

Williams, C.M.J. (1977). Gluing inflorescences increases yield in vegetable crops. Proc. Australian Seed Research Conference. Tam Worth, New South Wales.

Yalawkar, T.S. and Ram, H.H. (2004). Vegetable Crops of India, Vth edition. Agri-Horticultural Publishing House, Nagpur, 326 p.

□□□

5

Technology for Production of Fruit Crops

SANJAI K. DWIVEDI AND DEEPA H. DWIVEDI*
CEPTAM (DRDO), Metcalfe House, Delhi-110054
* Deptt. of Applied Plant Sciences, BBA University
Vidyavihar, Lucknow-226025 (UP)

Fruit industry in the Higher Himalayan region is very limited and still in its infancy. Although, abiotic stresses like poor soil conditions, low atmospheric humidity, poor irrigation facilities, short-growing period, low temperatures and isolation from the main land for about period of 7 months in a year, are major limitations for development of horticulture in the region, at the same time long day length with high light intensity, high day temperature followed by low night temperature and low atmospheric humidity make conditions congenial for quality production of certain fruits. The major fruit crops grown in the region are apricot (*Prunus armeniaca*), apple (*Malus domestica*), walnut (*Juglans regia*), Pear (*Pyrus communis*), peach (*Prunus persica*), plum (*Prunus domestica*), cherry (*Prunus avium*), mulberry (*Morus alba*), oleaster (*Elaeagnous hortensis*), grape (*Vitis vinifera*) and seabuckthorn (*Hippophae rhamnoides*). Recently strawberry (*Fragaria* sp.) has been successfully introduced.

Apricot is the major fruit crop grown in the region on commercial scale. However, the scattered plantation are found along the agricultural fields. Other fruit crops are very few in number and found in limited patches, especially in the altitude below 10,000 ft above MSL. Besides its use for table purposes, apricot is also consumed after drying with or without stone. The sweet kernel is used as dry fruit and bitter kernel varieties are used for extraction of oil. The cold arid regions of India have excellent genetic diversity in apricot.

Apple is the second major crop of the region only after apricot. It has been found growing at 12,000 ft above MSL where even apricot failed to grow successfully. The study conducted by Field Research Laboratory (DRDO), Leh has also highlighted the presence of valuable indigenous apple genotypes in the region. The laboratory has identified Tha, Mangol, Karkichoo, Shing, Mar, Nas, Yangma, Squirmo, Khara, Shamer kushu (apple) genotypes which have not been reported so far from Leh and Kargil districts of Ladakh. However, detailed studies on conservation and exploitation need to be initiated. The local varieties are generally small in size, highly juicy but have poor shelf life. However, their abiotic stress tolerance can be exploited.

Strawberry varieties are mostly short-day in nature and thus, cannot be grown successfully under the long-day conditions of the cold arid region. Introduction/development of long-day or day neutral varieties has opened up avenues for successful cultivation. Varieties recommended for Ladakh conditions are as under:

i. Long-day type: Addie, Confetura, Hera

ii. Day-neutral type: Brighton, Fern, Douglas, Selva, Tieoga

Strawberry can be grown mainly in the open and also in greenhouse or in pots for domestic consumption. Flowering generally occurs from May to June under open conditions while in the greenhouse it starts by end of March to April. Fruits are harvested 5-6 weeks after flowering and generally 3-4 pickings are available at weekly intervals.

1. Seabuckthorn: An Emerging Fruit

Seabuckthorn, a naturally growing thorny, deciduous bush, indigenous to the region, an unexploited fruit crop of the region, is fast emerging as an important crop due to its nutritional and medicinal properties. Seabuckthorn (*Hippophae rhamnoides*) belonging to the family Elaeagnaceae is a dwarf to tall (3-15 ft), much branched, thorny, dioecious shrub which grows naturally in approximately 30,000 ha area in Ladakh . It grows in a wide range of soil conditions from rocky, sandy and marshy to saline soils. The plant is able to tolerate moisture as well as temperature stress (-30° C). Flowering generally occurs in early June and ripening sets in by mid august. The fruits are small, round to oval in shape with a diameter of 6-8 mm, yellow to scarlet in colour and sour to highly acidic in taste. Fruits of seabuckthorn are a rich source of vitamins having vitamin C (800-1600 mg/100 g), vitamin E (162-255 mg/100 g), vitamin A (11 mg/100 g), vitamin B_1 (0.04 mg/100 g), vitamin B_2 (0.56 mg/100 g) and organic acid (2-4%). They also contain lysine and microelements such as iron, cobalt, selenium and molybdenum. Presence of antioxidants and flavonoids makes it ideal for health. However, because of the perishable nature and highly acidic taste of fruits they can not be consumed fresh (Dwivedi *et al.*, 2006).

2. Genetic Diversity

The indigenous fruit wealth of the Ladakh possess wide variability in tree and fruits parameters. Out of various germplasm collected, valuable morpho types have been identified. The most important among all are apricot (84), seabuckthorn (19) apple (9), mulberry (5) (Table 1, 2, 3, 4 & 5) and Oleaster (3), on the basis of physico-chemical properties of fruits. The collected, and evaluated genetic diversity of fruit crops is conserved in various field gene banks located in various regions/altitudes of Ladakh. The need of the hour is to develop modern methods of propagation of these crops/varieties for their quick multiplication and exploit the genetic wealth for its inherent potential under fruit breeding programmes.

Table 1: Physico-chemical characteristics of 84 genotypes of apricot from cold arid region of Ladakh

Parameter	Range		Mean
	Min	Max	
Moisture (%)	70.4	82.5	75.0
Fruit weight (g)	8.7	35.5	19.04
Fruit volume (ml)	9.3	34.6	20.4
Specific gravity	1.02	1.14	1.10
Fruit length (mm)	26.5	44.0	34.00
Fruit width (mm)	24.9	41.8	33.20
Cheek diam (mm)	22.9	38.8	31.00
T.S.S. (°B)	12.0	33.4	19.35
Pulp weight (g)	2.5	31.2	16.97
Stone weight (g)	1.0	20.0	2.37
Pulp:stone ratio	1.5	17.86	9.42
Kernel oil (%)	27.0	45.0	35.0
Av. kernel weight (g)	0.3	0.5	-

Table 2: Physico-Chemical parameters of fruits morphotypes of Seabuckthorn (*Hippophae rhamnoides* ssp. *turkestanica*).

Parameters	Range	Mean value	SEm
Fruit length (mm)	5.42-9.27	7.34	0.0889
Fruit width (mm)	4.67-7.16	5.91	0.0736
Weight of 100 fruits (g)	7.61-20.54	14.07	10.69
Volume of 100 fruits (ml)	7.9-24.1	16.0	0.314
TSS (°B)	10.2-18.4	14.3	0.189
pH	1.50-2.3	2.15	0.016
Acidity (%)	1.46-3.621	2.54	0.103
Sugar (%)	3.65-5.23	4.44	0.097
Vitamin C (mg/100g)	138-1400	769	26.36
Juice (%)	68.18-80.85	74.5	0.655
Seed (%)	4.44-8.65	6.54	0.377
Pomace (%)	11.4-27.5	19.45	0.585
Seed weight (g/100 seeds)	1.13-1.35	1.24	0.021

Table 3: Physico-chemical properties of apple germplasm of cold arid region of Ladakh

Parameter	Range		Mean
	Min	Max	
Fruit weight (g)	6.63	89.43	48.30
Fruit volume (ml)	5.90	77.00	44.55
Fruit length (cm)	3.79	5.37	4.36
Fruit width (cm)	4.55	5.93	5.18
T.S.S. (°B)	3.46	16.09	5.67
Acidity (%)	0.32	0.50	0.44
Vitamin C (mg/100 g)	50.00	50.00	50.00
Reducing sugar (%)	2.41	2.58	2.78
Non reducing sugar (%)	4.04	7.74	6.30
Total sugar (%)	6.62	10.90	9.08
Sugar/acid ratio	20.69	23.91	20.95

Table 4: Physico-chemical characteristics of fruits of cream coloured *Morus* sp. in cold arid Ladakh

Characters	Range	Mean	SEm
Length (mm)	21.35-24.94	23.33	1.0527
Width (mm)	11.38-12.52	12.02	0.3379
TSS (°B)	21.8-22.4	22.13	0.1763
Weight of 100 fruits (g)	160.98-173.50	171.13	0.1186
Volume of 100 fruits (ml)	198.00-201.00	199.76	0.0881
Acidity %	1.27-1.28	1.275	0.0081
pH	4.0- 4.1	4.066	0.0331
Vitamin C (mg/100 g)	30.98-31.12	31.05	0.0404

Table 5: Physico-chemical characteristics of fruits of purple-red *Morus* sp. in cold arid Ladakh

Characters	Range	Mean	SEm
Length (mm)	20.73-22.85	21.21	0.838
Width (mm)	11.83-12.26	12.00	0.132
TSS (°B)	15.6-17.4	16.73	0.569
Weight of 100 fruits (g)	149.80-158.60	155.33	0.278
Volume of 100 fruits (ml)	179.00-181.00	180.00	0.0577
Acidity %	0.56-0.65	0.61	0.028
pH	3.8- 3.9	3.866	0.0331
Vitamin C (mg/100 g)	34.10-34.80	34.46	0.202

3. Propagation and Nursery Development

Non-availability of genuine planting material is a major problem since there is no operational nursery which can meet the demand of local orchardists. To meet this demand the State Horticulture Department, imports about 30-40 thousand apple plants from Shimla and Srinagar which are air-lifted to Ladakh, since transport by road is restricted in winter months. This enterprise has not proved very successful since survivability of the plants is very poor, perhaps because of the fluctuations in temperature during the transit. One major problem is that in the hills of Uttaranchal, Himachal etc. generally, temperate fruit trees are planted before spring, *i.e.*, February, for which these are uprooted in January. However, the optimum time for planting in Ladakh is mid March-April when the ambient temperature becomes conducive for growth. In the interim period, the plants are either in transit or they are stored in cellar houses. Thus, the time gap between procurement and planting is a major problem which affects survival of these trees.

Seabuckthorn has been growing wild under natural conditions, has been planted by local farmers or by Forest Department for protection of flood/ soil erosion on river banks, sand dunes fixation or for fencing the new plantation or agricultural fields where fruit yield and its quality has no importance. However, for commercial cultivation, it is imperative that the crop be multiplied from the best material which has been selected for its fruit yield and other desirable parameters.

4. Standard Propagation Techniques

Traditionally seedling plants of apricot, apple, walnut *etc.* are found in Ladakh. No vegetative propagation was practiced by the locals which often resulted in long gestation period and wide variation in fruit quality. It was felt that in order to overcome this problem vegetative propagation techniques suited to the cold arid region needed to be standardized. Thus, vegetative propagation techniques in apricot were standardised, using seedling rootstock. It was concluded that different methods of vegetative propagation, *viz.*, chip budding (Dwivedi and Singh, 2001a), t budding (Dwivedi *et al.*, 2000b) and cleft grafting (Dwivedi *et al.*, 2000b) are successful at different periods. However, chip budding was found to be the best method (Dwivedi *et al.*, 2002c). Old seedling apricot plants were converted to productive plants by top working using the shield method of budding which was performed during mid June-early July. This has shown 90.0% success (Table 6-9) (Dwivedi and Singh, 1999). This technique can also be used for germplasm conservation by adopting the orchard on a tree concept.

Seabuckthorn can be propagated successfully by seed which are viable for a period of two years. May is the best time for sowing of seeds in nursery beds. Before sowing the mature seeds should be stratified for about 25 days by keeping in a box in alternate layers of moist sand. These boxes are kept under cool and shady condition for about 25 days. This method gives over 85% germination (Dwivedi *et al.*, 2001b). Seabuckthorn has profuse suckering habit and has been reported to yield 13-40 suckers per plant. These suckers are true to the type which can be separated along with the roots from the mother plants and planted directly at the planting site. The best time for separation of suckers and planting is in March before bud break (Dwivedi *et al.*, 2006). However, the major disadvantage is difficulty in operation due to thorny plant structure.

Seabuckthorn can be propagated successfully by stem cuttings and this method has been recommended for commercial propagation. One to two year old stem cuttings having pencil thickness are ideal for its propagation. The stem cuttings are taken from identified male and female plant to get elite planting material. Seabuckthorn cuttings are not easy to root on their own, thus indole butyric acid (IBA) is applied @ 500 ppm before planting it in the nursery beds. The best time for planting of cuttings has been found to be the month of April. About 85% success has been recorded in rooting of stem cuttings under Ladakh conditions. These cuttings are ready for plantation in the main field in March of the following year (Dwivedi *et al.*, 2006b). DRDO has successfully standardized asexual method of propagation of seabuckthorn through soft wood cuttings using 200-300 ppm NAA. The technique has yielded over 90% rootability in cuttings with excellent root quality under mist condition having coarse, sand rooting media. The plantlets get ready for transplantation in poly bags within 90-100 days after planting of cuttings (Dwivedi *et al.*, 2007b).

5. Orchard Management

Local farmers do not follow orchard management practises, *viz.*, manuring, fertilizer application and irrigation. Since the fruit trees are generally planted around the agriculture fields and receive nutrients and water from field crops, it is essential to highlight that no scientific study has been undertaken to work out the nutrient and water requirement of fruit crops for the cold arid region. The practice of pruning and training of fruit crops is also uncommon. Trees are allowed to grow freely which is why substantial fruit damage occurs during harvesting. An effort should be made to evaluate modern training systems for adopting the high density planting system. The training method which ensures maximum exposure of fruits to sunlight and enables the branches to bear maximum crop load should be preferred, since there are no hail storms in the region.

Table 6 : Physico-chemical characteristics of the some common apricot cultivars of Ladakh

Cultivars	Length (cm)	Breadth (cm)	Chick diameter (cm)	Fruit weight (g)	Pulp weight (g)	Stone weight (g)	Pulp- stone ratio	TSS (°b)	Titratable acidity (%)
Afghani	2.63	2.64	2.63	13.2	12.0	1.2	10.0	19.6	-
Australian	4.15	3.86	3.67	33.2	30.8	2.0	15.4	15.0	0.92
Charmagz	3.44	3.76	3.42	24.8	23.0	1.8	12.8	19.0	-
Halman	2.43	2.43	2.39	11.6	10.2	1.2	8.5	18.4	0.67
Raktsey karpo	3.85	3.82	3.63	34.6	32.0	2.4	13.3	21.2	0.61
Safeda	3.35	3.17	3.42	19.5	18.1	1.4	12.9	18.0	0.64
Shakanda	3.05	2.89	2.78	11.5	10.0	1.5	6.6	19.3	0.95
Shakarpara	3.73	3.58	3.64	28.0	25.3	2.7	9.3	12.0	0.13
Tokpopa	3.30	3.34	3.39	14.5	12.8	1.7	7.5	12.2	1.00
CD (0.05)	1.51	1.41	1.15	-	-	-	-	3.93	0.09

Table 7 : Physico –chemical properties of major apple genotpes of Ladakh

Genotypes	Fruit length (cm)	Fruit diam (cm)	Fruit weight (cm)	Fruit volume (cm)	T.S.S.	Acidity (%)	Vit-C (mg/100g)	Reducing Sugars (%)	Non Reducing sugars (%)	Total Sugars (%)	
Mongol	5.37	5.93	89.42	77.00	3.46	0.16	50	2.28	1.66	3.94	-
Tha	4.29	4.95	48.06	54.17	6.66	0.28	50	3.53	2.05	5.24	-
Bong Kushu	4.63	5.42	60.47	75.00	5.20	0.24	50	3.62	3.29	6.91	-
Shing Kushu	3.79	4.55	36.25	37.74	5.50	0.32	50	4.04	2.58	6.62	-
Mar Kushu	4.17	4.84	43.88	53.33	5.33	0.16	50	1.88	0.23	2.11	-
Khara Kushu	4.40	5.70	62.50	54.00	5.88	0.46	50	1.24	0.63	1.87	-
Squirmo	4.07	5.60	66.25	56.70	6.10-	0.46	50.0	6.53	2.36	8.89	-
Karkichoo	4.44	5.09	51.20	49.76	7.30	0.16	50	2.99	2.66	5.66	-
Phamer Kushu	4.34	4.70	40.67	41.67	12.21	0.46	50.0	3.60	2.89	6.49	-
Sanker Kushu	4.14	5.04	44.33	46.67	16.09	0.41	50	4.33	2.68	7.01	-

Table 8 : Physico-chemical properties of fruits of important strawberry cultivars grown in cold arid conditions of Ladakh.

Cultivars	Fruit length (cm)	Fruit diam (cm)	Av fruit weight (g)	Fruit volume (ml)	T.S.S. (°B)	Titrable acidity (%)	No. of fruits/ plant	Yield/ plant (g)	Estimated fruit Yield (t/ha)
Addie	3.32	2.98	9.09	11.03	16.76	0.69	11.6	100.54	7.45
Brighton	2.76	2.33	6.37	8.81	10.03	0.56	15.3	97.46	7.22
Chandler	3.22	2.48	5.88	7.98	17.12	0.73	18.4	108.19	8.01
Confitura	2.79	2.32	5.76	7.65	10.28	0.44	20.3	116.93	8.66
Douglas	2.80	2.57	6.22	8.03	9.91	0.56	13.7	85.21	6.31
Gorella	2.87	2.26	7.47	8.58	8.98	0.56	17.9	133.71	9.90
Fern	3.52	2.86	9.09	11.50	18.68	0.49	13.1	119.08	8.82
Pajaro	3.26	2.84	7.14	9.15	17.32	0.61	12.8	91.39	6.77
Selva	3.16	2.84	7.69	8.95	18.68	0.60	13.9	106.79	7.91
CD at 0.05 P	**0.38**	**0.15**	**1.07**	**1.30**	**2.41**	**0.09**	**3.54**	**10.67**	**0.76**

Table 9 : Performance of different methods of vegetative propagation in apricot in cold arid region of Ladakh.

Method/ Techniques	Time	Sprouting (%)	Bud take (%)	Days taken in sprouting	Saleable plant (%)	Linear growth (cm)	Radial growth (cm)
Grafting - tongue/cleft	First week of February	95.0	90.0	73.5	82.5	164.0	0.55
Budding - chip	Fourth week of February	90.0	85.0	71.4	77.5	28.4	0.58
Budding - shield	Second fortnight of august	88.3	81.6	29.5	76.5	17.0	0.45
Top working - shield	Second fortnight of July	93.3	80.0	31.5	-	76.5	0.80

Seabuckthorn should be pruned regularly for ease in harvesting and exposure of plant to sun for production of quality fruits. The best time for pruning of seabuckthorn in Ladakh is the month of March, before sprouting of the plant in spring (Dwivedi *et al.*, 2006). Intensity of pruning depends upon the age and growth of plant and soil conditions. However, 50% growth of 3-4 year old shoots should be removed. Seabuckthorn, being a thorny bush, should be restricted to a plant height of less than 2 m for ease in harvesting. Normally, properly managed seabuckthorn plants develop a crown of 2-3 m diameter at the age of four years after planting.

6. Insect-Pest/Diseases and Disorders

Incidence of pests and diseases in the cold arid region is negligible due to low temperature and very low relative humidity. This makes the region ideal for producing organic fruits which has rising in the market world over.

Codling moth (*Cydia pomonella*) is a major threat to the fruit industry of Ladakh. Apple is the primary host for the insect which destroys the fruit by feeding on the fruit pulp. The pest has shifted even to apricot, pear and walnut. In India, the problem of codling moth is restricted to the Ladakh region, while rest of the apple growing belts like Kashmir, Himachal Pradesh, Uttaranchal, etc. have not reported significant incidence of the pest. It is for this reason that the government has imposed restrictions under SRO 397 dated 8th September 1981, issued under the Jammu and Kashmir Plant and Disease Act, 1973, on the movement of apple and apricot from Ladakh. Other pests reported in fruit crops are aphids, leaf minor, pin hole borer, white grubs, blossom thrips, two spotted spider mite, gypsy moth, leaf curling aphid etc. However, their incidences are almost negligible and are restricted to certain areas in the region. Diseases and disorders like blight, gummosis and sunscald etc., in apricot have been observed which need to be studied in detail. Bird damage of ripe fruits of apricot is one of the major threats, which needs to be managed properly (Dwivedi *et al.*, 2007a).

7. Post-Harvest Management

Development of techniques for post harvest handling of fruit crops is the immediate requirement for the cold arid region. Lack of knowledge regarding maturity indices, harvesting, grading, packaging methods and poor storage facilities result in 40-60% loss in apricot alone. Apricot is generally preserved by open-sun drying which affects the fruit quality. Solar driers, *viz.*, polyhouse, tent and tunnel driers have been developed for hygienic dehydration of surplus fruits and vegetables in the region. Methods for value addition, *viz.*, osmotic dehydration, intermediate moisture, apricot bar, hurdle technology, squash, nectar, jam etc., have also been developed. Another potential area is extraction and processing of apricot kernel oil which is of medicinal and cosmetic use and is found in abundance in the region.

Studies on value addition and processing of seabuckthorn have also been initiated. A multi-vitamin herbal beverage as squash and nectar, without using any artificial colour, flavour and stabilizers etc., has been developed in which maximum vitamins and nutrients have been conserved in their natural form (Dwivedi *et al.*, 2002a). This technology has been patented and subsequently transferred for manufacture on commercial scale. The beverage has also been inducted in the Indian army as special ration for Siachen troops since it has special ability to help in acclimatisation in the high altitudes. The process for preparation of seabuckthorn jam, sauce, pickle and *chyavanprash* has also been standardised. Studies have been undertaken to develop procedures for extraction and evaluation of seabuckthorn oil having immense medicinal value for its use in the cosmetic industry.

8. Problems of Fruit Industry in Cold Arid Zone

The climatic conditions prevailing in the region, inaccessibility and lack of technical know how are the major bottlenecks in the development of the fruit industry. Despite the efforts being made, the existing problems are listed below:

- Non availability of quality planting materials through local nurseries,
- Lack of location specific scientific techniques,
- Improved cultivars of fruit crops for various purposes/locations,
- Crisis of irrigation water and modern scientific irrigation methods,
- Lack of post-harvest technology and processing methods of fruit crops,
- Poor knowledge of packaging, grading, transportation and marketing,
- Problem of transportation, storage and marketing facilities.

9. Greenhouses in Fruit Production

Ladakh remains cut off from rest of the nation from November to June every year due to heavy snowfall which causes blockage of roads. Sub-zero temperature prevails from December to February throughout Ladakh whereas, zero degree is experienced during the rest of the winter months (Singh, 1995b). Various greenhouse structures, *viz.* glasshouse, polyhouse, *Ladakhi* polyhouse, trench and tunnel for cultivation were studied. Of these, the trench has been found to be the most suitable and economic structure for the region followed by *Ladakhi* polyhouse (Singh *et al.*, 1998b). This is because it retains the increased day temperatures even during night. The trench, as proposed by Field Research Laboratory, is an underground structure of convenient variable dimensions covered with UV stabilized transparent polyethylene (400 gauges) sheet. At night another covering of black polyethylene is given to harness the soil temperature which increases during the day and to minimise the heat loss during the night (Singh, 1995a). Thus, greenhouses suited to local needs were developed and are called '*Ladakhi* polyhouses'. Three sides of this structure are made of unbaked bricks and the fourth side, which harvests sunlight, is made of polythene. This has been found to be very economical and functions efficiently even in the winters.

However, another structure especially suited for the region is the underground trench. This is an underground structure 30′ x 12′ and 3′ deep covered on the top with transparent polythene with the support of wooden poles. The polythene is removed on bright sunny days and the trench is covered when there is snowfall. This has been found to be the most effective structure for propagation of plants through runners (strawberry), cuttings (seabuckthorn, grape, mulberry, oleaster), grafting/budding (apricot, apple) and the production of crops like strawberry, grapes etc. since it regulates the microclimate inside, the trench most effectively as compared to any other protected structure (Singh *et al.*, 2000b). These structures can be effectively used for winter cultivation, raising quality planting material, conservation of valuable germplasm and extending the period of growth of crops.

10. Mulching

Water scarcity is a major limitation in the area, and use of mulches has been found to enhance water retention by the soils by decreasing loss of soil moisture through evapotranspiration under high light intensity and low environmental humidity. Besides, this mulching also regulates the soil temperature which subsequently improves plant growth. It decreases weed population in the vicinity of the main crop thereby decreasing competition for nutrients and water and thus, indirectly productivity is enhanced. It has been observed that use of organic as well as synthetic mulches, *viz.*, straw, hay, dried leaves and paper, pebbles, polythene, asphalt sheets has immense advantages. Organic mulches improve soil organic carbon levels and also enhance water holding capacity of the soil. Black polythene of 150 gauge thicknesses has been found to enhance yield and also reduces frequency of irrigation by 50% and completely negates the need for weeding (Singh, 1995a; Singh *et al.*, 1998b). Among the fruit crops mulching has obvious benefits. It enhances establishment of plants, accelerates plant growth and reduces the need for frequent irrigation in crops like apricot, seabuckthorn and apple (Dwivedi *et al.*, 2000d). In strawberry, mulching has been found effective to improve the runner production and fruit yield and its quality. Use of dry grasses and hay has also been found to increase fruit yield and summer production in strawberry.

11. Organic Farming

The cold arid region in India is alienated from the rest of the country because of its geographical topography. One major advantage of this is that the farming society here remained untouched by the use of inorganic fertilizers, pesticides, fungicides etc. This part of the world still follows traditional methods of organic farming and manures such as night soil, farm yard manure etc. for enriching soil. Legumes are included in crop rotation. Another feature is that the land remains fallow for 5-6 months in the year during winter when it is able to rejuvenate itself. Low atmospheric humidity and temperature are effective in preventing disease and insect-pest infestation in most of the crops. This ensures minimum use of pesticides in the region. The natural advantage of the region can be economically exploited by cultivation of organically grown fruits on a commercial scale and their export to foreign countries as well as to the domestic market.

12. Scope of Fruit Production

- Development of organic farming of fruits
- Production of Export quality fruits meeting international standards

- Off season production of crops like strawberry under natural conditions
- Exploiting fruit crop (apricot, seabuckthorn) for oil production
- Conservation and utilization of highly valuable germplasm of fruits
- Identification of gene responsible for plant growth under extremely stress conditions due to low temperature, moisture stress, high altitude
- Production of organic fruit pulp/value added products for domestic/global markets

13. Problems of Fruit Industry

The climatic conditions prevailing in the region, inaccessibility and lack of technical know-how are the major bottlenecks in the development of the fruit industry. Despite the maiden efforts being made the existing problems are listed below:

- Non-availability of quality planting materials through local nurseries
- Lack of location specific scientific cultivation/orchard management techniques
- Improved cultivars of fruit crops for various purposes/locations
- Crisis of irrigation water and modern scientific irrigation methods
- Lack of post harvest technology and processing methods of fruit crops
- Poor knowledge/infrastructure for packaging, grading, storage, transportation, and marketing etc.
- Neglection of fruit crops by policy makers/local administration/scientists/research institutions/government agencies working in the region.

References

Anonymous (2002). Annual report 2001. Field Research Laboratory (DRDO). C/o 56APO.

Dwivedi, Deepa H.; Kareem, A.; Dwivedi, S.K. and Mir, A.A. (2002). Physico-chemical characteristics of important apricot (*Prunus armeniaca* L.) cultivars of the cold arid region of Ladakh. Indian J. Hort. 59(2): 118-121.

Dwivedi, S.K. and Singh, B. (1999). Topworking of apricot in Ladakh. Prog. Hort. 31(1-2): 29-31.

Dwivedi, S.K..; Singh, B. and Paljor, Eli (2000d). Effect of mulch on establisment and growth of apples saplings in cold arid conditions of Ladakh. Prog. Hort. 32: 42-45.

Dwivedi, S.K. and Singh, B. (2001a). Studies on chip budding in apricot (*Prunus armeniaca*) arid conditions of Ladakh. Indian J. Hort. 58(3): 215-217.

Dwivedi, S.K.; Singh, B. and Attrey, D.P. (2001b). Studies on distribution, propagation and ultilisation of seabuckthorn in Ladakh. Seabuckthorn-a resource for health and environment in 21st century. Eds. Singh, V. and Khosala, P. K. Proceedings of international workshop on seabuckthorn, Feb. 18-21, 2001. CSKHPKVV-Palampur, India. pp. 20-24.

Dwivedi, S.K.; Attrey, D.P. and Kareem, Abdul (2002a). Post harvest management of seabuckthorn (*Hippophae* spp.) in cold arid regions of Ladakh. In: National workshop on post-harvest management of horticultural produce organized by National Productivity Council, 7-8 Feb. 2002 Chandigarh. pp. 89-98.

Dwivedi, S.K.; Attrey, D.P.; Paljor, Eli and Singh, B. (2002c). Studies on conservation of genetic diversity in apricot in cold arid region of india. In: 26th international horticultural congress, Toronto (Canada), 11-17 August, 2002. (Abstract) . pp. 524-25.

Dwivedi, S.K.; Paljor Eli; Attrey, D.P. and Singh, B. (2002d). Propagation of common seabuckthorn through hard wood cutting in ladakh. Indian J. Agri. Sci. 72(4): 228-229.

Dwivedi, S.K.; Singh, Ranjit and Ahmed, Z. (2006). The Seabuckthorn. Field Research Laboratory, DRDO, Leh, Ladakh (J&K) India.

Dwivedi S.K.; Singh, R and Ahmed, Z. (2007a). Apricot in Ladakh. Field Research Laboratory, DRDO, Leh, Ladakh (J&K) India.

Dwivedi, S.K.; Hemraj and Ahmed, Z. (2007b). Standardisation of a fast method of vegetative propagation of Sea buckthorn through soft wood cutting in India. Paper presented in the 3rd International Seabuckthorn Association conference held at Quebec, Canada, August, 12-16, 2007. Paper No. 83.

Singh, B. (1995a). Vegetable production in Ladakh. Field Research Laboratory, C/o 56 APO.

Singh, B. (1995b). Green house development in Leh. Paper presented in National workshop on commercialization of tissue culture and greenhouse Technology, Banglore, India 5-6 May 1995.

Singh, B.; Dwivedi, S.K.; Singh, N. and Paljor, Eli (1998a). Sustainable Horticulture Practices for cold arid areas. Paper presented in International symposium on sustainable agriculture in hill areas. 29-31 oct. 1998. HPKVV Palampur. (Abstract), pp. 35-36.

Singh, B.; Chaurasia, O.P.; Dwivedi, S.K. and Paljor, Eli (1998b). Effect of plastic mulch on vegetable production in Ladakh. In: National symposium on emerging scenario in vegetable R&D. PDVR, Varanasi (UP), 12-14 Dec.98. (Abstract) p. 137.

Singh, B.; Dwivedi, S.K. and Sharma, J.P. (2000a). Greenhouse technology for winter vegetable cultivation in cold arid zones. In: Dynamics of cold arid agriculture. Sharma, J.P. and A.A. Mir, Kalyani Publishers-Ludhiana, pp. 279-293.

Singh, B.; Dwivedi, S.K. and Paljor, Eli (2000b). Studies on suitability of various structures for winter vegetable production at sub zero temperatures. Acta Hort. 517: 309-314.

❑❑❑

6

Floriculture: An Untapped Potential Venture

DEEPA H. DWIVEDI AND SANJAI K. DWIVEDI*
Department of Applied Plant Sciences, B B A University
Vidya Vihar, Lucknow-226025 (UP)
* CEPTAM(DRDO), Metcalfe House,Delhi-110054

A sizable area in the country is located at an altitude between 2900-4000 m above MSL which is characterized by uneven land, barren mountains, arctic desert, extremely cold weather and very low precipitation and thus it is called the cold arid region. These regions in India are located in the high altitude zone which is described as a treeless world and an environment of low atmospheric pressure. Leh and Kargil districts in Jammu and Kashmir, Lahul-Spiti in Himachal Pradesh, Nelang, Niti and Mana in Uttarakhand and certain patches in North Eastern states are the major representative cold arid regions in India. The higher Himalayan regions of our country offers unique opportunities for growing of ornamental crops with a very attractive coloured good quality produce during summer/ rainy season, to trap the off season domestic market when there are very high demand of these crops in the national market. The qualities of produce are such that they can meet the international standards also and export possibilities can also be explored. The scope and the technology of growing various ornamental crops have been reviewed and discussed in the paper.

1. Annual Flowers

Botanically, annuals are the plants which normally complete their life cycle within one year. Horticulturally, annuals are the ornamental herbaceous plants grown in different seasons of the year for aesthetic purposes. Annuals complete their life cycle within a particular season. Some herbaceous biennials and perennials which flower in a particular season are also treated as annuals by the gardeners. At present, there are a large number of annuals grown in different seasons and are used as ornamentals. The present day annuals, particularly grown in winter season, were introduced in India from Africa, Europe, America, China, Japan and other countries, mainly by the Muslim rulers, Britishers and other Europeans who were very fond of gardens and flowers. Most of the annuals are free flowering and available in a wide array of dazzling flower colours. Their ease of cultivation and multiplication through seeds has made them most popular in Indian gardens. Annuals are like jewels which enliven other permanent plantings and features of the garden. No other plants in the garden can influence the general landscape such an incredible way by virtue of their mass array of colours and quick performance.

In Ladakh, flowers are grown on a very limited scale in window boxes, in pots and in small beds at the entrance of houses. These are generally grown during summer months from March-September when there is adequate sunshine and temperature is high. It is easier to grow annual flowers propagated from fresh seed sown each season which flower, set seed and die within a year and thus do not have to survive the sub zero winter temperature types. Perennial flowers, which once planted, continue to flower for a number of years are not able to survive due to the cold winters. Hence, it is obvious that the choice of practical gardener is annuals (Table 1).

Classification

Annuals have been used for varied purposes and have, thus been classified in different ways, *viz.*, growing habit, size, uses *etc.* Annuals which are commonly cultivated as ornamentals have different growth habits. Some are tall while others are medium, bushy, dwarf and prostrate. Flowers of annuals also vary in wide range of colours. Annuals may be used as edgings along paths, as ground covers for posts and containers in rock gardens etc. It is essential that the height and colour of the flowers chosen is in consonance with the purpose for which they area being used. A broad classification of annual flowers is given below:

According to Temperature Requirement

In the plains the annuals have been classified into three according to the season in which they are sown and that in which they flower, *viz.*, summer annuals, annuals and sunny season annuals. It is important to note that this classification does not hold good in Ladakh because of the limited period is available for the cultivation of flowers. Thus, all the three types are sown only during summer, from March-September.

The cultivation of flowers requires very little competence. However, certain operations, *viz.*, raising of nursery, transplanting *etc.* need special attention and care. On the basis of temperature sensitivity the annuals may be classified as follows:

Low temperature tolerant : There are large numbers of annuals grown in winter season which are successfully grown in Higher Himalayas. They may be sown early and may also be grown till the onset of autumn. The important ones are: Antirrhinum, Arctotis, Aster, Brachycome, Calendula, Dianthus, Helichrysum, Hollyhock, Larkspur, Lupin, Marigold, Nasturtium, Pansy, Phlox, Poppy, Sweet Pea, Verbena *etc.*

Sensitive to low temperatures : The choice for summer season annuals is limited as compared to the winter season. The commonly grown summer season annuals are: Gaillardia, Kochia, Portulaca, Balsam, Gomphrena, Marigold, Sunflower and Zinnia.

According to Plant Height

Tall (above 0.75 m)- African Marigold, Clarkia, Cleome, Coreopsis, Cornflower, Cosmos, Helichrysum, Hollyhock, Lady's Lace, Larkspur, Sunflower, Sweet Pea, Sweet Sultan *etc.*

Medium (0.45 m to 0.75 m)- Ageratum, Antirrhinum, Arctotis, Aster, Balsam, Brachycome, Candytuft, Carnation, Celosia, Cockscomb, Dianthus, Gamolepis, Gypsophila, Petunia, Poppy, Salvia, Wallflower, *etc.*

Dwarf (below 0.45 m)- Calendula, Cineraria, French Marigold, Gaillardia, Gazania, Hymenantherum, Mesembryanthemum, Nasturtium, Nemesia, Pansy, Portulaca, Sweet Alyssum *etc.*

According to Flower Colour

Different annuals are used in informal beds as borders in front hedge or along paths. The flowers are massed irregularly according to height but grouped in suitable colour schemes to produce a natural, harmonous and pleasing effect. This is extensively necessary to choose the annuals according to their colour in order to utilize them in the best possible manner in creating a beautiful landscape.

White Flowered- Annuals China Aster, Dianthus, Gypsophila, Lady's Lace, Phlox, Petunia, Stock, Sweet Alyssum, Zinnia etc.

Yellow Flowered- Annuals Calendula, Dimorphotheca, Gamolepis, Hymenantherum, Marigold, Zinnia *etc.*

Blue and Purple Flowered- Annuals Ageratum, Anchusa, Browallia, China Aster, Cornflower, Cineraria, Larkspur, Lobelia, Lupin, Pansy, Phlox, Sweet pea, Verbena *etc.*

Pink and Red Flowered- Annuals Clarkia, Dianthus, Phlox, Poppy, Salvia, Saponaria, Verbena, Zinnia *etc.*

According to Maturity Time

Early Blooming- Balsam, Candytuft, Celosia, Coreopsis, Dianthus, Gaillardia, Gomphrena, Phlox, Salvia, Sweet Alyssum *etc.*

Late Blooming- Antirrhinum, Arctotis, Petunia, Hollyhock, Portulaca, Rudbeckia, Sweet William *etc.*

According to Growth Habit

Carpet Bedding- Mesembryanthemum, Pansy, Portulaca, Sweet Alyssum, Verbena

Edging- French Marigold (Dwarf), Hymenantherum, Phlox, Pansy, Sweet Alyssum *etc.*

Climbers- Canary Creeper, Morning Glory, Sweet Pea, Quamoclit

Rockery- Antirrhinum (Dwarf), Brachycome, Eschscholzia, Gazania, Godetia, Linaria, Mesembryanthemum, Nasturtium, Pansy, Petunia, Phlox, Portulaca, Sweet Alyssum, Verbena *etc.*

Window Sill- Aster, Calendula, English Daisy, French Marigold, Nasturtium, Petunia, Phlox, Portulaca, Sweet Alyssum, Verbena etc.

Shade Loving-Ageratum, Candytuft, Cineria, Clarika, Delphinium, Godetia, Lobelia, Lupin, Nicotiana, Salvia, Sweet Alyssum, Verbena *etc.*

For Hanging Baskets- Brachycome, Daisy, Lobelia, Nasturtium, Petunia, phlox, Portulaca and Sweet Alyssum

Annuals with Ornamental Foliage

Amaranth, Cabbage/Kale, Kochia, Nasturtium (Variegated)

Annuals Having Fragrant Flowers

Cleome, Dianthus, Heliotrope, Mignoette, Stock, Sweet Alyssum, Sweet Pea, Sweet Sultan and Sweet William

Annuals for Cut-Flowers

Antirrhinum, Aster, Calendula, Carnation, Cornflower, Gaillardia, Gomphrena, Gypsophila, Lady'lace Statice, Sweet Sultan, Sunflower, Zinnia etc.

Annuals used as Loose Flowers

Annual Chrysanthemum, Aster, Balsam, Cornflower, Gaillardia, Marigold, Sunflower and Zinnia

Procurement of Seed

The Primary requirement to ensure a good crop is good quality seed. Thus must be done from a standard company of repute. Due to the poor climate conditions and the low soil fertility in Ladakh, seed germination is poor. Hence, quality seed having a good germination per cent is the foremost requirement. Seeds of most annuals are sown in a nursery to raise seedlings which are transplanted later in the permanent flower beds in the desired fashion. Seeds may be sown in nursery beds, earthern seed pots in wooden seed trays according to convenience. Nursery beds of one meter width and convenient length are made raised above the ground level to facilitate proper drainage of water. However, since the soil in most areas is a sandy loam, one can do without raising of the raising beds. The best soil mixture for raising of a nursery is one which contains two parts loam, one part well rotten organic manure and one part coarse sand. Such a mixture is made fine before filing it in the pots or pan. Although the incidence of disease in this area is low, it is beneficial to treat the seeds with a suitable fungicide, *viz.*, Fytolan, Captan etc. before sowing as a prophylactic measure. Very small seeds such as those of petunia, Calendula, ornamental Kale *etc.* may be mixed with a small quantity of sand to facilitate uniform sowing. The Ladakh, low temperature continues till the month of April-May. Thus for sowing of seed under open field conditions preparation is possible only by mid-May.

Sowing of Seed

For sowing of the seed, straight rows, 6 cm apart and 0.3-0-6 cm deep may be made with the help of a wooden stick or pencil. Seeds are placed in these rows uniformly which are then covered with sand or leaf mould. Watering is done immediately taking care firstly, that the sand and leaf mould given as a cover is not washed off, thus exposing the seeds and secondly, that over watering is not done else seed will rot. The seed pan or pot may be covered with a glass sheet or newspaper and kept in partial shade for a few days. Once germination takes place, the pans may be kept in full sunshine to harvest healthy nursery and seedlings. In Ladakh, one can even go in for raising of nursery under cover, *i.e.*, in trenches or in polyhouse. The seed can thus be sown under cover by April and nursery can be obtained about one month in advance; this facilitates an early crop. Generally, seedlings are raised in nursery beds. Bold seeded flowers can be sown directly in the beds/pots. Seeds can be sown in beds or in pots under open conditions/greenhouse for raising seedlings. Greenhouse provides flower seedlings 15-20 days early as compared to open. However, these seedlings need hardening before planting in open to avoid high mortality. Best time for sowing of seeds is 3rd week of March (greenhouse) and 2nd week of April (open field). Seeds should be sown in lines 10-15 cm apart and should be covered with light soil/soil+FYM mixture which should not be more than three times thicker than the diameter of the seed. Mulching (covering) of seed bed with dry grasses/ polythene sheet helps in faster and better germination. Seeds normally take 8-12 days to germinate and after 3-4 weeks plants attain 4/5 leaf stage which is optimum for planting. Light irrigation should be given soon after sowing. Thereafter 2-3 irrigations per week should be given.

Transplanting of Seedlings

Seedlings are transplanted about a month after sowing, at the 3-4 leaf stage. Seedlings are hardened before transplanting by withholding water for a couple of days. It is preferred that seedlings be transplanted on a cloudy day or in the evening to protect the harsh sun and thus help in their establishment. The soil around the neck of the seedlings should be pressed firmly after transplanting and watering should be done immediately. A friable, well drained sandy loam, rich in humus, is considered suitable soil for planting of flowers. The bed is prepared well by adding a bucket full of well decomposed cow dung or farmyard manure and about 45 g super phosphate or bonemeal per square meter of bed. For preparation of flower beds, deep ploughing or digging (20-30 cm) should be done to make a fine tilth. Stones, pebbles, dried roots etc. should be removed. Well rotten farm yard manure (FYM) @ 10 kg/sqm should be added in the soil atleast 30 days before planning /sowing. Fresh or undecomposed FYM/poultry manure in the flower beds especially in nursery beds and pots should be awarded.

Tips for Growing Annual Flowers in Pots

The best potting mixture is fertile Soil: FYM: Sand in the ratio of 3:2:1. Ensure drainage holes in the bottom of the pot and cover them with stone/pebbles to avoid its blockage. This helps proper drainage of excess water. Leave about 1/2" space on the top of the posts for watering. A layer of pebble should be kept at the bottom of the pot so that water stagnation could be prevented. A layer of pebble should be kept at the bottom of the post so that water stagnation could be prevented. Pots kept indoors should be put in the sun at a weekly interval. Ensure light and frequent watering and avoid excess watering in pots.

After Care

Regular weeding of flower beds is necessary to ensure a healthy crop, height; and regular watering at frequent intervals ensures a luscious growth since the plants cannot withstand heavy watering. In such annuals as sweet pea and carnation staking is required since they have long tender stem. Seeds of Pansy, Antirrhinum, Petunia and Ice plant, Portulaca etc. are very tiny and difficult to germinate. These seeds should be mixed in sand before sowing in nursery beds for even and equal placement of seeds. Seeds sown in the open field should be covered with polythene and must be protected from cold waves and snowfall. Care should be taken while uprooting the seedlings to avoid root damage. Irrigation should be stopped 4-5 days before uprooting the seedlings which crates water stress and helps harden the plant for better establishment. Very small or over mature seedlings should not be transplanted. Time gap between uprooting and transplanting should be minimized and daily light irrigation should be given for initial 5-7 days. Avoid heavy watering at the time of flowering to prolong the flowering period. Staking and support is required for tall plants and climbers.

Disease and Insect Pests

The cold dry climate of Ladakh is not at all conducive for the survival of disease causing microbes and insect pests. However, a close check should be kept on the disease and pests and prompt measures for their control should be adopted. A prophylactic spray of Bavistin and Dithane-M-45 could be given at the initial stage of plant growth by the more cautions grower. The incidence of viral diseases has been noticed in Zinnia, Aster, Marigold, Petunia, Dahlia *etc.* The whole plant is affected and killed. The only means of control is to uproot the plant and burn it off. Rust is another disease which often shows its presence. It is caused by fungi and occurs in Marigold, Hollyhock, Stock, Antirrhinum *etc.* Rust resistant varieties should be grown to control this problem. Among the insect pests, aphids and thrips are found in abundance. These are all sucking type of insects and attack a variety of species viz. Dahlia, Antirrhinum, Chrysanthemum, Sweet Pea, Larkspur, Petunia *etc.* Thrips are found in gladioulus, lily *etc.* These are sucking tube of insects which feed on the cell sap. They can be controlled by spraying or dusting the plant with Malathion, Rogar or Metasystose.

2. Cultivation of Gladiolus

Gladiolus cultivation, as a flower crop, has been very successful in India since it is easily grown in a wide range of climate, soils and other conditions. It has a beautiful spike with a large number of florets, which gives a long vase life. It is also available in a number of colours and can be used in all kinds of flower arrangements. There is huge demand of the flower in domestic as well as international market. Gladiolus growers not only earn profit by selling the flowers but also by producing the corms, which are used for its multiplication. The main advantage which Ladakh has over other parts of the country is the availability of gladioli in Ladakh from July to September, when there is slump in plain markets of the country due to rainy season. Thus, the crop grown in Ladakh can fetch handsome price in plain markets. Scientific cultivation package of gladiolus for Ladakh has been reported by Dwivedi and Attrey (2001) and Dwivedi and Singh (2000).

Climate and Soil

The desired climate for Gladiolus cultivation is easily met in Ladakh from May to September. Hence the crop can be easily grown during this period in Ladakh. Gladiolus is a heavy feeder and requires highly fertile conditions for its cultivation. In general fields, used for vegetable cultivation, are also suitable for gladiolus cultivation. Soils which are poor in organic matter, marshy or water logged and saline soils, are unfit for growing Gladiolus. Soil pH between 6-7 is recommended for its cultivation. A well drained, sandy loam soil, with high organic matter content and free from weeds, gravel and stones, is recommended for commercial cultivation of Gladiolus. Gladiolus is a sun loving plant. Thus, the field selected for gladiolus should be in open sunny location. Sites susceptible to strong winds should be avoided. Gladiolus requires fine tilths. A deep ploughing (20-25 cm) of field becomes necessary. Well rotten Farm Yard Manure @ 60-80 tonnes/ha (3-4 tonnes/kanal) should be mixed in the top soil, atleast 30-45 days before planting. DAP fertilizer @ 120 kg/ha (6 kg per kanal) should also be mixed before planting. A light irrigation, a week before planting, helps to facilitate better germination.

Varieties

There are over five thousand varieties of Gladiolus world wide. In India about five hundred varieties have been developed/introduced in various regions. Gladiolus varieties are generally selected on the basis of their colour. While selecting the variety preference of colour of concerned market should be taken into account. Out of twenty varieties evaluated at Field Research Laboratory, Leh, five varieties, namely, American Beauty, Friendship, Princess Margret Rose, White Prosperity and Souvenir, have been found to be the most suitable for commercial cultivation in Ladakh (Table 2) (Dwivedi *et al.*, 2002). These varieties should be preferred over other varieties. However, new varieties can also be selected for commercial cultivation, provided their colour and other features are known.

Planting

In Ladakh, Gladiolus takes on an average 110-120 days for flowering. Quality flowering has been recorded when the crop was planted in 4th week of April to 1st week of May. To get continuous supply of spikes the corms should be planted at weekly intervals. Very early or late planting reduces the yield and quality of flowers. Gladiolus can be planted on raised bed, ridges or on flat bed. In light soils of Ladakh, the corms are planted in flat bed in rows made at a distance of 30 cms each with 10 cms distance between corms. This would mean that 30-35 corms are planted per sq.m. The corms are planted 6-8 cm deep because a shallow planting may result in lodging of the crop in the light soils of Ladakh. Corms having 8-10 cm diameter are considered to be the best for commercial flower production. However, smaller corms having at least 4-6 cm diameter can also be used for obtaining flowers. Uniform corm size should be selected for cultivation. Variation in corm size may result in non-uniformity in days to flowering. Damaged, cut or diseased corm should not be taken for planting. Planting of corms should be done only after completion of dormancy. The best indication of completion of dormancy is the visibility of root initials at the base of the corm and sprouting of the terminal bud.

Watering

Gladiolus requires frequent and light irrigation throughout its growing period. However, excess of watering is detrimental to the crop as it causes rotting of the corms. Moisture stress at the time of flowering, *i.e.,* 7-8 leaf stage and at cormel development stage may affect adversely the success of the crop. Thus, care must be taken to provide sufficient moisture in the soil at these stages.

Other Management

To ensure a healthy crop, regular removal of weeds becomes necessary. Earthing up is an important operation to provide support to the plant to stand erect against the strong winds. It also removes the weeds and conserve the moisture. Staking is done with stick to give support to the plant. It is done just before flowering. Care should be taken that underground corm is not damaged during these operations.

Harvesting

The stage of harvesting of spike is crucial and depends upon the purpose of its use. For local markets, the spikes are harvested when the last (lower) bud of the spike starts opening. For distance marketing, harvesting is done slightly earlier, *i.e.,* when last bud begins to show colour. Spike should be harvested in the morning with a sharp knife. The spike should be cut at the base, atleast 15-20 cm below the flowers, alongwith 2-3 leaves. The harvested spikes should not be exposed to the direct sun and should be placed in a bucket of water in shade. Spikes are then tied in a bundle having 12 to 24 spikes and then packed in cardboard boxes or in hasian cloth for further despatch to the market.

3. Cultivation of Asiatic Lilium

Lilium is among the top ten cut flowers traded in world market. Its attractive shape, size, wide range of colour, long vase life and exotic appearance makes it very appealing. The genus Lilium is widely used in floriculture industry as cut flower as well as potted plant in home gardens where primarily three types are common, *viz.,* Asiatic, Oriental and Longiflorum. In addition hybrids of Longiflorum x Asiatic are also commonly cultivated both commercially and by amateurs. Possiblities of cultivation of Asiatic Lilium in Ladakh have been reported by Dwivedi *et al.* (2002a) and Dwivedi and Attrey (2004).

Selection of Site and Soil

Site selection is perhaps the most important factor for successful cultivation of Lilium in Ladakh. It prefers to grow under semi shady conditions. 50% shade net is effective for reducing light intensity under open conditions. Sites which are susceptible to water logging and strong winds are not suitable for cultivation of Lilium. It requires sandy loam soil, rich in organic matter, free from weeds, stones and salinity with good texture and suitable irrigation and drainage facility. Lilium requires soil of fine tilth. Thus, deep ploughing of field (20-30 cm deep) becomes necessary. Well rotten FYM should be mixed @ 60-80 tonns per hectare (3-4 tonns per kanal) at least 30 days before planting. Lilium needs maximum nutrition 3 weeks after planting. Thus, two dose of Calcium ammonium nitrate should be applied @ 1.10 kg per 100 m^2 three weeks after planting and again at active vegetative growth stage.

Varieties

The Lilium varieties are generally of three types, namely, Asiatic, Oriental and Longiflorum hybrids which are generally selected on the basis of their flower colour. On the basis of trials conducted at Field Research Laboratory (DRDO), Leh, a number of Lilium varieties have been found suitable for cultivation in cold arid conditions of Ladakh (Table 3).

a) Asiatic Lilium: Alaska (White), Amsterdam (Orange red), Elite (Orange), Gran Paradiso (Orange red), Jessica (Dark Orange), Kinks (Orange), Nove Centro (Yellow), Pasadena (Dark Yellow), Yellow Giant (Yellow)

b) Oriental Hybrid Lilium: Alhambra, Amanda, Atlantis, Cascade, Kissproof, Mediterannee, Marco Polo, Star Fighter, Star Gazer, White Merostar

c)Longiflorum Hybrid: Avita, Snow Queen, White Fox, White Satin

Planting

Lilium generally takes an average of 75 to 90 days for flowering depending upon the variety. May to mid September is ideal period for its cultivation and hence, Lilium should be planted in May-June, so that crop is harvested before mid September, *i.e.,* onset of autumn in the region. To get a regular supply of flowers, lilium should be planted at interval of 10-15 days. Lilium can be planted on raised beds, rides or even on flat beds. In light soils of Ladakh the bulbs are planted 15 cm apart in flat beds in rows made at distance of 30 cm. Approximately 18-20 bulbs are planted per sq. m. For development of proper root system they should be planted at a depth of 10-15 cm. Shallow planting may result in improper root development and lodging of the crop.

Crop Management

Sufficient soil moisture must be maintained after planting the bulbs. Since stem and root development requires sufficient soil moisture, it is essential that top 30 cm soil should be kept moist continuously. However, water stagnation at any stage should be avoided. Thus, frequent and light irrigation should be given throughout growing period. Staking is another important operation to keep the plant erect. This is done with the help of sticks for providing support to the individual plants. Netting can also be used for staking which should be gradually raised as the plant grows. To ensure healthy crop, regular removal of weeds becomes necessary. Care should be taken that underground bulbs are not damaged during these operations.

Harvesting

Lilium bulbs are generally harvested between 90-110 days after planting, depending on planting time and variety. As soon as first bud shows distinct colouration, lilium should be harvested. Premature harvesting may lead to improper development of flower bud whereas late harvesting (when flowers are fully open) may result in damage of flower during transit. Flowers should be harvested preferably in the morning hours with sharp knife. Cut stems should be placed in a bucket of cold water, in shade immediately after harvesting and should not be exposed to direct sun.

Grading, Storage and Vase life

Lilium flowers are generally graded on the basis of number of flower buds per steam and length of the stem. Hence, these factors should be kept in mind while harvesting and grading of Lilium. Freshly harvested flower can be stored at 25°C for a period of 7-10 days. Removal of leaves from the bottom of steam (10-15 cm) should also be done after harvesting to improve the keeping quality of the flowers. Addition of sucrose (5%) along with Hydroxy quinoline sulphate (200 ppm) in water has been found to improve the vase life of the flowers significantly.

4. Cultivation of Carnation

Carnation is one of the top ten flowers traded in the world market. Commercial cultivation of Carnation has taken deep roots in hills of India due to its attractive flower, long flower stalk, better vase life, hardy crop and high market demand. Carnations are excellent for cut flowers, bedding, pots, border, edging and rock gardens. It has immense potential to be grown in open and under protected conditions in hills of India. (Table 4) has shown significant variations in fifteen varieties of carnation evaluated under polyhouse conditions of Ladakh with regards to plant height, days to flower, flower stalk length, number of flowers per plant and duration of flowering. Average plant height varied from 47.65 cm to 57.66 cm in New Espana and Arthur Sim, respectively. Flowering in carnation took 115.8 days in Corpalmor while Red Carmo took maximum 130.66 days to flower. Flower stalk length is important trait in cut flowers. Varieties evaluated have shown minimum stalk length in Cabaret variety (19.65 cm) whereas, maximum length was observed in Dunty Pink variety (27.00 cm). Number of flowers per plant has also shown considerable variation from 4.3 flowers per plant (New Espana) to maximum 6.3 flowers per plant (Red Carmo). Size of flowers varied from 5.0 cm to 5.9 cm in Candy/White Candy variety and Corplamor/Shocking Pink respectively. Calyx splitting is a serious malady affecting the quality of flowers in Carnation. As the bud opens and petals approach their full size, the calyx may split down due to non formation of cylindrical calyx tube which supports the base of the petal. The low calyx splitting, therefore, is a desirable trait in carnation flowers. In this study, inspite of more number of petals and large size of the flower, no calyx splitting was observed. The longest duration of flowering (34.3 days) was observed in variety Flair followed by 33.0 days in Etora. Red Carmo variety has shown shortest duration (26.65 days) of flowering. The varieties evaluated under the investigation have also shown better and attractive flower colour and shape. The cooler temperature at flower development stage and longer photoperiods during entire growing season may have attributed in production of attractive flower colour and shape. These varieties have not shown any symptom of disease and pest incidence under Ladakh conditions. Low atmospheric humidity and low temperature during growing period could be possible reason for no infestation of pest and diseases.

Among the fifteen varieties evaluated Flair, Etora, Red Carmo and Corpalmor are some important varieties of Carnations which have performed better under cold arid conditions of Ladakh since Carnation is a long day plant and requires cools season for its growth and development and such growing conditions are existing in the region. These varieties therefore, can be taken for commercial cultivation under Ladakh conditions (Dwivedi and Kareem, 2004). Carnation being one of the most important commercial flowers of the world has been successfully introduced under cold arid conditions.

5. Potential Wild Flora as Ornamental

The region has rich floral diversity with special tolerance to abiotic stresses. Several explorations have been conducted and about 200 taxa have been repotted from the Higher Himalayan regions which have excellent ability to withstand ultra low temperature during winter and complete its life cycle in a very short span of time (100 to 150 days) besides having the excellent ability to tolerate soil, water and nutrient stresses. The valuable floras have the potential for domestication and scientifically exploited for their potential as medicinal, aromatic, food, fodder, forestry, ornamental, dye, fence purposes etc. to revolutionize the economy of the tribal region. About 24 wild flora of the region have been identified as potential ornamental plants. Initial studies have shown that these indigenous plants can be domesticated and used for ornamental purposes (Chaurasia *et al.*, 2006). Brief description (namely, botanical name, family, local / English name, habitat, morphological features, ornamental use and propagation method) of the potential ornamental plants have been given in the (Table 5).

Table 1 : Performance of important annual flowers in ladakh

Name of flower (1)	Avg. plant height (cm) (2)	Row Spacing (cm)(3)	Flowering time (days after sowing)(4)	Flowering duration (in days) (5)	Uses/Remarks
Antirrhinum**	60-65	15-20	75-80	30-35	Potted plant, Bedding, Cut flower
Aster	50-60		90-100	30-35	Bedding, Cut flower, transplant carefully
Bells of Ireland	70-80	20-25	80-90	50-60	Bedding Potted plant, Dry decor
Calendula	35-45	20-25	35-45	20-25	Bedding, Borders, Potted plant
Californian poppy*	40-50		80-85	25-30	Bedding, Direct seed sowing
Candytuft	30-40	25-30	90-100	35-40	Bedding, potted plant
Coreopsis	40-45	20-25	70-75	25-30	Borders ,along the path
Chrysanthemum	115-125	50-60	95-105	15-20	Border, Bedding, Cut flower, tolerate moisture stress.
Clarkia	70-80	40-50	95-105	15-20	Potted plant, Cut flower
Cleome	135-145	50-60	90-110	20-25	Borders, Bedding
Cornflower	85-95	20-25	90-100	20-25	Potted plant, Mix border
Cosmos	135-145	40-45	70-80	30-40	Bedding, Borders
Dahlia	110-120	40-50	100-120	20-25	Bedding , Potted plant, requires staking
Dianthus	25-35	30-40	100-110	20-25	Cut flower, Potted plant, propagation by cutting also
Dimorphotheca	50-60	25-30	75-85	20-30	Bedding Potted plant
Gazania	50-60	30-35	80-90	35-40	Bedding Potted plant
Gaillardia**	35-45	30-40	80-90	35-40	Bedding , Perennial, sunny locations
Godetia	50-60	30-40	70-80	30-40	Potted plant, Bedding exotic look
Gypsophila	70-80	35-45	70-80	20-25	Bedding, Cut flower
Hollyhock**	160-180	50-60	100-120	30-40	Borders, Background
Iceplant	20-25	15-20	80-90	25-30	Edging, Rockery, Hanging basket, sunny locations

Contd...

Kochia	60-80	40-45	Bush in 60 days	45-60	Borders, Potted plants, grows in shade also
Larkspur	60-70	30-35	80-90	25-30	Bedding, Potted plants, sun loving
Lin	70-80	20-30	80-90	20-25	Bedding
Linaria	60-70	20-25	70-80	20-25	Bedding Potted Plant
Marigold(African)	80-90	30-40	80-90	25-30	Fragrant flower for worship , garland
Marigold (French)	40-50	20-25	80-90	25-30	Bedding , Potted plant, Garland
Morning glory*	130-160	20-25	60-70	20-25	Climber
Nasturtium*	30-40	25-30	60-70	25-30	Hanging baskets, sunny locations
Ornamental cabbage	25-30	30-40	70-80	40-45	Potted plant
Ornamental chilli	20-25	40-45	90-100	40-45	Potted plant
Paper flower	50-60	30-40	90-100	40-45	Bedding , Potted plant, cut flower
Pansy	20-25	15-20	80-90	25-30	Good for pots, difficult to grow
Petunia	60-70	30-35	80-90	50-60	Bedding, Hanging Baskets, Potted plant, sow carefully
Phlox	35-40	20-25	70-80	45-50	Borders, Edging, cut flower
Poppy*	60-80	20-25	80-90	20-25	Bedding ,Borders
Portulaca	20-25	15-20	90-100	20-25	Borders, Hanging Basket
Salvia	60-75	30-40	100-120	25-30	Good for sun and shade
Stock	45-60	25-30	100-120	20-25	Cut flower, Night scented
Sunflower*	120-150	30-60	100-110	30-40	Background, needs staking
Sweet alyssum	20-25	15-20	40-50	50-60	Sweet scented flowers, Edging, Rockery
Sweet Pea*	60-90	15-20	60-70	25-30	Annual climber, scented flower
Sweet Sultan	70-80	30-35	80-90	20-30	Bedding, Cut flowers, sunny places
Sweet William	60-70	30-40	110-120	30-40	Cut flowers, Bedding Perennial
Verbena	25-30	20-25	45-50	20-25	Rockery, Hanging Basket

(Anon, 2006)

* Bold seeded flowers, can be sown directly.

** Flowers perennial under Ladakh conditions.

Table 2 : Performance of gladiolus varieties under Ladakh conditions

S.No.	Variety	Plant Height (cm)	No. of leaves / per plant	Days to flowering	Spike length (cm)	Florets/spike	Cormels / plant
1.	American Beauty	101.9	9.3	118.0	98.9	16.4	42.9
2.	Cambourne Butterfly	114.3	8.9	120.4	82.9	17.3	33.6
3.	Eurovision	119.4	9.1	122.6	82.4	15.2	48.8
4.	Friendship	117.8	9.1	98.2	102.5	19.6	38.4
5.	Ice Gold	105.8	8.9	113.8	96.4	18.4	36.9
6.	Novalux	107.1	9.2	100.9	79.6	15.3	18.7
7.	Novelty	108.1	9.1	103.5	91.4	18.9	48.8
8.	Princess Margaret Rose	136.6	8.9	124.5	111.0	18.4	45.4
9.	Priscilla	104.6	9.4	91.9	80.8	16.8	18.9
10.	Rose Delight	103.6	9.1	109.0	88.5	16.9	38.6
11.	Spic and Span	120.7	9.1	115.0	96.4	18.4	46.5
12.	Souvenier	104.5	9.4	107.4	95.4	20.7	45.6
13.	Tiger Flame	114.5	8.9	120.2	83.0	16.2	36.7
14.	Tropic Seas	104.3	8.7	117.9	79.1	14.7	22.2
15.	Venetia	119.5	9.4	103.7	96.2	16.9	31.6
16.	Victor Borge	112.6	8.7	108.0	88.0	15.8	55.7
17.	White Prosperity	125.9	9.7	117.8	96.2	21.1	50.2
18.	Wind Song	118.6	9.7	112.5	98.3	15.8	42.3
19.	Wings Sensation	121.7	8.9	114.7	91.7	17.3	40.4
20.	Winter Goddess	117.5	9.3	118.3	89.7	16.5	41.9
	CD at O.05P	15.6	NS	14.6	8.3	3.7	5.4

(Dwivedi *et. al.*, 2002)

Table 3 : Performance of Asiatic Lilium varieties under Ladakh conditions

Variety	Colour	Plant height(cm)	No. of leaves Per plant	Days to flower	No. of buds/plants	Flowering period(days)	Flower spread (cm)
Alaska	White	47.8	45.1	97.0	7.13	9.0	12.1
Amsterdam	Orange, red	65.0	61.0	99.0	6.30	8.6	11.6
Elite	Orange	42.2	34.5	96.0	5.33	7.3	10.3
Gram Paradiso	Orange red	59.6	57.1	92.0	4.33	8.1	10.6
Jessica	Dark orange	62.0	59.0	98.6	6.40	7.0	12.2
Kinks	Orange	56.4	52.0	101.5	5.50	6.9	11.2
Nove Cento	Yellow	47.6	38.6	103.8	6.30	6.7	12.2
Pasadena	Orange	52.0	49.6	98.7	4.80	8.1	12.3
Pollyana	Dark yellow	48.7	46.0	94.0	4.13	6.7	10.7
Yellow Giant	Yellow	43.8	42.5	95.3	5.80	9.0	11.9

(Dwivedi *et. al.*, 2002)

Table 4 : Performance of carnation varieties in cold arid regions of Ladakh (2000-2001)

Variety	Plant height (cm)	Days to flower	Flower stalk length (cm)	No. of flowers per plant	Size of flowers (cm)	Duration of flowering (days)
Arthur Sim	57.66	125.17	25.65	5.0	5.8	29.27
Cabaret	48.65	117.66	19.65	5.5	5.2	28.33
Candy	51.25	127.33	23.65	6.1	5.0	29.25
Corpalmor	51.60	115.80	19.80	4.9	5.9	32.25
Dunty Pink	54.65	124.00	27.00	5.6	5.3	28.35
Espana	50.75	119.75	20.75	5.0	5.3	32.55
Etora	58.10	118.33	22.50	4.8	5.7	33.00
Flair	54.30	120.00	21.80	6.1	5.2	34.30
Irma	49.75	120.25	21.75	5.5	5.2	28.15
Managio	51.80	126.25	23.45	5.9	5.5	30.45
New Espana	47.65	123.50	25.25	4.3	5.8	30.85
Pink Pirra	53.40	121.50	22.00	4.9	5.6	29.15
Red Carmo	55.66	130.66	23.70	6.3	5.2	26.65
Shocking Pink	56.15	123.00	23.45	5.2	5.9	32.30
White Candy	52.20	127.15	26.25	5.2	5.0	31.75
CD (p=0.05)	**1.40**	**4.47**	**1.60**	**0.34**	**NS**	**1.80**

(Dwivedi *et. al.*, 2002)

Table 5 : Indigenous Plant species of Higher Himalaya having potential to be used for ornamental purposes

Botanical Name (Family)	Local/English Name	Habitat	Plant Morphology	Ornamental uses	Propagation
Acantholimon lycopodioidies (Girard.)Bois.(Plumbaginaceae)	Longze/ Acantholimon	Changthang, Higher Peaks	A tufted spiny shrub, with white flowers	Whole plant as dry decoration	Suckers
Allium Prezewalskianum Regel. (Amaryllidaceae)	Skotche/Wild garlic	Partapur, 2895m. (Nubra valley)	Bulbous herb, with pink-purple flowers	Pink flowers for pots/fflower beds	Seeds/bulbs
Aquilegia fragrans Benth. (Ranunculaceae)	Fragrans/Colmbine	Sangri,2880m(Suru valley)	A perennial herb, with cream-coloured flowers	Scented flowers fit for pots beds	Seeds
Aster tibeticus Hook.f. (Asteraceae)	Skalzangmentok/Wild aster	Tsoltak, 4,720 m.(Changthang valley)	A herb with attractive flowers	Flowers for beds	Seeds
Chorispora sabulsa Camb (Brassicaceae)	Imbuk/Chorispora	Zingral, 4,710m (Indus valley)	A perennial herb, with pink flowers	Scented flowers for pots	Seeds
Cremanthodium plantagineum Maxim . forma illisii (Hook.f.) Kitamura (Asteraceae)	Lukumentok/ Cremanthodium	Penzila, 4,260m (Zanskar valley)	A perennial herb with yellow flowers	Flowers	Seeds
Dactylorhiza hatagirea (Don) Soo (Orchidaceae)	Angmolapa/Spotted heart orchid	Partapur, 2,850m. (Nubra valley)	A perennial herb, with purple spikes	Flower-spikes as cut flowers	Roots
Delphinium brunonianum Royle (Ranunculaceae)	Lunde-kaown/Himalayan Larkspur	South Pullu, 4,735 m. (Indus valley)	A perennial herb, with purple spikes	Suitable for pots	Roots
Delphinium cashmerianum Royle (Ranunculaceae)	Lunde-kaown/Himalayan Larkspur	Tsoltak, 4,750m (Changthang valley)	A prostrate herb, with Purple spikes	Suitable for pots	-----
Dracocephalum heterophyllum Benth. (Lemiaceae)	Zinkzer/Dracocephalum	Zingral, 4.710m. (Indus valley)	A spiny herb,with Cream-Coloured flowers	Suitable for beds	Runners/seeds
Echinops cornigerus DC (Asteraceae)	Eckzema/Echinops	Leh City, 3,505m. (Indus Valley)	A Spiny herb, with white flowers	Flowers for dry decoration	Seeds
Epilobium angustifolium L (Onagraceae)	Willow herb	Randum,3780 m. (Zanskar valley)	A branched herb, with blue flowers	As ornamental hedge and cut-flower	Suckers/seeds

Contd...

Indula obtusifolia (Asteraccae)	Kerner Rupmak/Inula	South pullu, 4,780 m(Indus valley)	A herb with yellow flowers	Flowers for pots	Seeds
Iris lactea Pallas (Iridaceae)	Tesmamentok/Wild Khalsi, iris	3,820m (Indus valley)	A bulbous herb, with purple flowers	Flowers for beds	Bulb
Juniperus macropoda Boiss. (Juniperaceae)	Shuppa/Juniper	Dras, 3,120 m. (Suru valley)	An evergreen perennial shrub/tree	Ornamental foliage	Seed
Meconpsis aculeate Royle (Papaveraceae)	Achatsarnmm/Blue poppy	Perzila 4,260 m. (Zanskar valley)	A herb with blue flowers	Suitable for pots	---------
Papaver nudicaule L (Papaveraceae)	Ice land Poppy	Tscltak, 4,715m. (Changthang valley	A herb with onange flowers	Suitable for beds	--------
Primula rosea Royle (Primulaceae)	Sulumentok/Prime rose	Penzila 4,210 m. (Zanskar valley)	A bulbous herb, with purple flowers	Scented flowers for beds/cut-flowers	Bulb Seeds
Rosa eglentaria L (Rosaceae)	Siah/Wild rose	Kargl, 2,590m. (Suru valley)	A shrub with yellow flowers	Ornamental shrub/cut-flowers	Suckers/cuttings
Rose webbiana Wall. Royle (Rosaceae)	Siah/Wild rose	Panamic, 3,110 m.(Nubra valley)	A shrub with pink flowers	Ornamental shrub/cut-flowers	Suckers/stem cuttings
Saussurea obvallata (DC) Edgew. (Asteraceaaae)	Pangchi/Brahma Kamal	South pullu, 4,715 m. (Indus valley)	A herb with attractive flowers	Attractive flower-spike	Seeds
Sedum ewersii Ledeb(Crassiulaceae)	Gomnii/Stone crop	Stock, 3,825 m. (Indus valley)	A succulent herb	Suitable for hanging baskets	Suckers
Sedum heterodonta Hook.f.&Thoms.(Crassulaceae)	Gomnii/Stone crop/Wild rose	Panamic, 3,110 m. (Nubra valley)	A herb with yellow flowers	Suitable for hanging baskets	Root cuttings/Seeds
Thylacospermum caespitosum (Cambess) Schischkin (Caryophyllaceae)	Tegarcon/Thylacospermum	Tanglanla, 5,330 m. (Changthang valley)	A tufted herb	As a dry decoration	----------

(Chaurasia *et. al.*, 2006)

6. Challenges and Opportunities

Low humidity, long day length, bright sun shines and pest/disease free environment offers unique condition for the production of high quality Gladiolus, Lillum and Carnation for the purpose of cut flower although Floricultural crops are not yet commercially cultivated a wide range of annuals can be grown successfully in beds and pots as well. These regions can also be exploited for production of quality seeds of annual flower crops and disease free planting material of crops like Gladiolus, Lilium, Dahlia *etc.* Poor soil, low atmospheric humidity, extreme diurnal temperature, very low precipitation, short growing period, inaccessibility, poor technical awareness, lack of standardized agro-techniques, poor post harvest management, transportation/ marketing network, poor logistic support *etc.* are the major challenges which may limit the performance of the floriculture industry of the region. The region however, boasts of a disease and pest free environment, superior quality/highly attractive produce, possibilities of growing rare/unique/exotic flower crops, potentials of flower seed production, organic farming practices, off season availability of crops which ensures better prices in export/distant markets *etc.* The region does not have proper transport facilities because of which the floriculture may face tremendous problems in marketing of their produce. The most important part of flower cultivation is the technical know-how which a grower normally lacks. Non availability of quality seeds and other planting materials and varieties specific for the region are other challenges before the industry. The region has rich floral diversity with special tolerance to abiotic stresses. A number of wild floras of the region have the potential to be used as ornamental plants. Initial studies have shown that these indigenous plants can be domesticated and used for ornamental purposes. Future research efforts need to target introduction and evaluation of new crops/ varieties suited to the region, their scientific production technologies and utilization/value addition, alongwith awareness programmes for farmers to ensure sustainable development of the floriculture industry in the region.

References

Anonnymous (2006). Identify your Garden Flowers and Growing tips for Annual Flowers in Ladakh (Flower Poster). Field Research Laboratory (DRDO) - Leh, India.

Chaurasia O.P.; Dwivedi S.K.; Attrey D.P.; Singh B. and Ballabh B (2006). Wild flora of Ladakh Himalaya -Potential ornamental plants. J. Econ. Toxon Bot. 30(1): 79-82.

Dwivedi, S.K. and Attrey, D.P. (2004). Lilium Cultivation in Ladakh, Field Research Laboratory (DRDO) Leh, India.

Dwivedi S.K. and Kareem, Abdul (2004). Introduction and evaluation of carnation varieties in cold arid region of India. J. Orn. Hort. 7(2):207-209.

Dwivedi, S.K.; Paljor, Eli. and Attrey, D.P. (2002). Gladiolus in Ladakh. Field Research Laboratory, DRDO, Leh, India.

Dwivedi S.K.; Attrey, D.P.; Paljor, Eli and Kareem, Abdul (2002a). Introduction and evaluation of Asiatic Lilium in cold arid conditions of Ladakh. In: National Symposium on Indian Floriculture in the New Millenium. Bangalore, India on 25-27 Feb. 2002. *In: Floriculture Research trend in India.* (eds.) R.L. Mishra and Sanyat Mishra. pp. 293-294.

Dwivedi, S.K. and Attrey, D.P. (2001). Gladiolus Cultivation in Ladakh. Field Research Laboratory (DRDO), Leh, India.

Dwivedi, S.K. and Singh, B. (2000). Gladiolus-A Potential Crop in Ladakh. Exploring the gladiolus in India. NBRI, Lucknow. pp. 75-80.

□□□

7

Woody Elements of Plant Wealth of Trans-Himalayas

ANUP RAJ
Regional Agricultural Research Station, S.K. University of Agriculture Science & Technology, Leh-194 101, Ladakh

Of the total trans Himalayan region of 2.6 million sq km, a small portion of approximately 0.18 million sq km is in India. The rest of it is in Tibetan autonomous region in China. Besides Ladakh, abutting areas of Lahul and Spiti and Pooh sub division of Kinnaur in Himachal Pradesh, small areas in Uttaranchal and in Sikkim, North of Kanchen Junga in the Eastern Himalayas constitute the Indian trans-Himalayas, one of the ten bio-geographical zones identified in the country. Ladakh is lying across four mountain ranges, which run in South East-North West direction, and delimits its main sub regions.

i. The Suru and Zanskar river valleys are situated between the Great Himalayan Range, which marks Ladakh's Southern boundary and the Zanskar Range to the North.

ii. The Indus valley between Zanskar and Ladakh Ranges.

iii. The Nubra valley between the Ladakh Range and the Karakoram, which forms the Northern boundary of Ladakh.

In the South East, the regularity of range and valley dissolves in a complex of high altitude ridges and plateaus, many of them forming enclosed basins, dotted with lakes and wetlands. Known as Ladakh's Changthang, this region is considered as an extension of Tibetan Chang Thang (Northern Plains) into India (Anon., 1997). Whole of Ladakh is administratively divided into two districts namely Leh and Kargil. Leh district is comprising of Indus and Nubra valleys, and the Changthang. The area of Kargil district is comprising of Zanskar, Suru and Drass valleys.

General Features of Vegetation

The xeric environment of Ladakh is further accentuated by the dominance of xerophytic and semi xerophytic plants like *Echinopsis cornigerus, Capparis spinosa* and *Crepis flexuosa.* On ultimate analysis of vegetation Kachroo *et al.* (1977) grouped the plant of Ladakh in different plant types and found that therophytres (annual plants) dominated with 33% of share, chaemoephytes 27%, geophytes 21%, hemicryptophytes 10% and planerophytes (trees and bushes) 7%; a life from spectrum typical of semi desert environment. In spite of that other plant types are not excluded. Numerous species are extremely delicate herbaceous plants with unbelievable tender leaves and flowers that wilt rapidly within a few minutes of being uprooted (*Oxyria dignya*). There is in fact nearly every gradation from totally hygrophile to arid steppe type: delicate hygrobiont annual herbs, delicate hygrophile herbs which have capacity of reviving even after drying and continuing growth on being moistened again, semi hygrophiles (mesophytes) on grassy meadows and xerophiles and xerobionts.

Aerial parts of the plant of high altitude ecosystem are mostly annuals (*Peganum harmala, Artemisia* spp., *Lepidium latifolium*) with rare exceptions. These plant have perennial underground stems, rhizomes, and root stock which sprout forth annually during the abbreviated growing season in summer into prostrate (*Capparis spinosa),* decumbent or erect but dwarf (*Peganum harmala, Crepis flexuosa)* aerial shoots that wither away at the end of growing season after having flowered and set seeds. In some species the aerial tufted matted or dome shaped growth do not succumb to the approach of winter but persist as hard schlerophyll, much branched aerial perennials. The remarkable hemispherical *Caragana* in the sand river plain of Changthang, the dome shaped *Thylacospermum* and flat and stunted *Acantholimon* on moist hilly slopes in Kargil are examples of this group.

Some of these like *Hippophae* and *Myricaria* are characteristically stunted, stout, much branched dwarf bushes. These plants remain partially or completely buried under snow during winter. When the snow melts in the following summer reawakened from their dormancy, the buds of the plant unfold and short branches grow. Trees grow when they are irrigated, or in a rare spot where they can find water naturally, but they form a very small part of the covering of the area.

General Floristics

Despite tough, inaccessible and hostile terrain Ladakh has attracted many botanists from early times. Information gathered by such professional as well as non-professional botanists, are gleaned by Stewart (1917) from the scattered documents available at that time. He was the first botanist who conducted the vegetation survey in Ladakh and documented the floristic of the region in a systematic manner. During his expeditions, though, he did not cover Nubra and Zanskar valleys, which are within the geographical boundaries of present Ladakh. The species enlisted by him include those mentioned by the earlier botanist besides his own collection, which obviously constituted the major chunk. The plants found in Ladakh were classified into three majors groups by him: alpine, desert and oasis. Alpine group is largely confined to narrow belt below the melting snow and along the upper coarses of streams and does not spread into the valley (Stewart, 1917). The oasis are confined to wet lands (riverine as well as lacustrine) and along the cultivated fields and irrigation channels. Most of the Ladakh is covered by open desert steppe characterized by low plant diversity and abundance, and high level of adaptation to high altitude and aridity as well. Across Ladakh's landscape, the vegetation changes gradually from alpine meadows (*Kobresia, Cares, Potentilla, Nepeta* etc.) on North side of the Himalayan crest to steppe vegetation (*Caragana, Artemisia, Stachys, Ephedra, Stipa* etc.) to the North and East with shrub land (*Hippophae, Myricaria, Salix)* along river courses. The region is virtually tree less except for isolated patches of Junipers (*Juniperus macropada, J. recurva, J. communis)* and birch (*Betula utilis)* in some valleys and mostly cultivated varieties of poplars (*Populus* spp.) and willows (*Salix spp.*) along the major water courses (Fox *et al*., 1994).

Woody Species

Although there are no forests in Ladakh, there are frequently thickets along streams and on little islands in the rivers. These have a typical association of *Myricaria, Hippophae, Rosa and Clematis orientalis.* The commonest trees are willow, poplar, walnut, mulberrie, apple and apricot, mostly introduced. Small grooves of native junipers (*Juniperus macropoda)* are sometimes found (Stewart, 1917). Treelessness of the tract is explored by many workers (Mani, 1990; Fox *et al.,* 1994). In such a high altitude environment paucity of tree vegetation is a general rule. If the trees are present at all they change their habit and become bushy in nature. Even the tree species known to have strong apical dominance, *Juniperus macropoda* and *Populus nigra* for example, often found to have acquired shrubby nature in Ladakh. Inadequate moisture regime and severe cold during winter prevailing here cut short the growing season to approximately five months. In such the condition stems die back at apparently irregular intervals possibly because they failed to maintain a positive carbon balance and are replaced by epicormic shoots. For a species to reach tree size, the appropriated growing condition must prevail in the free air at the height of tree canopy (position of apical bud). In addition to that the growing season must be long and warm enough for trees to develop shoots to maturity and harden them against winter adversity. Trees must accumulate sufficient carbon to offset the winter period when negative carbon balance and loss of tissues through abscission and injury are likely and to provide a surplus of assimilates for reproduction (Wardle, 1993).

Notwithstanding these facts the diversity of woody plant species in Ladakh is fairly high (Anup Raj, 2004). There are many patches of natural forest still existing (Abbas, 2004) in remote unexplored valleys. Besides, there are many regions at relatively lower altitudes, Kargil and Nubra for example, where large stretches of tall trees of indigenous species are found growing in abundance. *J.macropoda, P.alba, P. balsemifera, Eleagnus hortensis, Morus alba* etc. are some of the indigenous species that grow to large dimensions. Many of these may reach up to 15 meter in height at favourable locations. In addition to these there are many exotic trees like *P. nigra, S. alba* and *Robinia psuedacacia* are performing very well. Trees of *P. nigra* growing to a height of 20 meter are not uncommon even at an elevation of 3,500 m and more in Leh. Shrubs like *Myricaria germanica, Tamaricara elegans, Tamarix gallica, Hippophae rhamnoides, Rosa webbiana,* and *Ephedra gerardiana* are some of the examples which dominate the natural landscape in many locations of the region including Changthang.

Forest Type

In the revised classification of forest type of India (Champion and Seth, 1968) the vegetation of Ladakh is spread over various forest types under three groups namely I) Himalayan Dry Temperate Forests (Group 13), ii) Sub-alpine Forest (Group 14) and iii) Dry Alpine Scrub (Group 16). Under group 13 these are West Himalayan dry Juniper forest (type 13/C5) and two serial types, namely *Hippophae-Myricaria* scrub (type13/1S1) and *Populus-Salix* forest (type13/1S2); under group 14, it is *Hippophae-Myricaria* scrub under general serial and degraded types (type14/1S1); and under group 16 it is dry Alpine scrub (type 16/C1). These forests and scrubs are confined to the riverine wetlands developed in the valley bottoms along Ladakh's main river courses and their tributaries. These habitats in the main valley are under intense biotic pressure and many of them have been replaced by permanent cultivations or plantations of exotic willow and poplar. In spite of that there are relatively remote areas where these vegetation types still remain intact. Inaccessibility of these thickets owing to unavailability of motorable road and insurmountable terrain has saved these ecologically significant biological formations. On the other hand it has prevented the scientists to gather information about their extent, species' composition and other phyto-sociological parameters.

Trees and Shrubs

Notwithstanding the extremes of climate prevailing, the region is surprisingly more diverse in tree and shrub species. In the following section the trees and shrubs of Ladakh are dealt with in detail. Many of these plants either growing in the arboratum raised at the Stakna farm near Leh or preserved in the form of herbarium of SKUAST-K. In describing the plants, the details related to systematics has been avoided. However, the information regarding their habit and habitat, which can help in identifying the plants *in situ*, are given. Much of the information is based on extensive survey of Kargil, Nubra and Changthang and intensive survey of Nubra valley. For rest of the area of the region, information from relevant literature and knowledge of the local people has been relied upon. A distinction has been made between the primary and secondary sources of information wherever necessary in the following section.

Tree Species

1. *Populus* species

Populus is one of the most prominent genera in this region. There are four species of *Populus* growing in this region, which are *P. nigra*,

P. basemifera. P. alba and *P. euphratica.* Of these *P. nigra* is exotic species and is widely planted in the farmers field and along streams. The timber is used for making rafters of the roof of local Ladakhi houses and is much valued in this timber deficient region.

P. balsemifera is the next most important timber species. This plant is reported to be growing indigenously in Nubra valley (Hooker, 1885). During survey, however, we could not find it to be growing in wild though it is commonly found in the farmers' field in the Nubra valley especially in Hunder and Diskit villages. Waving bole and resinous buds are characteristic of this species. Though the crookedness of the bole reduces its economic value as compared to that of *P. nigra* it is also preferred by the locals due to relatively greater strength of its wood.

P. alba and *P. euphratica* are the two indigenous poplars of this region. *P. alba* is distinguished from the other poplars by the silvery young shoots (due to white tomentose growth) and dorsal side of the leaves. The leaves are also lobed, and not entire, like other poplars. This species is found growing around Tirit, Turtuk and Thang villages of Nubra valley. Best size trees are present in Thang village, the last Indian villages on the Indian side of LOC with POK. One tree of this species is also growing in Alchi monastery campus near Leh and is reserved by the monks. Common people are not allowed to sever even the branches of the trees except for religious purpose.

P. euphratica is very interesting with respect to its morphology as well as its habitat. This is found along the riverbank of Shyok. At least two patches of this species are encountered when one traverse along its left bank. First patch is found at Degar near Rongjuk in Nubra valley where number of individuals are less about 20-25. However, the trees have single stem (not bushy) and about 3-4 m in height. Bark is white in colour and trees are old age. This patch is under high biotic pressure and no regeneration is observed in this area.

Near Changmar, down stream Shyok another patch is present on sandy tract in a comparatively larger area. Here *P. eupharatica* is bushy and associated with *Tamarix gallica* and *Berberis ulcina.* The whole area is under fencing by soil conservation department and the biotic interference is minimal. Scattered plants of *P. euphratica* can also be found along the riversides continually between these two patches. Polymorphism in the leaves of this species are expressed to the extreme with atleast four different leaf shapes can be easily recognized on a single tree.

One more species of *Populus* is also reported in literature (Kachroo *et al.*, 1977) which we have not been able to locate is *P. ciliate.* This species is grouped together with *P. balsemifera* as both have resinous buds. However, these two can be distinguished by two major characters. The leaves of *P. ciliata* are white on its dorsal surface and the male flowers are petioled.

2. *Salix* Species

The region is very rich in species diversity in wild willow as well as cultivated. As we find the identification difficult, we could not reach to final conclusion about botanical name of many of these. In that case we have used the common names.

i. *Malchang* (*Salix alba)*: It is the most commonly grown willow in Leh and nearby areas. The tree is pollarded at a height of 2 meter approximately and is managed for harvesting tallu (joist) in 3-4 years rotation. The tallu of this species is not much valued due to its wavering nature. However, the leaves are of much fodder value and the wood of main stem is used for making chogtse (a small table like structure) when the stem grows thick enough. If not pollarded it grows as a large tree with a bole up to 8 m in height. Trees near Zorawar fort in Leh city are one of the best examples.

ii.S *elchang*: The bark of young shoot is bright yellow green that becomes dark yellow to orange colour with maturity. The tallu is highly prized because of its straightness and comparatively uniform thickness through out. This species is free from willow leaf gall and scale insect unlike the other two commonly grown ones.

iii. *Shoa*: It is another commonly growing willow on the farmers' field along the streams for tallu purposes. The young shoots are reddish brown in colour.

iv. *Other Species*: Besides these 3 common species of *salix* four others are also growing in the arboretum of SKUAST(K) at Stakna farm. These plants are collected from Nubra valley and most of them are growing in wild. The willow species from Ladakh are reported by different workers are mentioned.

3. *Juniperus* spp.

The common juniper found in Ladakh is *Juniperus macropoda.* One can encounter this species very frequently on the moist hilly slopes between Kargil and Drass in Kargil district and in Dha-Hanoo belt in Leh district. As one moves eastward in Ladakh proper, this species becomes rare. A few small patches are reported from Hemis Shukpachan and Tia-Khasa phoo (Sharma and Anup Raj, 2004). Besides this patches of few scattered trees are also found in Nubra and Leh valley. Most of these trees are bushy in nature. Others are mature and over mature. The natural regeneration in these areas are very scanty and its probable reasons are discussed by Sharma and Anup Raj (2004).

Two others species, *viz., J. wallichiana and J. comnvnis* are also listed in the flora of Ladakh and western Tibet by Stewart (1917). Kachroo *et al.* (1977) have also mentioned three Junipers from this area: *J. recurva* from Kargil, *J. communis* from Zojila and *J. macropoda* from Nubra. *J. communis* is reported from Chalunkha village in Nubra by Chaurasia and Singh (1996). However, we have not seen *J. communis* (a small shrub, with all leaves linear and acicular and not dimorphic leaves as in *J. macropoda*) in Nubra or in Leh valley. Whatever may be the status of other junipers in Ladakh, *J. macropoda* is the most common juniper, which is highly revered by the Buddhish community and also under high biotic pressure mainly due to its demand in their religious rituals.

4. *Betula utilis*

Betula utilis is conspicuously present at Zojila and its presence was well noted by Stewart (1917) and Kachroo *et al.* (1977). However in Ladakh it has shrunkn to a few patches (Abbas, 2004) in Zanskar valley of Allam Tibet. Though we have not personally seen it in Ladakh proper.

5. *Ulmus waltichiana*

Yombuk, as it is popularly know in Ladakh is present in one patch only in village Tegar in Nubra valley and around the Samathanling Gonpa. In early days local people used its bark as shampoo to wash hairs. But as the modern day technology and its product reached to the interior of this otherwise in accessible area, the importance of this tree has reduced to such an extent that it is now treated as unwanted tree there. The seedlings of this plant, which comes up profusely near the few existing mature trees, is uprooted and thrown as weed and this species has now become a threatened species in Ladakh (Anup Raj, 2004).

6. *Eleagnus hortensis*

This species is grown in farmers field and yields edible fruits. This species is very common in Kargil where at least five phenotypes, namely, Bee sersing, Chapacha, Balti, Marpo and Ringmo are easily identified by the local people. However, its status in systematic should be determined. This species is also found in Leh and Nubra valley as scattered trees. However, it is not uncommon in these localities. Another species *E. umbellata* is also found in the region.

7. *Morus alba*

It is very commonly grown in Kargil district in the farmers' field for its fruits. At least two morphoforms of this species are found in Kargil. Its position in the taxonomy should be investigated. This species is not seen growing in Leh or Nubra valley. However in literature it is also reported to be growing in Indus valley (Stewart, 1917) and Nubra valley (Chaurasia and Singh, 1998; and Kachroo *et al.*, 1977).

8. *Juglans regia*

It is also found in Nubra valley especially in Deskit, Hunder and Turtuk area. In Leh and nearby areas it is found growing in farmers field but growth is poor. In Dha-Hanoo belt it is very commonly grown.

9. *Rhododendron anthopogan*

Rhododendrons are one of the timberline species found growing with *Betula utilis* and *Juniperus* spp. in Himalayas. However, this species is conspicuously absent from Ladakh. None of the flora have reported it from Ladakh. There is one report (anonymous, 2001), nonetheless, from Zanskar where it is known as balu.

10. *Cratagus oxycantha*

This species is found in Baima village in Leh and in Turtuk in Nubra valley. This is present in the herberium of SKUAST-K at Leh. This is a small tree having lobed leaves and fruits generally small ovoid berry. It is called Shorot in Baima.

11. *Pinus wallichiana*

Literature is generally silent about its presence in Ladakh. However, a recent publication (Abbas, 2004) reported it from Shhrtsay Doh in Kargil district. Some local people also claimed to have seen this species growing near Wanla along Zanskar river. This area may have remained unreported by botanists due to its inaccessibility and remoteness.

12. *Robinia psuedacacia*

This one is the recent introduction in Ladakh and due to its xerophytic nature it is highly adapted to its new climate. It is a fast growing tree with leaves having highly nutritive fodder value. The wood of this tree also has high calorific value. Due to these characters, this is a very potential tree species for future forestry in this area. At present, however, due to its thorniness and some other reasons, it is still an alien to the local people.

13 *Prunus tomentosa (wild plum)*

This species, locally known as Skyurmo, is highly endangered wild tree species of Ladakh. Only 5-6 plants of this species are growing at Skurbuchan village in Leh valley. Due to neglegence this species is at the verge of extinction and hence demands immediate attention of the plant scientists. Local people use the twig of the plant for cleaning teeth as toothbrush (datoon).

Besides these trees, two commonly grown fruit trees, apricot (*Prunus armeniaca*) and apple (*Malus pumila*) are commonly grown all over Ladakh.

Shrubs and Bushes

1. Tamarix gallica

This beautiful tall robust shrub (up to 3-4 m tall) is conspicuously found growing in abundance in Nubra valley. It was first observed at Panamik from where downwards its abundance increases. Its best growth is found all along the Shyok river between Khalsar, where it is present in the river beds on channel bar and sand bars, and Changmar where it is present on sand dunes beside the river Shyok along with *Populus euphratica.* It is locally known as gyalta. Its wood is hard and used for making agricultural implements.

2. Myricaria germanica

It is found growing gregariously all along the streams where slope is gentle, in Leh and Nubra valleys and also in Changthang. It is highly branched shrub upto 2 m in height with pink terminal inflorescence. The bark is reddish brown and the cambium of the plant is used in traditional Tibetan medicines. The leaves are very small, scaly and adpressed. Wherever there is gentle flowing water for at least growing season the species grows.

3. *Tamaricaria elegans*

Earlier this species was known as *Myricaria elegans* and is very similar in appearance to the *M. germanica*, excepting that it is more vigorous and taller and its leaves are bigger and not scaly or adpressed. The closely stacked stems of Umbo, as it is locally known, is used for decoration above the entry porch in monasteries. Its habitat is similar to that of *M. germanica* and is common in the thickets along Indus and its tributaries.

4. *Hippophae rhamnoides*

It is the most dominant plant species in Leh and Nubra valleys of Ladakh. The much-branched spiny plant is famous for its fruit pulp and fixed oil of its seeds. At higher altitudes (in Changthang) it is 1-2 meter tall shrub, having bushy appearance and at lower attitudes (in Turtuk) it becomes tall (upto 5-6 meter) and possesses tree like habit. It is found all along the sandy river and streamsides. The best growth is found in Nubra valley where its branches are used for fencing and wood is used for burning as well as for charcoal making. The potential of its fruit products is under-utilized yet.

One more species of this genus are reported to be growing in Ladakh, *viz., H. tibetica* in Changthang.

5. *Rosa webbiana*

The most common five petelled pink rose is found along the sandy beds of streams. It is a handsome tall (upto 2 m) erect shrub whose best growth is found in the upper reaches of Nubra river. The name of the glacier Siachen from which this river originates is in fact derived from this species Sia, as it is locally known. This species can be seen growing near Khardungla at an altitude as high as 4,800 m above mean sea level.

Yellow coloured wild rose (*R. eglantaria*) is the other species found commonly growing in Leh and Kargil area. However, this is confined to the farmers' field and gardens where it is grown for ornamental purpose. White rose is also found but is rare in Leh district.

6. *Berberis ulicina*

It is a true xerophytic species found growing in dry rock and sandy areas with no visible water source. It is a spiny small (up to 1 meter) shrub with thin stout branches. It is not seen in Leh valley. The fruit of the plant yields dark purple dye. The bark is silvery white in colour and flower is light purple. It is one of the dominant species on sand dunes in between Deskit and Hunder in Nubra valley. Another species *B. zabeliana* is also reported from Nubra (Chaurasia and Singh, 1998) and from Drass (Kachroo *et al.*, 1977).

7. *Ribes orientale*

It is common in kargil district and found along the streams in rocky slopes. A much spread small shrub with lobed leaves and glandular hairs on it. It is not common in Leh and Nubra valley. However, another species *R. alpestre* with prickles is found along the streams on way to Hunder Drok in Nubra valley.

8. *Lonicera spinosa*

This is one of the common species among many under *Lonicera* genera found in Ladakh. This species is known as Sheth and stem of this plant is used to make arrows for traditional archery events in Ladakh. This is very abundantly growing around Tiger village along with *Sophora moorcroftiana* on gentle moist slopes along the stream. Small ovoid berries are borne in pair joined together at the base.

9. *Ephedra gerardiana*

This species is found on dry rocky sites growing gregariously in Nubra and Leh valley. This is also found around Khardungla. Its best growth is found in Nubra valley where it becomes stout, woody, much branched shrubs upto 1 m in height. The stem is green, rough to touch, leaves only scaly and fruits ovoid with red fleshy bract. The dried stem is used to extract "ephedrine" which is used as drug against asthama.

Another species *E. intermedia* is a smaller shrub which is found on much drier rocky sites and is found in Kargil, and Markha valley of Leh district.

10. *Cotoneaster integerrima* (=*C. vulgaris* = *C. gilgitensis* = *C. falconeri*)

It is an erect shrub with dark red fruits. It is found on moist northern slopes along the Drass river in between Drass and Kargil. This species is locally known as hembi in kargil.

Another species *C. numularia* a prostrate but stout shrub, is also found near Hemiya along Indus on dry, rocky, almost leveled ground. In Leh this species is known as Kumchung.

11. *Colutea multiflora*

It is a tall (up to 2 m) shrub with inflated pods due to which it is also known as bladder senna. Young stem is grey white and smooth, when mature bark peels of in the form thread along the longitudinal axis. It grows on sandy and dry rocky sites and is confined to Tirit nallah and near Chalunka just before Bogdang in Nubra valley.

12. *Cicer microphyllum*

This wild chickpea is found growing at higher elevations on moist slopes. It is quite frequent on slopes just below South Pullu on way to Khardungla. It is also found in Hemis Shukpachan village and around many high altitude passes. It is very nutritive fodder plant.

13. *Sophora moorcroftiana*

It is a large (up to 2 m height), much branched spiny shrub with moniliform pods with 4-5 seeds in each pod. The ovoid seed is black, smooth and hard. This species is confined to Tiger village in Nubra valley. It is found growing on moist gentle slopes on well-developed soil along with *Lonicera spinosa* and also along the road leading to Samsthanling Gonpa.

Another species *S. alopecuroides* from Ladakh is also reported in literature (near Bodhkharbu, Kachroo *et al.*, 1977; in Zanskar, Chaurasia and Singh, 1998).

14. *Caragana versicolor (=C. pygmaea)*

A much-branched spiny shrub forming a hemispherical canopy is very commonly found on the wide sandy plains along Indus river in Changthang. It is also found near Khardung village on way to Nubra valley where its growth is not as vigorous as in Changthang. In winter local people use the uprooted plant as fuel.

Another species *C. cuneata* is reported from Drass by Kachroo *et al.* (1977). Besides, there are many species of *Astragalus* found growing throughout Ladakh, some of which are woody in nature.

Research Need

The horizon of forestry research in the region is wide open. As the climate and hence the natural vegetation is inherently different from the other ecological zones of India, the problems and the approach to solve them are also different from those applied anywhere else. The field of forestry research is comparatively new. The problems to be addressed therefore are of basic and fundamental in nature. Some of these that need immediate attention are discussed in the following section:

I. Vegetation survey and study of autecology of indigenous species

Though the region has received attention of botanists from early 19th century (Stewart, 1917), many of the natural forests and thickets are still uninvestigated mainly due to inaccessibility of such areas. Some of these natural grooves are mentioned by Abbas (2004), however little information is available about them. Even the revised classification of forest type of India (Champion and Seth, 1968) has dealt with the natural vegetation of the region very superficially. Further, the ecological study of the individual species, for example the agencies governing their propagation in nature, their microclimate requirements, biotic pressure acting on them etc. is wanting.

II. Establishment of Propagation Techniques of Threatened Species

Many of the indigenous trees and shrubs of the region have become threatened (Anup Raj, 2004) due to various reasons. Poor regeneration *in situ* is one of these. *Juniperus macropoda* is one of such tree species, which has now shrunk in to one or two stands of a few individuals in Leh district. The stand structure of these stands is strongly right hand skewed indicating predominance of mature and overmature trees and lack of regeneration (Sharma and Anup Raj, 2004). Efforts to raise this species *ex situ* has also not resulted in any success yet. However, one Konchok Thoba, from Hanoo Thang village in Leh district has succeeded, though to a limited extend, to raise it through to a limited extent to raise it through seeds collected from the excreta of a bird locally known as *Chhubi*. His experiences can be utilized by refining the method scientifically to raise the seedlings on a larger scale. Similarly, the regeneration method of *Populus euphratica* and other indigenous species also need to be standardized.

III. Bioreclaimation of Salt Affected Soils

There are some patches of salt affected soils along the Indus river in the recent alluvial flood plains (Anon., 2003). High water table is one of the main reason for salt accumulation on the surface. These lands are thus not suitable for crop cultivation. Tree and shrub species should be screened for such land that can serve as biodrainage of the land. Increased evapotraspiration results in lowering down of water table and hence leaching of salt from the surface. After few rotation the land came be bioreclaimed and become suitable for crops.

IV. Afforestation

The region is a timber deficient, which depends, to a large extent, on Kashmir valley for its wood requirement. Afforestation in the region can alleviate this condition to some extent. But most of the area is either barren rocks or covered by perennial snow and hence unavailable for plantation activities. In spite of that, there are large extent of land lying waste mainly as fans and piedmont (Anon., 2003). Most of the land in its present form not suitable for agriculture due to poor soil characteristics and hence most prospective sites for afforestation. The principal constraint in developing these barren lands in green belt is unavailability of water. With proper choice of species and better management of available runoff of snow melt supplemented with ground water resources, which is largely untapped, plantation activities, however, can be extended to these areas.

The plant species that can withstand poor soil fertility, soil and atmospheric aridity and severe cold in winter can succeed in this environment. Indigenous trees and shrubs that are product of natural selection operating in these conditions are well adapted to different habitat in this region. A list of indigenous woody plant species suitable for such habitat is given in Table. The expansion of these in their respective microclimates can be achieved by encouraging their natural regeneration supplemented with artificial regeneration wherever needed.

Another approach to afforestation of the region may be through introduction of exotics. Though introduction of exotics is a controversial issue and it may replace the natural forest which is already under intense biotic pressure, its benefit can not be undermined. People are already harvesting economic benefits of exotic poplars and willows in the region. However, these species are confined to villages and nearby areas as a component of agro-forestry and farm forestry. For resource poor sites away from villages, species base has to be broadened. *Robinia psuedacacia* is one of the recent successful introductions suitable for such sites. Other potential species which are able to fix atmospheric nitrogen in the roots and are drought and frost tolerant are listed in Table. These species need proper testing (through scientific trials) before being introduced. In addition to that it must be ensured that their introduction be confined to barren lands and should not replace the existing natural forests.

Table 1: Tree diversity in Ladakh

Local/ common name	Botanical name	Family	Reference[x]
Malchang	*Salix alba*[*]	Salicaceae	-
Byerpa	*Populus alba*	Salicaceae	-
Hotong	*P. euphratica*	Salicaceae	-
Himalayan Poplar	*P. ciliata*[+]	Salicaceae	C&S, 1996[a]; RRS, 1917; Kachroo *et al.*, 1977
Poplar	*P. deltoides*[0]	Salicaceae	C&S, 1996[a]; Kachroo *et al.*, 1977
Yurlad	*P. nigra*	Salicaceae	-
Yurpa	*P. balsemifera*	Salicaceae	Brandis; *P. candicans* by RRS, 1917
Shugpa	*Juniperus macropoda*	Cupressaceae	-
Shugpa	*J. communis*[+]	Cupressaceae	C&S, 1996[b]; Kachroo *et al.*, 1977; RRS, 1917
Shugpa	*J. recurva*[+]	Cupressaceae	Kachroo *et al.*, 1977
Shugpa	*J. wallichiana*[+]	Cupressaceae	RRS, 1917
Kail	*Pinus wallichiana*	Pinaceae	Abbas, 2004
Stakpa	*Betula utilis*	Betuleaceae	-
Starga	*Juglans regia*	Juglandaceae	-
Yumbok	*Ulmus wallichiana*	Ulmaceae	-
Sersing	*Eleagnus hortensis*	Eleagnaceae	-
Sersing	*E. umbellate*[@]	Eleagnaceae	-
Otse	*Morus alba*	Moraceae	-
Shorot	*Cratagus oxycantha*[@]	Rosaceae	-
Skurmo/ wild plum	*Prunus tomentosa*[@]	Rosaceae	RRS, 1917
Chuli	*P. armeniaca*	Rosaceae	-
Kushu	*Malus pumila*	Rosaceae	-
Quince	*Cydonia obloga*[@]	Rosaceae	-
Black locust	*Robinia psuedacacia*	Fabaeceae	-
Balu	*Rhododendron anthopogan*		Anonymous, 2001

* Others species of salix are described in Table 3
\+ Presence in Ladakh doubtful
x References are given for those species that we have not found in Ladakh
0 Some of the introduced trees are growing in Ladakh but growth is very poor
@ Not reported from this region earlier
P. sylvestris is another species of *Populus* mentioned by C&S, 1996[a] however the literature is silent about it.
C&S= Chaurasia and Singh RRS= R.R. Stewart

Table 2: Shrub diversity in Ladakh

Local/ common name	Botanical name	Family	Reference[x]
Khitchermo	*Berberis ulicina*	Berberidaceae	-
Khitchermo	*B. zabelina*	Berberidaceae	C&S 1996[b]; Kachroo *et al.*, 1977; Anon., 2001
	B. vulgaris	Berberidaceae	RRS, 1917
Umbu	*Myricaria germanica*	Tamaricaceae	-
Umbu	*Tamaricaria elegans*	Tamaricaceae	-
Gyalta	*Tamarix gallica*	Tamaricaceae	-
Shingtik	*Sophora moorcroftiana*	Febaceae	-
Shingtik	*S. alopecuroides*	Febaceae	RRS, 1917; Kachroo *et al.*, 1977; C&S, 1996
Sari	*Cicer microphyllum*	Febaceae	-
Bladder senna	*Colutea multiflora*	Febaceae	-
Trama	*Caragana versicolor*	Febaceae	-
Zomosing	*C. girardiana*	Febaceae	Anon., 2001
	C. cuneata	Febaceae	RRS, 1917; Kachroo *et al.*, 1977; C&S, 1998
Zomosing	*Astragalus rhizanthus*	Febaceae	RRS, 1917; C&S, 1998
Serma	*A. macropterous*	Febaceae	RRS, 1917; Kachroo *et al.* 1977, C&S, 1998
	A. zanskarensis	Febaceae	RRS* 1917; Kachroo *et al.* 1977, C&S, 1998
Lakhman	*A. ademiaefolius*	Febaceae	RRS, 1917; Kachroo *et al.* 1977, C&S, 1998
	A. coluteocarpus	Febaceae	RRS, 1917; Kachroo *et al.*, 1977
	A. cicerifolius	Febaceae	RRS, 1917; Kachroo *et al.* 1977
Supalo	*Indigofera hetrantha*	Febaceae	Kachroo *et al.* 1977; C&S, 1998
Hembi	*Cotoneaster numularia*	Rosaceae	RRS, 1917; Kachroo *et al.*, 1977
KumchungC	*.falconeri*	Rosaceae	RRS, 1917; Kachroo *et al.*, 1977; C&S, 1998
	C. acuminata	Rosaceae	RRS, 1917

Contd...

Sia marpo	*Rosa webbiena*	Rosaceae	-
Sia serpo	*Rosa eglantaria*	Rosaceae	-
Sia karpo	*R. moschata*	Rosaceae	Kachroo *et al.*, 1977
Sia serpo	*R. ecae*	Rosaceae	C&S, 1998
Cyan rose (White)	*R. sericea*	Rosaceae	Anon., 2001
Emong karvo.	*Potentilla salesoviana*	Rosaceae	-
Ascuta	*Ribes orientale*	Capparidaceae	-
Ascuta	*R. alpestre*	Capparidaceae	-
Shet	*Lonicera spinosa*	Rubiacea	-
Fumarpe	*L. heterophylla*	Rubiacea	RRS, 1917; C&S, 1998
Lamora	*L. glauca*	Rubiacea	RRS, 1917; C&S, 1998
Lamora	*L. rupicola*	Rubiacea	RRS, 1917; C&S, 1998
Lamora	*L. asperifolia*	Rubiacea	RRS, 1917; C&S, 1998
Tsermang	*Hippophae rhamnoids*	Eleagnacea	-
Tsermang	*H. salicifolia*	Eleagnacea	RRS, 1917
Tsermang	*H. tibetana*	Eleagnacea	Blaize, 2004
Chhapat	*Ephedra gerardiana*	Epedraceae	-
Chhapat	*E. intermedia*	Eleagnacea	-
Thuja	*Thuja orientalis*	Cupressaceae	-
Longze	*Acantholimon lycopodioides*	Plumbaginaceae	-

C&S= Chaurasia and Singh RRS= R.R. Stewart

x References are given for those species that we have not found in Ladakh

Table 3: Willows of Ladakh in literature

Species	RRS	C&S	Kachroo *et al.*, 1977
Salix elegans	Drass valley near snow	Leh	Drass, Leh
S. sclerophylla	Drass	Leh	Ladakh
S. fragilis	Cultivated in western Tibet	Leh	Nubra
S. hastate	North side Zoji	Leh	Drass, Leh
S. dephnoides	Sapi river ?	Leh	Drass, Leh
S. flabillaris	Cultivated in Ladakh	-	Ladakh
S. pychnostachya	Cultivated in Ladakh	-	Ladakh, Zanskar
S. angustifolia	Cultivated in Khaltse	Leh	Ladakh
S. divergense	Zanskar	-	-
S. tetrasperma	Drass valley	Leh	Leh, Nubra, Drass
S. oxycarpa	Kargil	-	Leh
S. bebilonica	-	Leh	-
S. caesia	-	Leh	Nubra, Zanskar

C&S= Chaurasia and Singh RRS= R.R. Stewart

Table 4: List of herbarium sheet of willow from Ladakh at the northern circle, Botanical Survey of India, Dehra Dun

Species	Location	Collector's name
Salix oxycarpa	Mulbekh (3310 m)	Wadhwa, 1976
	Deskit (3200 m)	Wadhwa, 1976
	Igoo nalla (3400 m)	Banerjee, 1970
	Chushul (4300 m)	Banerjee, 1970
S. pychnostachya	Drass, Kargil (3100 m)	Wadhwa, 1976
	Phyang	Viswanathan 1975
	Saspole	Wadhwa, 1976
S. angustifolia	Langsdung village	Wadhwa, 1976
S. flabillaris	Matayan (3330 m)	Wadhwa, 1976
	Kargil (3570 m)	Wadhwa, 1976
S. acmophylla	Thiksey (3505 m)	Wadhwa, 1976
S. divergence	Gomburong Zang (4500 m)	Banerjee, 1973
	Panikha (3300 m)	Wadhwa, 1976

Table 5: List of indigenous trees and shrubs having potential for afforestation in different habitat

Habitat type	Plant species	Tree/ shrubs	Local/common name
Dry sand with gravel	*Capparis spinosa*	Shrub	Kabra
	Colutea muliflora	Shrub	Bladder senna
	Berberis ulicina	Shrub	Khichermo
	Rosa webbiana	Shrub	Sia marpo
Dry rocky sites	*Ephedra gerardiana*	Shrub	Chhapat
	Berberis uliciana	Shrub	Khicharmo
	Capparis spinosa	Shrub	Kabra
	Roca webbiana	Shrub	Sia marpo
Sand dunes	*Populus euphratica*	Tree	Hotong
	Tamarix gallica	Shrub	Gyalta
	Berberis ulicina	Shrub	Khicharmo
Riverine site	*Salix* spp.	Tree	Changma
	Populus balsemifera	Tree	Yarpa
	Hippophae rhamnoides	Shrub	Tsermang
	Tamarix gallica	Shrub	Gyalta
Along streams	*Salix* spp.*Eleagnushortensis*	Tree	Drokchang
	Juniperus macropoda	Tree	Sersing
	Prunus armeniaca	Tree	Shugpa
	Myricaria germanica	Tree	Chuli
	Tamaricaria elegans	Shrub	Umbu
	Hippophae rhamnoides	Shrub	Umbu
	Sophora moorcroftiana	Shrub	Tsermang
	Lonicera spinosa	Shrub	Shingtik
	Rosa webbiana	Shrub	Shet
	Ribes orientale	Shrub	Sia marpo
		Shrub	Ascuta

Table 6 : Exotic nitrogen-fixing trees and shrubs suitable for Ladakh conditions

Name	Family	Origin	Use	Charecterstices
Acacia aneura	Mimoceae	Australia	Fuel wood, ornamental	Shrub or tree to 12m, high drought tolerance (to 200mm), wide range of frost tolerance
Afarnesiana	Mimoceae	Australia	Fuel wood, ornamental	Shrub 2-7m; forms thickets, stipular spine, frost tolerant, rapid growth, tolerant to heavy clay
A. nerifolia (Silver wattle)	Mimoceae	E. Australia	Wind breaks, ornamental, fuel wood	Shrub to 8m or tree to12m; drought toleranc (to 300mm) and frost tolerant, fast growth
A. salicina (willow wattle)	Mimoceae	Australia	Wind break, timber, fodder (Pods & dried leaves)	Tall shrub or tree to 20m, sub-humid, sub tropic, drought and frost tolerant
Chamaecytisus palmensis (tree Lucerne)	Fabaceae	New Zeeland	-	Shrub to 6m; temperate, frost tolerant to -10° C, drought tolerant
Prosopis alba	Mimoceae	S. America	Fire wood, timber, fodder (pods)	Tree to 15m, tolerant to 100mm up to 3000m in Argentina
Purshia spp. (Bitter brush)	Rosaceae	N. America	Browse fodder for game, soil erosion control	Shrub to 3m; temperate, high cold tolerant
Shepherdia argentea (Buffalo berry)	Eleagnacae	N. America	Fodder (fruits, foliage), soil erosion control	Shrub to 5m; temperate, high drought and cold tolerant
Ongeinia oojeinensis	Fabaceae	N. India	Wood for implements, fodder	Tree to14m, frost & drought tolerant
Ceanothus spp. (Buckthorn)	Rhamnaceae	America	Fodder, soil amelioration	Shrub to 5m; drought hardy, diverse habitats

References

Abbas, M. (2004). Natural forest areas of Ladakh and their vegetation. In: Biodiversity of Ladakh: Strategy and Action Plan. (Humbert Droz, B. and Dawa, S. eds.). Sampark, New Delhi, pp. 161-164.

Anonymous (1997). Biodiversity of Jammu and Kashmir, WWF - India, New Delhi.

Anonymous (2001). Conserving Biodiversity in the Trans-Himalayas: New Initiations of Field Conservation in Ladakh. First Annual Technical Report (1979-2000). Wildlife Institute of India, International Snow Leopard. Trust and US Fish and Wildlife Service.

Anonymous (2003). Natural Resources Management, Ladakh Region: A remote sensing based study. Regional Remote Sensing Service Centre, ISRO, Dehradun, pp. 192.

Anup Raj (2004). Status of Indigenous Trees and Shrubs of Ladakh and Recommendation for their conservation.

Champion, H.G. and Seth, S.K. (1968). A Revised survey of Forest types in India, pp. 321-342.

Chaurasia, O.P. and Singh, B. (1996-2001). Cold Desert Plants Vol - V, FRL, DRDO, Leh.

Fox, J.L.; Nurbu, C.; Bhatt, S. and Chandola, A. (1994). Wildlife conservation and land use change in the Trans-Himalayan region of Ladakh, India. Mountain Research and Devel. 14(1): 39-60.

Hooker, J.D. (1885). The Flora of British India. L. Reeve and co. Ltd London. Vol V.

Kachroo, P.; Sapru, B.L. and Dhar, U. (1977). Flora of Ladakh: An Ecological and Taxonomic Appraisal. Bishen Singh Mahendra Pal Singh, Dehradun, 172 p.

Mani, M.S. (1990). Fundamentals of High Altitude Biology. 2nd Ed. Oxford and IBM/Publishing Co. New Delhi.

Sharma, B.M. and Anup Raj (2004). Status of Natural Regeneration of Juniperus Macropoda and Boiss in Ladakh, the cold arid region of Western trans-Himalayas. Indian J. For. 27(3): 237-240.

Sharma, J.P. and Mir, A.A. (2000). Dynamics of Cold arid Agriculture. Kalyani Publishers, Ludhiana, 338 p.

Stewart, R.R. (1917). The flora of Ladakh Western Tibet. Bull Torr. Bot. Club: 571-590.

Wardle, P. (1993). Causes of Alpine Timberline: A review of the Hyphthesis. In : Forest Development in cold climate John Alden, J.Louise Mastrantonia and Soren Odum (ed.). p. 89-104.

❑❑❑

8

Soil and Plant Nutrition

V.K. SHARMA*
RARS, SKUAST, Housing Colony, Leh-194108
* Presently at Central Institute of Temperate Horticulture (ICAR), Old Airport, Srinagar (J&K)

The farmers need suitable fertilizer/manure of good quality and appropriate application information in order to correct nutrient deficiency. The farmers are interested in a fertilizer which is agronomically effective (increase yield), economically attractive (increase profit) and practically convenient to handle and apply. Nutrients are continuously removed from the soil by crops and lost by leaching and erosion. It is therefore, imperative that sound soil and crop management practices are adopted to improve and maintain soil fertility and soil physical condition. The soils of the region are coarse textured, shallow, sandy, derived from weathered debris of rocks, subjected to severe wind erosion, have high permeability, low water holding capacity, low organic carbon and poor capacity of nutrients reserve. Where barley, wheat and alfalfa are major crops, this chapter will help the farmers of cold arid region for maintaining soil fertility by means of balanced fertilization with respect to major as well as micro-nutrients.

1. Nutrients Management

Soil fertility and agricultural production can be maintained only by efficient and judicious management of nutrients, their addition to the soil from external sources. The two sources most widely used all over the world are (i) organic source (organic manures) simply called manures and (ii) chemical fertilizer simply called fertilizers.

1.1 Manures

Manures are the organic materials derived from animal, human and plant residues, which contain nutrients in complex forms.

1.1.1 Farm Yard Manure (FYM)

This is prepared from a combination of cattle dung and urine, waste feed and fodder, material used as liter etc. On being allowed to decompose under favourable condition with the help of microbes already present in excreta, both aerobic and anaerobic reactions take place almost simultaneously. The decomposition is carried out in pits or trenches, about one meter in depth, the length and breadth depending upon the quantity of raw materials to be decomposed. The cattle shed wastes are dumped directly in to the trenches/pits till the level of heaped materials is 30 to 40 cm above the ground levels. The top is plastered with slurry of cattle dung and soil. After about two months, the materials, in the pits are ready for use as manures (Table 1).

1.1.2 Compost

Composting, as in the case of FYM, is essentially a microbiological decomposition of organic residues, collected from the rural area. The well rotten manure form a rural source is called rural compost. In compost making, wide variety of carbonaceous and nitrogenous organic waste material, besides cattle dung and urine are utilized. In the preparation of rural compost, any farm waste, weeds etc. are used as raw material. The decomposition is carried out in trenches/pits. At first, a layer of refuse is dumped, followed by a layer of animal dung or night soil. The process of dumping in alternate layer is continued till the level of dumping is above the ground. The top most layer should be a layer of refuse and this should be covered with a layer of soil. In about four months compost is ready for use as manures.

1.1.3 Green manure

Green leaf manuring is a practice in which green leaves from shrubs and trees are collected from outside and turned in to soil.

Green leaf manuring is possible because leguminous crops are not grown in Ladakh region and there is mono-crop season. Moreover, for *in situ* decomposition, adequate soil moisture, either through rainfall or by irrigation, is essential.

Table 1: Major nutrient contents present in some manure

Manures	N (%)	P_2O_5 (%)	K_2O (%)
Farm yard manure	0.5	0.3	0.5
Compost rural	0.5	0.5	1.5
Green manure	0.4	0.2	0.6
Horse dung	2.5	1.5	1.5
Sheep and Goat dung	3.0	1.0	2.0
Cow dung	1.5	0.5	0.5
Night soil	5.0	3.0	2.0

Source: Biswas and Mukherjee (1989)

1.2 Fertilizer

These are materials either natural and manufactured, containing nutrients essential for normal growth and development of plants. It may also be said that any material or substance intended to be used as a nutrient carrier in soil or crop for boosting crop yield is called as fertilizer. Presently fertilizers have become an integral part of agricultural economy as they increase the fertility of soils and enable them to support high yields (Table 2). About 50 per cent of the increase in crop production during recent times has been attributed to the fertilizer use, though the fertilizer use efficiency is very poor. The production and usage of fertilizer are mainly aimed at fulfilling needs of N, P and K. Most of materials which enter in fertilizer trade contains one or more primary nutrients. The source of nutrients in most of the fertilizers are:

Source of nitrogen	:	Air and organic matter
Source of phosphorous	:	Phosphate rocks (mined)
Source of pottassium	:	Potash minerals (mined)
Source of sulphur	:	Mined sulphur or sulphur recoverd in soil refinieries

The following method can be used to calculate the amount of fertilizer required to provide the required amount of nutrients:

Table 2: Nutrient contents in commercial fertilizers

Fertilizers	N (%)	P_2O_5 (%)	K_2O (%)	Micronutrient (%)
A. Straight fertilizers				
1. Nitrogenous fertilizers				
Urea	46.0	_	_	-
Ammonium sulpate	20.6	_	_	-
Ammonium chloride	25.0	_	_	-
2. Phosphate fertilizers				
Single super phosphate	_	16.0	_	-
Triple super phosphate	_	46.0	_	-
Di-calcium phosphate	_	34.0	_	-
3. Potassic fertilizers				
Muraita of potash	_	_	58-60	-
Sulphate of potash	_	_	48.0	-
B. Complex fertilizers				
Di-ammonium phosphate	18.0	46.0	-	-
Ammonium phosphate sulphate	20.0	20.0	-	-
Urea ammonium phosphate	28.0	28.0	-	-
Nitro-phosphate with potash	15.0	15.0	15.0	-

Formula

$$\text{Fertilizer (kg) needed} = \frac{\text{Nutrient needed (kg)}}{\text{\% Nutrient in fertilizer}}$$

For example

i) Urea (kg) needed to get 5 kg nitrogen = $\frac{5}{46} \times 100$

= 10.85 kg urea

ii) DAP (kg) needed to get 5 kg phosphorous = $\frac{5}{46} \times 100$

= 10.85 kg DAP

∴ 10.85 kg DAP will get how much amount of nitrogen

100 kg DAP .... 18 kg nitrogen.

1 kg DAP .... $\frac{18}{100}$

10.85 kg DAP .... $\frac{18}{100} \times 10.85$

= 1.944 kg nitrogen from 10.85 kg DAP

1.2.1 Factor Affecting Fertilizer Use

a. Soil factor: The most important factors are physical condition of soil, soil fertility and reaction. On soils with a poor physical condition, plant growth will be poor due to impeded drainage restricted aeration and unfavorable soil temperature, and fertilizers nutrients will evidently not be utilized efficiently. The coarse-textured soils are usually poor in available nutrients than fine-textured soils. The higher the fertility of soil, the lower is the response of crop to fertilizers. Soil reaction is an important consideration in the selection of right type of phosphatic fertilizers. In general, the higher the organic matter status of soils, the more efficient is the fertilizer utilization by the crop.

b. Climatic factor: These factors are temperature, rainfall or frequent irrigation, evaporation, length of day and growing season. The rate of nitrification is slower in cooler climate than in a warmer climate, more ammonical nitrogenous fertilizers will therefore be needed in cool climate. Higher fertilizer requirements are expected in regions of high rainfall or frequently applied irrigation because of leaching loss of soil nutrients. For short-duration crops, early application of fertilizers is more effective than in long-duration crops, which take up nutrients over a long period. Often higher uptake occurs at later stage of growth. Light intensity and length of day increase/decrease fertilizer requirements of crop. The higher the light intensity, the better is the fertilizers utilization.

c. Crop factor: One important characteristics of the plant which is related to fertilizer responsiveness is the cation exchange capacity of the roots. The difference among crops ability to absorb nutrients from the same soil depends upon the size of the root system. The pattern of nutrients uptake and development of roots are important factors in timing and placement of fertilizer for their efficient use. With alfalfa in the crop sequence, nitrogen requirement may be less.

d. Agronomic factor: These factors include fertilizer responsiveness of crops and their varieties, timely sowing, proper spacing, proper dose, time and methods of fertilizer application. These parameters are key to efficient use of fertilizers for crop production. The nature of crop, its growth stages and nutrient requirement, soil condition and nature of fertilizers determine the time of application which may be before sowing, at the time of sowing and after sowing. Method of application may be broadcasting, placement and foliar application, depending upon the overall situation. Nitrogen and phosphorus are the two major components for their efficient use. Leaching as nitrate and voltilisation as ammonia may lose nitrogen from nitrogenous fertilizers. The problem with phosphorus is reversion and fixation on clay particle. Attempts have been made to increase efficiency of fertilizers nitrogen through the use of slow release nitrogenous fertilizers and to apply in three split dose of nitrogen at critical stages such as basal dose, at tillering stage and heading stage (ear).

Based on the content of available nutrients, soil test values (N, P and K) the soils are grouped into classes such as low, medium and high (Table 3). The recommendations can be used on the basis of soil test value (Table 4).

Table 3: Rating chart of soil test values and nutrients recommendation

Organic carbon %	Applied dose (kg N ha^{-1})	Avail. phosphorus (kg ha^{-1})	Applied dose (kg P_2O_5 ha^{-1})	Avail. potassium (kg ha^{-1})	Applied dose (kg KO ha^{-1})
Low					
Below 0.2	160**	Below 10	80	Below 100	60
0.2 - 0.4	140	10 - 20	60	101 - 150	50
Medium					
0.41 - 0.60	120	21 - 30	45	151 - 200	40
0.60 - 0.80	90	31 - 40	30	201 - 250	30
High					
0.81 - 1.00	75	41 - 55	15	251 - 300	20
Above 1.00	60	Above 55	10	Above 300	0

Source: Biswas and Mukherjee (1989)
In this case use of FYM is necessary
Available phosphorus by Olsen's method
Available potassium by neutral ammonium acetate

Table 4: Suitable commercial crops and their normal recommended doses of nutrients for Ladakh

S.No.	Crop	N (kg ha^{-1})	P_2O_5 (kg ha^{-1})	K_2O (kg ha^{-1})
1.	Amarnath	50	50	50
2.	Artichoke	60	60	60
3.	Beet root	50	40	50
4.	Brinjal	80	40	50
5.	Broccoli	100	50	50
6.	Broad bean	30	80	50
7.	Bottal Guard	60	40	40
8.	Carrot	50	40	50
9.	Cauliflower	100	50	50
10.	Cabbage	100	50	50
11.	Cucumber	80	40	40
12.	Capsicum	80	40	50
13.	Celery	100	40	80
14.	Chinese cabbage	50	40	50
15.	Coriander	40	20	20
16.	French bean	30	80	50
17.	Fenugreek	20	40	20
18.	Knol khol	80	50	50
19.	Karam sag	80	50	50
20.	Ladys finger	60	40	40
21.	Garlic	100	60	120
22.	Long melon	50	30	30
23.	Lettuce	60	50	50
24.	Onion	80	50	60
25.	Potato	120	60	80
26.	Peas	30	60	50
27.	Pumpkin	80	30	30
28.	Palak	60	30	30
29.	Parsley	60	30	30
30.	Radish	50	40	50
31.	Summer squash	60	30	30
32.	Spinach	60	30	30
33.	Turnip	50	40	50
34.	Tomato	100	60	60
35.	Veg. Mustard	50	40	50
36.	Water melon	50	30	30
37.	Spring wheat	140	60	80
38.	Spring barley	110	35	70
39.	Maize	150	60	30
40.	Pulses	50	20	40
41.	Rape seed	150	70	150
42.	Fodder plant	350	90	90

Source : FRL, DRDO, Leh

1.2.2 Critical Status of Nutrients in Plants and Soils

Ladakh agriculture has also moved from an era of scattered single nutrient deficiency to more complex multiple nutrient deficiencies. At least five essential nutrients are already of widespread practical importance in terms of the need for application. These are nitrogen, phosphorus, potassium, sulphur and zinc. Thus, balanced fertilizer use, and balanced crop nutrition must now include the application and management of all those nutrients, which are deficient or not available to the crop in adequate amounts. If a nutrient is deficient in the soil and plant, it is of primary importance in farming. The critical limits of nutrients in plant and soil are given in Table 5.

Table 5: Critical status of nutrients in plants and soils

S.No.	Nutrients	Plant	Soil limit (Available form)	Class	Soil test methods
1.	Nitrogen	1.50	< 280 kg/ha	Low	Alkaline permagnet method
			280-560	Medium	
			> 560	High	
2.	Phosphorous	< 0.2%	< 10 kg/ha	Low	Olsen's method
			10-25	Medium	
			> 25	High	
3.	Potassium	< 1.5%	< 108 kg/ha	Low	Neutral Ammonium acetate extractable
		1.5 - 3%	108-280	Medium	
		> 3%			
			> 280	High	
4.	Sulphur	< 0.10-0.40%	< 10 mg/kg	Deficient	$CaCl_2$ solution extractable
5.	Zinc	< 15-20 ppm	< 0.6 mg/kg	Deficient	DTPA extractable
6.	Iron	< 50 ppm (25-80)	< 4.5 mg/kg	Deficient	DTPA extractable
7.	Manganese	< 20 ppm (10-30)	< 2.00 mg/kg	Deficient	DTPA extractable
8.	Copper	< 4 ppm (3-10)	< 0.2 mg/kg	Deficient	DTPA extractable
9.	Boron	< 20 ppm	< 0.5 mg/kg	Deficient	Hot water
10.	Molybdenum	< 0.1 ppm	< 0.2 mg/kg	Deficient	Ammonium oxalate

Source: Tondan (1991); Mohr *et al.* (1965); Rama Rao and Shekhon (1988) and Bhogal (2005)

1.2.3 Available Nutrients Status in Soils of Leh District of Ladakh Region

Data on the different characteristics of soils of all blocks of Leh district are presented in Table 6. The soils of Leh district were neutral to alkaline in reaction, pH ranging from 7.03 to 10.1, EC varied from

0.05–156 dsm^{-1} and organic carbon ranged from 0.7 to 14.1 g kg^{-1} in the soils of Leh district. Leh block contained highest mean value 5.79 g kg^{-1} of organic carbon, which is similar to Khaltse block (5.77 g kg^{-1}). The soils of Durbuk block has lower value (2.90 g kg^{-1}). The soils of Nubra block contain highest mean value of Phosphorous (11.59 kg ha^{-1}) while the soils of Leh block showed minimum value (7.45 kg ha^{-1}). Available potassium in Khaltse block showed the highest mean value (346.87 kg ha^{-1}) and the lowest mean value (211.94 kg ha^{-1}) has found in Durbuk block.

Table 6: Physico-chemical properties and DTPA- extractable micronutrients in soils of different blocks of Leh district of Ladakh

Soil characteristics	Blocks of Leh district						
	Leh (29)*	Kharu (10)	Nubra (17)	Nyoma (12)	Khaltse (3)	Durbuk (5)	Over all (76)
pH (1: 2) Range	7.81-8.40	7.00-8.39	7.39-10.12	7.43-9.13	7.75-8.08	7.03-8.72	7.00-10.12
EC (dsm^{-1}) Range	0.12-0.77	0.18-0.90	0.05-1.41	0.07-1.28	0.18-1.56	0.22-0.33	0.05-1.56
Mean	0.342	0.416	0.374	0.291	0.796	0.267	0.414
Organic carbon (%)							
Range	0.07-1.41	0.08-0.83	0.19-1.26	0.11-0.91	0.40-0.81	0.15-0.61	0.07-1.41
Mean	0.579	0.505	0.559	0.434	0.577	0.290	0.490
Available Nutrients (kgha^{-1})							
Available P							
Range	4.00-16.50	4.40-18.62	7.43-16.05	4.20-22.05	6.41-16.30	6.00-15.70	4.00-22.05
Mean	7.45	7.96	10.87	11.59	10.59	9.63	9.68
Available K							
Range	23.5-366.0	23.5-384.0	43.7-395.5	42.6-496.2	295.7-434.0	39.2-311.2	23.5-496.2
Mean	262.99	261.75	277.63	263.00	346.87	211.94	270.69
DTPA-Extractable micronutrients (mg kg^{-1})							
Zinc							
Range	0.06-3.42	0.08-1.80	0.09-16.30	0.06-1.80	0.30-0.47	0.08-0.50	0.06-16.30
Mean	0.451	0.524	3.340	0.440	0.370	0.274	0.900
% deficient	79.3	60.0	17.65	75.0	100.0	100.0	71.99
Copper							
Range	0.46-2.39	0.43-0.84	0.43-3.13	0.46-0.95	0.50-0.85	0.50-0.70	0.43-3.13
Mean	0.62	0.55	1.71	0.60	0.63	0.60	0.785
% deficient	Nil	Nil	Nil	Nil	Nil	Nil	Nil
Manganese							
Range	0.42-2.30	0.50-1.30	0.70-32.17	0.40-2.30	0.50-1.20	0.20-1.00	0.20-32.17
Mean	1.23	0.87	13.44	1.17	0.93	0.80	3.07
% deficient	96.60	100.00	23.53	100.00	100.00	100.00	86.69
Iron							
Range	0.72-2.16	0.80-2.21	0.54-33.79	1.28-4.20	1.20-1.84	0.80-1.46	0.54-33.79
Mean	1.83	1.54	13.07	2.02	1.42	1.26	3.52
% deficient	100.00	100.00	23.53	100.00	100.00	100.00	87.26

* Data in parenthesis indicate number of villages; Source: Sharma *et al.* (2005a)

Hundred per cent soils of Durbuk and Khaltse blocks and minimum 70.65 per cent soils of the Nubra block under present position are classified as deficient in available zinc. None of the soils were classified as deficient in available copper in all blocks of Leh district. The content of DTPA-Mn in these soils varied from 0.20-32.17 mg kg^{-1}. The soils of Nubra block has highest mean value of DTPA-Mn (13.44 mg kg^{-1}) followed by 1.23, 1.17, 0.93, 0.87 and 0.80 mg kg^{-1} in the soils of Leh, Nyoma, Khaltse, Kharu and Durbuk blocks, respectively. It was observed that 100 per cent soils were deficient in Kharu, Nyoma, Khaltse and Durbuk blocks with respect to DTPA-Mn, while 23.53 and 96.6 per cent soils in DTPA-Mn of Nubra and Leh blocks, respectively, were found to be deficient. The amount of DTPA-Fe ranged from 0.54 – 33.79 mg kg^{-1} in soils of Leh district. There was 100 per cent Fe deficiency in the soils of Leh, Kharu, Nyoma, Khaltse and Durbuk blocks where as only 13.07 per cent soils of Nubra block were deficient with respect to available Fe. Based on the percent deficient soils for micro- nutrients in the different blocks followed the order as Khaltsi = Durbuk > Nyoma > Kharu > Leh > Nubra. The soils of Durbuk and Khaltse blocks showed similar deficiency for micro-nutrients. However, Khaltse block is superior to Durbuk block with respect to mean value for organic carbon, available phosphorous and potassium.

1.2.4 Available Nutrient Status in Soils of Kargil District of Ladakh Region

The soils of Kargil district are slightly acidic to high alkaline in reaction (pH 6.57 to 9.47). The electrical conductivity ranged from 0.08 to 1.55 dsm^{-1} and organic carbon from 1.40 to 35.8 g kg^{-1}. The available phosphorus and potassium contain in the soils of Kargil district varied from 2.5 to 137.40 kg ha^{-1} and 103.5 to 861.3 kg ha^{-1} with the mean values of 24.62 and 313.13 kg ha^{-1}, respectively. And 51 and 4 per cent soils were deficient in available phosphorus and potassium. The DTPA-Zn, Cu, Mn and Fe in the soils of Kargil district varied from 0.14 to 5.17, 0.48 to 3.41, 0.34 to 4.74 and 0.20 to 38.01 mg kg^{-1} with the mean values 1.17, 0.82, 0.99 and 7.32 mg kg^{-1}, respectively. The per cent deficient soils for DTPA-Zn, Mn and Fe were 48, 98 and 66, respectively. The available Cu was classified as sufficient in all soils of Kargil district (Table 7).

Table 7: Status of nutrients in Kargil soils

Soil characteristics	Range	Mean	% deficient
pH (1:2.5)	6.57 - 9.47	7.74	-
EC (1:2.5) (dsm^{-1})	0.08 - 1.55	0.44	-
Organic carbon (g/kg)	1.40 - 35.80	8.42	-
Available nutrients			
Available P (kg/ha)	2.52 - 137.40	24.62	48
Available K (kg/hg)	103.50 - 861.30	313.13	4
Available Zn (mg/kg)	0.14 - 5.17	1.17	48
Available Cu (mg/kg)	0.48 - 3.41	0.82	Nil
Available Fe (mg/kg)	0.20 - 38.01	7.32	66
Available Mn (mg/kg)	0.34 - 47.4	0.99	98

Source: Sharma *et al.* (2005b).

1.3 Nutrient Deficiency

When plants are starving for any particular nutrient, characteristics symptoms usually appear on these plants. If crops are not vigorous and healthy, it is important to know and understand the cause. In case of nutrient deficiency, fertilizer may be applied as top dressing in the soil or on the foliage as a spray in time to save the crop. It should be pointed out that these deficiency symptoms are not always easily diagnosed. But information concerning these symptoms has been accumulating rapidly and have become a valuable aid in determining the need for certain nutrients.

1.3.1 Deficiency Symptoms

a. Nitrogen

The need for more nitrogen is indicated by a light green to yellow appearance of the leaves. The older bottom leaves start turning light green, then turn yellow at the tip. In the corn plant, the yellowing extends up to the midrib of the leaf, with the outer edges remaining green. A deficiency of nitrogen may cause the kernel of cereals to become shriveled and light in weight. In fruit trees, the early shedding of leaves, death of lateral buds, poor set of fruit and development of unusually colored fruits are indications of lack of nitrogen. Extreme nitrogen deficiency is most likely to occur in sandy soils, and water-logging soils usually need additional nitrogen.

b. Phosphorous

If phosphorous is deficient, cell division in plants is retarded and plant growth is stunted. A dark green color associated with a purplish coloration in the seeding stage of growth is symptom of phosphorus deficiency. Bronze or purple leaves sometimes are observed at the top of new shoots of phosphorous deficient apple trees. In the absence of sufficient phosphorous, general maturity of the crop and seed formation is delayed. The most characteristic symptom of phosphorous deficiency, among plant in general, is the stunted growth.

c. Potassium

A deficiency of potassium usually shows up as a 'leaf scorch' in most plants. This leaf scorch is frequently mistaken for 'burning' or firing. Corn indicates a need for potassium by a yellowing of the tips and margins of the lower leaves. This coloration does not move up the mid rib as with a nitrogen deficiency but gradually spread upward and inward from the leaf tips and edges. Where in potassium deficient soils, alfalfa frequently develops a series of white spots near the margins of the older leaves. The edges of the leaves finally dry up. Potato plants indicate a potassium deficiency by a marginal scorch of the lower leaves.

d. Sulphur

Sulphur deficiency decreases size of leaves and also red or purple pigmentation. In some plants, young leaves are more sensitive than older ones. Leaves will become yellow with necrotic margin.

e. Zinc

Characteristic symptoms of zinc deficiency in different crops are given below:

Wheat: Chlorotic blotches appear on the upper side of the older leaves from an irregular network. They first turn white, then brown and the whole leaf dies. Younger leaves appear tube like and are small.

Maize: Leaves become pale green, the young leaves become pale yellow or almost white. This symptom is named as white bud.

Potato: Stunted growth, deformation in leaves (fern leaf).

Tomato: Chloroplast of leaves lack normal lamellar structure. Decrease auxin productions.

Onion: The leaf tips chlorotic and misshapen and the leaves are twisted.

Apple: Narrow leaves (little leaf), rosette owing to inhibition of inter nodal growth.

f. Copper

Copper is induced by heavy liming and excessive nitrogen and phosphorus application

Wheat: Failure of grain formation, panicale does not come out fully, flag leaf become purple.

Peas: Pods may develop but seeds are aborted.

Cabbage: Inhibition of growth, wilting of plant, fail to develop head.

Onion: Leaves become chlorotic, thin skins, remains pale yellow or greyish. Onions are soft and perish quickly.

Tomato: Leaves show marginal wilting.

Apple: Plants shows 'S' shaped leaves.

g. Iron

Iron deficiency effects in different crops are given below:

Maize: Inter-veinal yellowing.

Wheat: The inter vein areas are most affected and the veins retain a darker color (Iron chlorosis).

Cabbage: Newly emerging leaves show deep purple colouration, plants fail to head.

h. Manganese

Manganese deficiency causes marsh spots of peas, grey speckled of oats and speckled yellows of beet. In potato, black rounded necrotic spots appear along midribs and major veins. In tomato, necrosis appears as small brown or orange tinted spots.

i. Boron

Boron deficiency is aggravated by liming and heavy application of nitrogenous fertilizers. Many physiological changes in plants, such as stem crack of celery, water core of turnip, yellows of alfalfa, corky core, internal cork or draught spots of apple fruit, hen and chicken of grape and bark measles of apple stem, are associated with a deficiency of boron.

Characteristics of boron deficiency in different crops are given below:

Corn: Corn cobs with reduced and deformed grain formation.

Cauliflower: Deformed foliage, brown curd, hollows stem disease.

Carrot and Alfalfa: Leaves may be purple, red orange and yellow mainly in the expending leaves.

Potato: Short internodes, plant is bushy in appearance tuber discolor rapidly when cut.

Tomato: Leaves turn orange or yellow, plants become stunted in growth. Wing of corky splits or scars in a longitudinal plain around calyx end of young fruits.

Apple: Poor flower set, leaves small, thick and brittle. Cracking of fruit, often small and misshapen, die back of shoots.

Grape: Die back of shoots, poor fruit set, shot berries (different size of grape on the same bunch). Hen and chicken symptoms are caused by parthenocarpic fruit development.

j. Molybdenum

Molybdenum deficiency in cauliflower causes a cupping of the leaves. It is owing to the reduced rate of expansion near the leaf margin compared to that in the center of the leaf, which is known as "Whip tail" formation. Growth of leaf lamina is severely restricted. Peas and beans are very sensitive to molybdenum deficiency. Leaves become pale, green, growth and yield are reduced. Most of the vegetables crops are prone to molybdenum deficiency.

Cauliflower: Yellow chlorotic blotches between the veins of the pale green, dull grayish or bluish green leaves, which become slender known as "whip tail".

Tomato: Chlorotic mottling of older leaves, inward curling of leaf margin.

Radish: Terminal growing tip dies, leaves are discolored and distorted.

1.3.2 Confusing Nutrient Deficiency Symptoms

Some nutrients have related roles in the plants with the result that their deficiency symptoms can also be confused. Some examples based on Singh (1984) are provided below:

i) N deficiency can be confused with S deficiency
ii) Ca deficiency can be confused with Bo deficiency
iii) Fe deficiency can be confused with Mn deficiency
iv) Leaf stripe diseases of oat can be confused with Mn deficiency
v) Effect of viruses, little leaf etc. can be confused with Zn or Bo deficiency
vi) Brown streak disease of Rice can be confused with Zn deficiency

1.3.3 Future Guidelines

On the basis of nutrients status in the soils of Ladakh region, the major issues that need to be addressed are:

i. The unique feature of the soils of Ladakh region is slow carbon sequestration due to low precipitation, low temperature and low microbial load. Therefore, use of FYM 10 to 20 t ha^{-1} in cultivated field is important critical input lynchpin. It stabilized soil structure with concomitant increase in water holding capacity.

ii. Due to scantly vegetation in this environment the available nitrogen content in the soil is very low as far as the nitrogen mineralisation in concerned. Nitrogen is building block for plants and play pivotal role in the qualitative and quantitative parameter for the crops. Hence use of N-fertilization in conjunction with FYM is very important for sustainable soil productivity.

iii. Due to coarse textured soil the percolation rate is very high in the region. Hence as nitrate leached down from the root zone, while nitrogen as ammonia form loss through voltilization due to high light intensity, therefore split application of nitrogen fertilizers in three split dose at critical stages needs to be applied in Ladakh region.

iv. Available zinc, iron and manganese are deficient in the soils of Ladakh region. Hence basal dose of recommended micronutrient fertilizers are to be applied in the Ladakh region, where soils are deficient in micronutrients and it will depend on amount of available micronutrient in the soils. This problem is further aggravated by crop removal. Hence, application of micronutrient fertilizers every 3 year is very important.

2. Future Thrust for Research

i. Study on integrated nutrient management to sustainable soil productivity and crop yield.

ii.S tudy on irrigation standardization because irrigation is very important as water is the main limiting factor for potential crop yield under Ladakh region.

iii. Study on reclamation and management of salt affected soils.

iv. With reduced dependence on chemical fertilizer and increased important of organic farming, nutrient stress requires to be managed by balanced fertilizer with N, P, K, S and micronutrient along with FYM, compost, bio-fertilizers and green manure.

References

Anonymous, (2001). Annual Report, FRL, DRDO, Leh

Biswas, T.D. and Mukherjee, S.K. (1989). Soil fertility and fertilizer use. In Text book of Soil Science. *pp. 191* and *199.*

Bhogal, Narindar Singh (2005). Role of sulphur on the productivity of rapeseed-mustard. In: Winter school on advances in rapeseed-mustard research technology for sustainable production of oil seed, Decmber 15, 2004- January 04, 2005, pp. 332-340.

Mohr, G.R.; Datta, N.P.; Sankara Subramoney, H. and Lelay, V.K. (1965). Soil Testing in India, 2nd edition: U.S. Agency for International Development in India, New Delhi.

Rama Rao, N. and Shekhon, G.S. (1988). Potassium concentration in rice at certain stages of growth as an indicator for the nutrient status of plants. J. Potassium Research. 4:1-11.

Sharma, V.K., Dwivedi, S.K. and Tripathi, Diwakar (2005a). Status of available and micronutrients in the soils of different blocks of Leh district of Cold arid region of Ladakh in relation to soil characteristic. J. Indian Soc. Soil Sci..

Sharma, V.K., Dwivedi, S.K. and Ahmad, Z. (2005b). Status of available major and micronutrients in soil of Kargil district of cold arid region of Ladakh. Annals Plant Soil Res.

Singh, Mahindra (1984). Nutrition disorders in crops and their remedy. Haryana Agri. Uni., Hissar, pp. 37.

Tandon, H.L.S. (1991). Nutrient status of soils and crops. In: Secondary and Micronutrients in Agriculture. pp. 15-19.

□□□

9

Principles for Breeding of Crops

VIJAY K. YADAV, VINOD K. SAH AND A.K. SINGH*
College of Forestry & Hill Agricultrue
G B Pant University of Agriculture & Technology,
Hill Campus, Ranichauri-249 199, Tehri Garhwal (Uttarakhand) India
*Deptt. of Horticulture, Banaras Hindu University, Varanasi (U.P.)

The growth and development of plants are the functions of Genotype (G), Environment (E) and G x E interaction impact. G x E interaction tends to reduce the correlation between the genotype and phenotype. Cold arid climate are complex situation where many environmental factors affect growth and development of plants and hinder the full expression of phenotype. Cold arid environment is highly unpredictable unlike the other agro-climatic regions, so high yielding varieties developed elsewhere do not yield up to the expectation levels under these conditions. Thus, it is imperative to design special breeding programs for cold arid regions, keeping in view the agro-climatic conditions and growers' preference. In cold arid region, wheat, barley and alfalfa are major field crops, while field pea and mustard are grown on limited area. In vegetables, beet palak, spinach, turnip, potato, cabbage, radish, cauliflower and tomato are important crops whereas, apricot and apple are important fruit crops of the region. Most of the *rabi* crops, which are grown during the winter months in other agro-climatic situations, are cultivated during the short period of April to September in these areas. Short cropping season, high light intensity with preponderance of UV radiation, low nutrients in soil, high and desiccating wind, soil salinity and prevalence of extreme temperatures are major constrains in successful cultivation of all the agricultural and horticultural crops. Therefore, any breeding strategy may not yield satisfactory results unless all the factors are put into consideration while making selections and developments of varieties for this region. To formulate an efficient breeding program for developing varieties which suite to the cold arid conditions, it is essential to understand the physiological and biochemical control of tolerance to these abiotic stresses.

1. Types of Stresses in Cold Arid Region

Various kind of stresses that plants are exposed in cold arid region are:

1.1 Biotic Stresses

Biotic stresses may be either in the form of diseases or insect-pests for which plants are exposed in the high altitude areas. Important diseases which are noticed across different crops are powdery mildew, rust, smuts, downy mildew and damping off in nurseries. However, important insects are cabbage butter fly, aphids, cut worm, diamondback moth, codling moth, fruit moth and gypsy moth which feed on crops. Generally, vegetable crops are more prone to different diseases and insect-pests attacks. However, diseases are more damaging in cereals and pulses as compared to insects. The extent of damage by these biotic stresses depend on crop cultivation conditions and predisposing factors available for spread of diseases or increasing population of insects and evolution of new biotypes of insects and races of pathogens.

1.2 Abiotic Stresses

In high altitude region abiotic stresses are major limiting factors in crop cultivation. As compared to biotic stresses, they are more important to be considered in varietal improvement programmes. The important abiotic stresses and their effects are given below:

1.2.1 Water Stress

Among all the stresses to plants, water stress is the most important in cold arid region. Scarcity of water during the growing period is a major limiting factor in crop cultivation. In combination with high insulation, desiccating effect of wind and low temperature, water deficit causes irreversible losses to plants growth and development. At PWP, stomata of most mesophytes lose their ability to close, causing unhindered transpiration losses and death by tissue desiccation. With high water loss due to high transpiration at water stress condition, vacuole shrinks and cytoplasm may be severed from the cell wall. Water stress conditions generally lead to reduction in leaf area and photosynthesis rate, increased respiration losses which ultimately reduces photosynthate accumulation in plants. A decrease in cell size occurs due to loss of turgor causing decrease in net photosynthesis area. No crop is possible in cold arid region without irrigation or supplemental water, therefore, all agricultural activities are confined to provision of water. A field/vegetable crop may need 10-20 irrigations depending upon soil conditions and nature of crop. Even short-term water stress may reduce productivity of crop. Drought avoidance (escape), drought resistance and drought tolerance are various mechanisms by which plants are able to perform under drought. Drought avoidance may be achieved by growing short duration crop during favourable soil moisture conditions, thus escaping drought. However, drought resistance is characterized by the ability of plants for better exploitation of soil water and ability of restriction of water loss through extensive root system, stomata control on transpiration, cuticular transpiration, smaller leaves, designing suitable ideotype of crop plants and higher root/shoot ratio etc. Drought tolerance is characterized by the ability of plants to maintain high plant water potential at a given level of water stress, through osmo-regulation. High level of solutes in the cell sap gives cells a low osmotic potential (high osmotic pressure), which keeps them turgid even under water stress, thereby avoiding dehydration. Tolerance to drought may therefore, be achieved by maintaining stomata open despite reduced water content, thereby avoiding CO_2 starvation. Under these conditions, transpiration is reduced by low cell osmotic potential but CO_2 exchange takes place.

Drought tolerance for a plant breeder would mean 'Yield' whereas, to an ecologist it would mean 'Survival'. Blum *et al.* (1983) has stated that the total drought resistance of genotypes can not yet be defined physiologically and most probably it does not exit as a unique plant trait. According to Levitt (1972) drought resistance, by its simplest definition, is the sum of drought avoidance and drought tolerance.

The drought resistance/tolerance by the plants results through coordination of different physiological and metabolic activities of plants which can only be achieved by accumulation of desired characteristics in genotypes by using appropriate breeding methodology. There are several traits contributing factors towards resistance of a genotype. In a genetic improvement programme their contribution towards water deficit conditions needs to be portioned and assessed by using appropriate simulation model and screening techniques.

1.2.2 Temperature Stress

Plants do not have the ability to control their temperature and tend to assume the temperature of their environment. Each plant species has its distinct cardinal and optimum temperature ranges for various growth and development processes. Active plant growth is generally confined to a temperature range of 10-40°C. Thermal extremes cause damage to biochemical processes in plants. The magnitude and effects of temporal variations may be visualized by the fact that in cold arid climate sunstroke and frostbite could be experienced in the same season. In cold arid region, both seasonal and diurnal temperature fluctuations are extremes. The variation in temperature provides to grow wheat, barley and other *rabi* crops in summer months which are normally grown during winters in plain.

In cold arid region, chilling injury occurs during germination and early stages of plant growth. A temperature range of 4-15°C is considered as chilling temperature in context to plant growth and development. During low soil temperature at germination stages in April may cause vernalization in wheat, chickpea and pea, due to that reduction in biomass and induction of early flowering was observed. Whereas frost injuries (freezing) are common during the last week of September and early October, this leads to rupturing of plasma membrane due to formation of ice crystals. Crop plants of sub-tropical origin and C_4 plants have very limited ability for cold tolerance. The prevalence of freezing injuries during early days of September forced farmers not to take second crop in the season and as a consequence most of the cultivated areas of the region are mono-cropped.

As in cold region temperature goes maximum up to 40° C during June and July that results in reduced growth, leaf tip burning, premature fall of leaves and poor and shriveled seed setting in wheat and other crops. High temperature in combination of low soil moisture, high radiation and high wind affects vegetable and fruit crops yield badly. While devising a proper strategy of crop improvement in these areas requires improving component characters related to increase tolerance for low temperature as well high temperature conditions.

1.2.3 Salt Stress

Unlike the halophytes, most of the glycophytes grow best with salt content below of 4 ds/m. According to Levitt (1980), if the salt concentration is high enough to lower the water potential appreciably (0.5-1.0 bar), the stress will be called as salt stress. Effects of salt stress on plants can be classified into the primary and secondary effects (Tal *et al.*, 1979). Primary salt injuries may include direct, specific toxic effects as well as indirect effects, such as metabolic disturbances and inhibition of growth and development. While secondary effects include nutrient deficiency and osmotic dehydration. Oretli (1968) concluded that salt injury in plants is due to ion accumulation in apoplast leading to physiological drought is the primary cause of salt damage. Salt stress depresses growth by affecting vital metabolic processes like photosynthesis or by altering the ionic relationships in plants and ultimately leads to deficiencies of essential nutrients like phosphate, nitrate or calcium. Different crops have varying limit of tolerance for sodicity and salinity (Table 1). In cold arid region where most of the soils are saline in nature with pH ranging from 7.5 to 9.5, that leads to poor growth of sensitive crops/varieties. Increase in soil salinity was also observed due to irrigations. Effects of salinity were observed on the growth and survival of crops. It also reduces the choice of crops that can be grown on such problematic soils. Salt tolerance mechanisms vary between different species. Some species develop high osmotic pressures in the tissue fluids to counteract the increased osmotic pressure, while others accumulate considerable quantities of salts in tissue fluids to regulate that accumulation. Different approaches of breeding leading to accumulation of desirable attributes in genotypes may be deployed for developing salt tolerance varieties. Pasternak (1987) suggested different strategies for developing salt tolerance varieties including chance selection of cultivars, breeding and selection for ion exclusion or inclusion, selection of vigorous cultivar, induction of polyploidy and hybridization involving salt sensitive and salt tolerant genotypes.

Table 1: Intra-crop variability of salt tolerance in some important crops

Crops	Sodicity (pH)	Salinity (ECe dsm^{-1})
Rice	9.2 -10.2	4- 10
Wheat	8.8 - 9.3	6.5 -11.0
Barley	9.0 -9.3	4.0 -6.5
Indian mustard	8.9 - 9.2	4.0 - 5.5
Toria	8.8 - 9.0	4.0 - 6.0
Taramira	8.7 - 8.9	4.0 - 6.0
Linseed	8.7 - 8.9	4.0- 6.0
Safflower	8.7 - 8.8	4.0 - 6.0
Sunflower	8.6 - 8.8	4.0 - 6.0
Sugarcane	8.8 - 9.0	-
Cowpea	8.6 - 8.8	4.0 - 5.0
Green gram	8.6- 8.8	-
Pigeon pea	8.3 - 8.6	-
Pea	8.6 - 8.8	4.0 - 5.0
Gram	8.6 - 8.8	4.0 - 5.0
Broad bean	8.6 -8.8	4.0 - 7.0
Sugarbeet	9.2 - 9.8	4.2 -10.0
Pearl millet	8.7 - 8.9	4.0 - 8.0

Adopted from: Mishra *et al.* (2008)

1.2.4 Nutrient Stress

The various kind of nutrient stress situation that plants face are deficiency of nutrients in soil, toxicity of nutrients in soil, nutrient imbalances, antagonistic effects among nutrients, enhanced nutrient losses and hampering of nutrient availability and uptake by water stress, salinity, sodicity, water logging and cold and heat stresses. Cold arid soils are generally poor in nitrogen and organic carbon content. Similarly, due to salinity, water stress, low temperature and nutrient loss the availability of nutrients are also major bottleneck in proper growth and development. Proper soil management and nutrient applications are required for taking higher productivity from these regions. However, by applying appropriate breeding approaches genotypes with high nutrient use efficiency can be developed.

1.2.5 Physical Impedance

Mechanical impedance refers to the physical resistance a soil offers to elongation of root and seedling emergence. The cold arid region soils are new in nature and contain high proportion of gravels and pebbles that hinder proper growth and development of roots. Similarly, land ploughing at the same depth every year may develop hard pans which show downward extension of roots. Such problems are acute for fruit, fodder and fuel yielding trees, wherein growth and development are affected adversely. Sub-soiling and management practices can be performed for increasing nutrient availability. While through genetic manipulation plants having vigorous and spreading root system can be developed.

2. Selection Criteria

One of the early selection criteria has been the number of embryonic roots, which has shown a fairly good correlation with survival and yield under stress conditions. The number and rate of appearance of embryonic roots was correlated with drought resistance. The other criterions which can be used for identification of suitable genotypes for the regions are:

2.1 Yield Stability Over Stress Sites

The cold arid environment is mostly unstable and unpredictable and keeps on changing year to year, which affects the performance of varieties to a great extent. Stability in phenotypes across varying environmental conditions can be utilized as useful criterion for the identification of varieties for cultivation in stress conditions. This procedure has the advantage of being easily integrated into an ongoing breeding program and of utilizing performance in natural stress environments as the selection criteria. Stable genotypes in their expression for the different growth and yield parameters can be selected by exposing them in different stress conditions. Further, the genotype x environment interaction for grain yield analysis will lead to improved yield testing programme. When selection is made in multiple trials for yield or additional traits, the identification of superior genotypes sometimes becomes complicated by an interaction between genotypes and environments, especially if genotypes re-ranking occurs. Yadav and Sharma (2008) used this strategy for selection of stable varieties for cold arid conditions of Ladakh in case of bread wheat by evaluating pre-release and breeding lines.

2.2 Stem Reserve Mobilization for Grain

Stem reserves from pre-anthesis plant assimilation are being increasingly recognized as an important source of carbon for grain filling when current photosynthesis in inhibited by drought, heat or disease stress during this stage (Blum, 1998). While root storage is important in some legumes and other species, there is no evidence that roots or leaves are as important as stem for reserve storage in small grained species. Gallaher *et al.* (1975) reported that 70 per cent contribution to grain filling in barley could be from assimilates stored prior to anthesis under stress. Stem reserve mobilization or the percentage of stem reserves in total grain mass, is affected by sink size, by the environment and cultivar. Thus, translocation of pre-anthesis assimilates to grain during stress is important for stable performance of a variety and offers scope for genetic improvement. Genotype differs in their capacity of reserve mobilization under stress and screening can be taken for selecting efficient genotype(s) from diverse germplasm pool.

2.3 Stomatal Behaviour

Regulation of water loss by the stomata is one of the important characters that have received a considerable attention from plant physiologists and breeders attempting to improve crop yields in stress conditions. Selection for stomatal characters potentially provides great scope for improvement in drought resistance not only because stomata provide main control of water loss and consequently of plant stress but also because of their ability to adapt to changing conditions, i.e., any change in aperture being reversible within a short time period.

The shape of the stomatal response curve to leaf water potential varies markedly with species, age and growth condition. In addition to the selection for particular values of maximum and minimum conductance, it is advantageous to select either for a high threshold potential for stomatal closure or for gradual closure over a wide range of water potentials. The former response (sensitive genotypes) would be particularly valuable for preventing the damaging effects of short stress periods while the latter (non-sensitive genotypes) might be of value in maintaining photosynthesis during drought of longer duration.

2.4 Root Growth and Root-Shoot Ratio

Increase in root length and weight under moisture stress indicates greater density of roots and or greater depths of root penetration; both of these are important morphological adaptations to moisture deficit condition and results in greater extraction of soil water (Hoogenboom *et al.*, 1987). Narayan and Mishra (1989) recorded deeper penetration of roots under moisture stress and concluded that depth of root penetration can provide a useful selection criterion for wheat breeding under drought stress. The root-shoot ratio of plants usually increases with moisture stress which reduces shoot growth in relation to root growth; but there are also evidences of absolute increase in root growth. Generally, the resistant lines had slower shoot and root growth rates, soil water extraction rates but higher shoot:root ratio than the susceptible lines (Matthews *et al.*, 1990).

2.5 Osmotic Adjustment

Osmotic adjustment results on account of accumulation of solutes within the cells in response to stress. Higher concentration of solutes in the cells enables plants to maintain turgor pressure and continue the turgor driven processes such as stomatal opening, transpiration, photosynthesis and growth at reduced rates. It may enhance biomass production by delaying leaf senescence and allocation of more assimilates to developing grain. Genotypes of wheat equipped with osmotic adjustment property produce more root biomass and greater root length, density; thus, have more access to soil water. Osmotic adjustment has been reported to maintain harvest index in wheat with several specific effects (Ludlow and Muchow, 1990). In moisture stress environments, the yield advantage by osmotic adjustments has been recorded upto 50% (Morgan *et al.*, 1986). The capacity of a genotype to adjust to stress via osmotic adjustment is genetically controlled by one or a few genes.

3. Physiological Screening Techniques for Abiotic Stresses

3.1 Screening Techniques for Drought Tolerance

The green pigments are thermo-sensitive. Degradation occurs when it is subjected to higher temperature. The chlorophyll destruction commence rapidly at critical temperature of 50-55°C. The Chlorophyll Stability Index (CSI) is a function of temperature. This property of chlorophyll was found to be correlated well with the drought resistance.

a. *Treated* : 250 mg of leaf sample → cut into bits → transfer it into test tube containing 10 ml of water → keep in water bath at 50-55°C for 30 minutes → mix it with 10 ml of 80% acetone and centrifuge (5000 rpm) → read at 652 nm.

b. *Control* : 250 mg of leaf sample → macerate with 10 ml of acetone → centrifuge at 5000 rpm → supernatant make up to 25 ml → read at 652 nm.

$$\text{Total chlorophyll content (mg g}^{-1}) = \frac{\text{OD at 652}}{34.5} \times \frac{V}{W}$$

The CSI can be calculated by chlorophyll content of the treated with untreated chlorophyll content treated

$$\text{CSI (\%)} = \frac{\text{Total chlorophyll content treated}}{\text{Total chlorophyll content untreated}} \times 100$$

3.2 Relative Water Content (RWC) by Relative Turgidity Technique (Barrs and Weatherley, 1962)

RWC is an important parameter, which indicates the ability of plant to maintain high water in the leaves under moisture stress condition. Also used as indices to determine drought tolerance of a genotype.

One gm of leaf discs (Fresh weight) in Petri dish → add 20 ml distilled water → keep it for one hour → take turgid weight → keep it in oven at 75-80°C → take dry weight

$$\text{RWC (\%)} = \frac{\text{Fresh weight (wt) -dry weight (wd)}}{\text{Turgid weight (wt) - dry weight (wd)}}$$

3.3 Plant Height Stress Index (PHSI)

High PHSI is a good indicator of drought tolerant genotypes and this can be calculated by:

$$\text{PHSI} = \frac{\text{Plant height under stress}}{\text{Plant height under non-stress}}$$

3.4 Dry Matter Stress Index (DMSI)

DMSI can be calculated for genotypes using the data of biological yield:

$$\text{DMSI} = \frac{\text{Total dry matter produced under stress}}{\text{Total dry matter produced under non-stress}}$$

3.5 Yield Stability Ratio (YS)

Yield stability Ratio (YS) can be calculated by taking the ratio of yield under stress and non-stress as suggested by Lewis (1954).

$$YS = \frac{\text{Yield under stress}}{\text{Yield under non-stress}} \times 100$$

3.6 Drought Injury Index (ID)

Drought injury index (ID) can be estimated by the method of Gebre and Kubas (1991).

$$ID = \frac{100\,(Rd-R0)}{(1-R0)}$$

Where Rd is the Ratio of solution conductivity from discs for stressed leaves before autoclaving to that of after autoclaving, R0 is the ratio of conductivity of control discs before autoclaving to that of after autoclaving.

3.7 Stomatal Index

Take a fresh clean leaf and remove all the dust and soil particles adhering to the surface. Apply small quantity of quick fix gum on the surface of the leaf smear evenly into a thin film layer. After 15 minutes of drying, remove the thin layer of the quick fix. This would have traced out the impression of the epidermal cells, stomata openings, oil glands and veins. Place the quick fix film on a glass side and observe under a microscope.

$$\text{Stomatal Index} = \frac{S}{E+S} \times 100$$

Where, S = Stomatal number/unit area, E = epidermal cells/unit area.

3.8 Stomatal Frequency

The number of stomata per unit area of leaf is described as stomatal frequency. Low stomata number per unit area is indicative of tolerance of the genotype.

4. Screening Techniques for High Temperature Tolerance

4.1 Triphenyl Tetrazolium Chloride (TTC) Reduction Activity or Availability Test

Leaves of temperature-stressed plant are washed with distilled water and cut into small discs. Ten leaf discs are weighed and placed in a test tube containing 3 ml of 0.06 per cent (W/V) TTC in 0.05 M phosphate buffer. The discs are then incubated at 30°C for 15 hours in dark. After incubation, discs are rinsed with distilled water and extracted with 7 ml of 95 per cent (V/V) ethanol in boiling water bath for 5 minutes. The extract is cooled, decanted and made to a volume of 10 ml with 95 per cent ethanol. Absorbance of the colour intensity read at 480 nm and 530 nm using spectrophotometer. The TTC reduction activity is expressed as the absorbance unit/g leaf. Also expressed, as the per cent decrease in TTC reduction compared to control. Thus, inhibition of TTC reduction is an indication of enzyme inactivation.

4.2 Pollen Viability Tests

Pollen grains from heat stressed plants are stained with acetocarmine and observed in microscope. Live pollen grains (well filled) are stained red and non-viable pollen grains remain unstained.

4.3 Chlorophyll Fluorescence and Chlorophyll Stability Index

Variable chlorophyll fluorescence (VCF) is indicative of the capacity for electrons flow through PS II. It decrease as leaves envisage heat injury and can be measured with chlorophyll fluorescence meter.

Chlorophyll stability index is inversely related to the heat resistance of chlorophyll. Chlorophyll stability index of stressed leaves (CSI) is calculated by:

$$\text{CSI}(\%) = \frac{\text{Total chlorophyll of stressed sample}}{\text{Total chlorophyll of unstressed sample}} \times 100$$

4.4 Heat Susceptibility Index

Susceptibility index (S) estimates (Fischer and Wood, 1979) for each cultivar the reduction in the given variable from the normal to the stress environment relative to the respective mean reduction for all cultivars is as follows:

$$S = (1 - Y/Yp/1 - X/Xp)$$

Where, Y is the yield under stress, Yp is yield without stress and x and Xp represent mean yield over all cultivars under stress and non-stress conditions, respectively. The term (1 – X / Xp) is defined as "stress intensity".

5. Screening Techniques for Frost Tolerance

5.1 Ratio of Leaf Dry Weight to Fresh Weight

This method is based upon the assumption that when tissue is frozen, a part of the tissue is killed and therefore per cent ratio of dry weight and fresh weight changes. Two sets of one gram each leaf/ seed samples are taken and ratio of dry weight and fresh weight of one set is calculated by drying is oven at 80°C for 3 days or till a consistent dry weight is gained. Another set is given freezing treatment and ratio of dry weight and fresh weight examined. Higher per cent reduction in the ratio would suggest that particular genotype to be more susceptible as compared to the one showing less per cent reduction.

5.2 Conductivity of Plant Extract (Flint *et al.*, 1967)

This method is very broad estimate of frost injury, particularly when the damage is localized to small structure. Take 2 sets of 1 g tissue each in 10 ml of double distilled de-ionized water. Freeze one set, whereas, boil other set and examine the conductivity of different types. Higher the percent change in conductivity in lechates, more susceptible would be a genotype to frost and vice versa.

6. Screening Technique for Salt Tolerance

This screening of genotypes can be performed in laboratory conditions by following the general protocol as given below:

6.1 Induce Seedling

Germinated seedling → induction treatment → Lethal stress → recovery in water
(6 h, 200 mM) (48 h) (400 mM) (48 h)

6.2 Non-induce Seedling

Germinated seedling → Lethal stress → recovery in water
(48 h) (48 h) (400 mM) (48 h)
(Sodium chloride)

After the completion of protocol the root length, shoot length can be measured at the end of induction, stress and recovery. From data, the growth during stress (GDS), growth during recovery (GDR) and growth during stress and recovery (GDS + GDR) can be calculated.

Growth during stress (GDS) = growth at the end of stress - growth at the end of induction.

Growth during recovery (GDR) = growth at the end of recovery - growth at the end of stress.

Growth during stress and recovery (GDS + GDR) = growth during stress + growth during recovery.

7. Breeding Approaches

Breeders look for new sources of variation when attempting to improve plants, but such empirical selection in the case of abiotic stresses is not possible. The reason for this is that any abiotic stress, *viz.*, drought, cold, salinity resistance is not a simple response, but mostly depend on a number of component responses that interact after each other and may differ from crop to crop. Similarly, selection under controlled environment rarely correlates with its performance in the field, because of yearly variability in the environmental conditions. Thus, improvement in a single trait or incorporation in single trait may not give expected results. However, importance of trait in improvement of resistance for the stress can be studied through isoline approach. Poor understanding of tolerance mechanism, genes controlling component trait and change in stress tolerance at different developmental stage further complicate the breeding programmes. To develop an efficient breeding programme for developing resistant genotype, it is essential to understand the mode of inheritance. Generally, most of the traits associated with enhance tolerance for abiotic stresses are quantitively inherited with a complex physiological reaction, thus its genetic basis has received limited attention and the development of varieties for complex situations like cold arid has been slow. However, appearances of these stresses are frequent in cold arid region, thus evaluation and screening of material can be easily done under field conditions. Following approaches can be deployed for selection and development of varieties for cold arid conditions.

7.1 Introduction and Evaluation of New Crops

Cold arid climates suited to only few crops for their growth and development. However, traditionally, few crops were being cultivated in this region. Recently, various crops were introduced and evaluated in different cold arid regions. In Ladakh region, most of vegetable crops, *viz.*, cauliflower, cabbage, tomato, brinjal, coriander, fenugreek etc. are recent introductions which were found suitable for cultivation. Similarly, various new crops, *viz.*, mustard, toria, chickpea, safflower, oats, sunflower was introduced by different workers in Ladakh region and were evaluated for their suitability. At SKUAST, Regional Research Sub-Station, Kargil, chickpea, both *kabuli* and *desi* type, was introduced from ICRISAT, Hyderabad to study the scope of their cultivation. *Desi* chickpea entry ICCV97031 yielded 33.24 q/ha and *kabuli* type ICCV00305 yielded 26.35 q/ha, which is comparable to other traditionally chickpea growing areas. Besides that, crop remained free from pod borer problem which is a major pest in chickpea growing areas (Yadav *et al.*, 2003). Similarly, in an attempt to increase the crop option for oilseed crops, Yadav *et al.* (2004) introduced sunflower population for their evaluation at SKUAST, Research Station, Leh. Crop performed well and per plant seed yield was up to 55.50 g, which was reasonably good for cultivation of crop in the region. These examples suggest that further efforts are needed for introduction and evaluation of new crops in the region.

7.2 Selection Under Target Environment

This is easiest and most commonly used approach by breeders for selection and development of variety for stress conditions. This approach is very suitable when the breeding program serves a target population under very diverse environments, where genotype by environment interactions is large (Ceccarelli and Grando, 1997). Under this method germplasm or segregating material of the crop are grown in the environment where it is naturally exposed to prevalent stresses during the duration of crop. The selection is not directed towards any particular stress but is based on the over all fitness of the genotype at site in target agro-climatic zone. The performance trend of genotype over the several crop seasons (4-5) would reveal its fitness. Screening of wheat germplasm regularly for five years at SKUAST, RARS, Leh, has resulted in the identification of several high yielding wheat genotypes possessing resistance to various abiotic stresses (Sharma and Mir, 1997). Selection of genotypes under stress environment was found more effective in developing varieties for stress conditions rather than selecting varieties in optimum or near optimum conditions (Singh *et al.*, 1993; Ceccarelli *et al.*, 1996).

7.3 Hybridization and Selection

Hybridization for transfer of stress tolerance is one of the methods for developing high yielding stress tolerant varieties. After hybridization segregating generations can be handled either by pedigree method or by backcrossing when the trait is being introgressed. Hybridization can also be employed for gene pyramiding for various stresses through successive hybridization. However, some special methods for handling of segregating generations for selecting genotypes wide adaptations used by different workers are given below:

7.3.1 Shuttle Breeding

Shuttle breeding has proved to be a model approach for developing abiotic stresses tolerant varieties for sustainable productivity in fragile ecosystems. This method has successfully been used for developing salt tolerant varieties in rice, wide adaptable varieties in wheat and other crops (Braun *et al.*, 1996). According to N.E. Borlaug, who invented this method "Through the use of this technique, we developed high-yielding, day-length-insensitive varieties with a wide range of ecological adaptation and a broad spectrum of disease resistance a new combination of uniquely valuable characters in wheat varieties". In this approach, crosses between desirable parents can be made in target environments and segregating generations can be evaluated in other environments entirely different from target environments. In crops like wheat, barley, lentil and other *rabi* crops, which is grown during summer months, two crops can be taken in one year by including centers where these are grown during winter months. For example, wheat is sown in April in Leh and harvested during September, while in Jammu, crop is sown in October-November and harvested in April. Thus, by selecting such locations for evaluation, the period involved in varietals development can be reduced, and on the other side, the adaptability of selected material will be high.

7.3.2 Selection for Salt Tolerance

The donors for salinity tolerance can be identified by screening germplasm under salted conditions. The donors can be involved in the hybridization programme for subsequent selection and release of varieties. The scheme for breeding a salt tolerant variety is shown in (Fig. 1). By using such approach, thousands of advanced lines from crosses between salt tolerant donors and elite breeding lines have been tested against salinity at IRRI and a few improved lines have been developed (Akbar *et al.*, 1987).

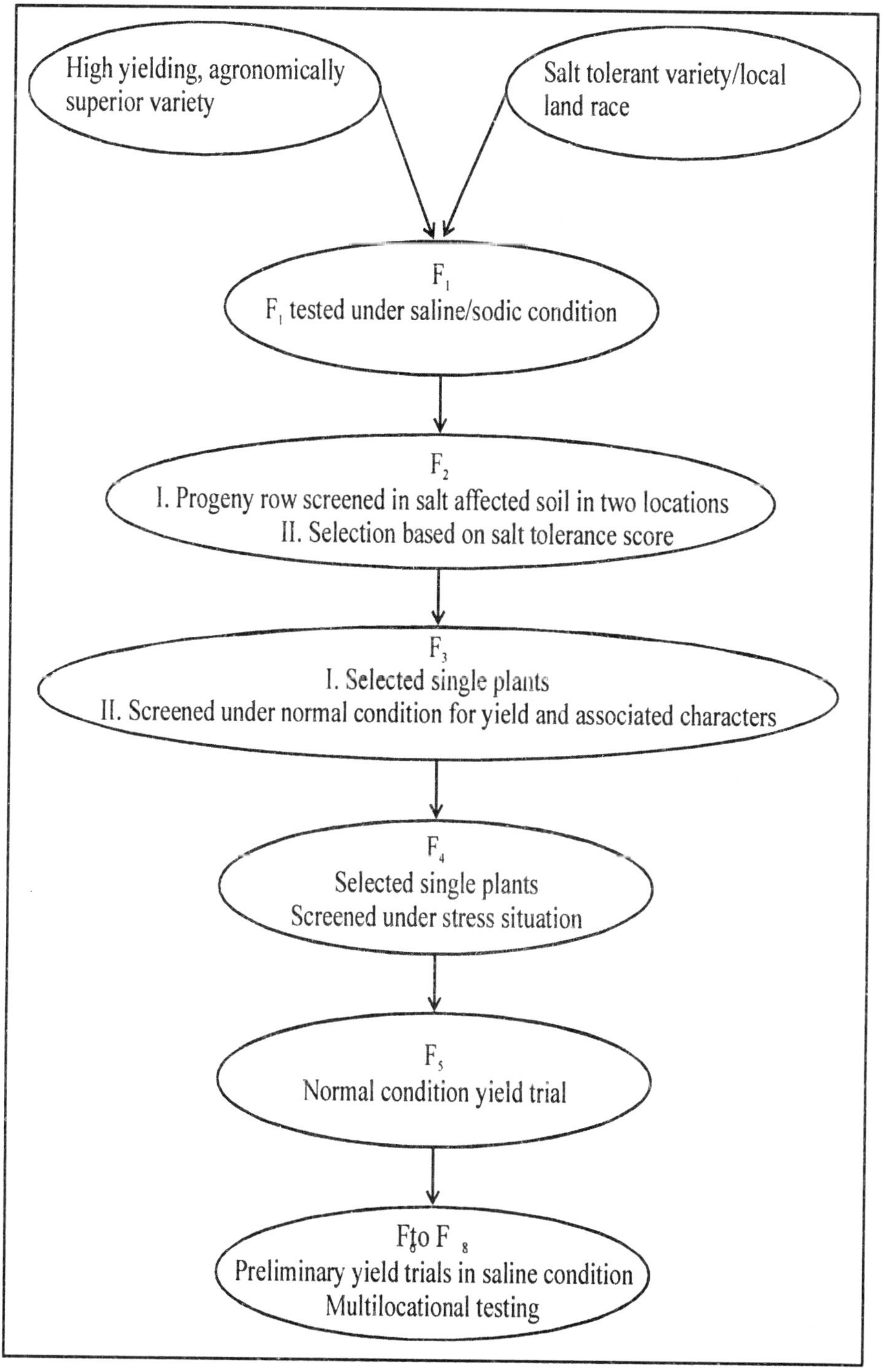

Fig. 1 : Breeding for salt tolerant variety

8. Biotechnological Approaches

The new technologies such as *in vitro* screening, somatic hybridization, transgenic development, marker added selections are some important biotechnological tools that can be applied for developing genotypes for abiotic stress conditions. These techniques are more precise and less time taking as compared to conventional methods. These techniques provide rapid mass selection and transfer of genes across the species and crops.

8.1 *In vitro* Screening

Tissue culture has been shown to produce somaclonal variation and that variation can be utilized for effecting selection. Selection for resistance is the most straight forward method for mutant selection, whereby resistant cells in a large population can be selected by their ability to grow in the presence of an inhibitor while the sensitive cells do not. The inhibitor compounds are put in the medium at a concentration such that some cell population survives and can be further grown on a selective medium. Different selection cycles are performed to get tolerant cell/callus cultures that are subsequently regenerated into plants. These plants are then *in vitro* screened against the inhibitor. A significant positive correlation between praline level and frost tolerance has been found in different crops. Chen *et al.* (1985) obtained cold tolerant wheat plants from suspension cultures and calli initiated from immature embryo of Norstar wheat after inducing cold hardening by low but non-freezing temperatures. Bouharmont *et al.* (1991) also selected plants tolerant to aluminum toxicity in rice through *in vitro* screening of calli to high strength of Almunium sulphate and cold tolerant plants by culturing calli for long term at sub-lethal temperatures. Similarly, screening calli in manitol or polyethylene glycol (PEG) at sub-lethal concentration could provide basis for selection for drought tolerance (Brar and Uchimiya, 1990).

8.2 Marker Added Selection (MAS)

In this method, linkages are sought between DNA markers and traits associated to abiotic stress tolerance. Instead of selecting for trait, the breeder can select for a marker that can be detected very easily in the selection scheme. Recent advances in plant genome mapping and molecular biology techniques offer a new opportunity for understanding the genetics of stress-resistance genes and their contribution to plant performance under stress. The availability of genetic map saturated with molecular markers allows a breeder to do selection for characters related to several abiotic stresses much more efficiently and effectively than the selection based on morphological traits. Strong root systems were reported to be associated with drought tolerance and better water use efficiency. Quantitative trait loci (QTLs) for characters have been identified in rice (Ray *et al.*, 1996), maize (Lebreton *et al.*, 1995) and in pearl millet (Yadav *et al.*, 1999) and increased capacities for osmotic adjustment have been reported to improve yield in drought affected plants (Tangspremsri *et al.*, 1995).

In order to improve yield under drought in common beans, studies showed that RAPD markers based MAS was efficient than the conventional selection in improving yield under moisture stress although these markers were population specific (Schneider *et al.*, 1997). In sorghum, stay green colour has been considered as drought tolerance mechanism and Haussmann *et al.* (2002) identified four QTLs for this trait by recombinant inbred populations. Thus these emerging technologies can be applied to complement the efforts to reduce the burden of selection to an extent, especially in the selection of parents, identification of desired genotypes in segregating generations and identifying the location of genes/alleles involved in determination and association with stress tolerance.

References

Akbar, M.; Jena, K.K. and Seshu D.V. (1987). Salt tolerant wild rice. IRRI Newsletter 12: 15.

Barrs, H.B. and Weatherly, P.E. (1962). A re-examination of the relative turgidity technique for estimating water deficits in leaves. Aust. J. Biol. Sci. 15: 413-428.

Blum, A.; Mayer, J. and Golan, G. (1983). Chemical desiccation of wheat plants as a simulator of post- anthesis stress. II. Relation to drought drought stress. Field Crops Res. 6: 149-155.

Blum, A. (1998). Improving heat grain filling under stress by stem reserve mobilization. *Euphytica* 100: 77-83.

Bouharmont, J.; Dekeyser, A.; Jansin Van, V. and Doge, Y.S. (1991). Application of somaclonal variation and *in vitro* selection to rice improvement. In : Rice Genetics II., IRRI, Philippines, pp. 271-277.

Brar, D.S. and Uchimiya, H. (1990). Genetic engineering of crop plants. In: Plant Tissue Culture: Application and Limitations. Elsevier Sci. Publ., Amsterdam, Netherlands, pp. 346-365.

Braun, H.J.; Rajaram, S. and Ginkel, M.V. (1996). CIMMYT's approach to breeding for wide adaptation. Euphytica 92: 175-183.

Ceccarelli, S. and Grando, S. (1997). Efficiency of empirical selection under stress conditions. J. Genet. Breed. 43: 25-31.

Ceccarelli, S.; Gramdo, S. and Booth, R.H. (1996). International breeding programmes and resource-poor farmers : Crop improvement in difficult environments. In: P. Eyzaguirre and M. Iwanaga (Eds.), Participatory Plant Breeding, pp. 96-116. Proceedings of a workshop on Participatory Plant Breeding, 26-29 July 1995, Wageningen, The Netherlands, IPGRI, Rome.

Chen, T.H.H.; Kartha, K.K. and Gusta, L.V. (1985), Cryopreservation of wheat suspension culture and regenerable callus. Plant Tiss. Organ Cult. 4: 101-109.

Fischer, K.S. and Wood (1979). In: Fisher, K.S., E.C. Jonson and G.O. Edmeades.1981. Breeding and selection for drought resistance in tropical maize. Proc. of Symposium on Principles and methods in crop improvement for drought resistance with emphasis on rice, held at IRRI, Philippines, May 4-8, 1981.

Flint, M.L.; Boyce, B.R. and Beattic, D.J. (1967). Index of injury- A useful expression of freezing injury to plant tissue as determined by electrolytic method. Can. J. Plant Sci. 47: 229.

Gallaher, J.N.; Biscoe, P.V. and Scott, R.K. (1975). Barley and its environment V. Stability of grain weight. J. Appl. Ecol. 12: 319-336.

Gebre, J.M. and Kubas, M.R. (1991). Seasonal and clonal variation in drought tolerance of *Populus deltoids*. Can. J. For. Res. 18: 519-521.

Haussmann, B.L.G.; Mahalakshmi, V., Reddy, B.V.S.; Seetharama, N; Hash, C.T. and Geigger, H.H. (2002). QTL mapping of stay-green in two sorghum recombinant inbred populations. Theor. Appl. Genet. 106: 133-142.

Hoogenboom, G.; Huck, M.G. and Peterson, C.M. (1987). Root growth rate of soybean ads affected by drought. Agron. J. 79: 607-614.

Lebreton, C.; Lazic'-Janic', V.; Steed, A.; Pekie', S. and Quarrie, S.A. (1995). Identification of QTLs for drought responses in maize and their use in testing casual relationship between traits. J. Exp. Bot. 46: 853-865.

Levitt, J. (1980). Responses of Plants to Environments Stresses, Chilling, Freezing and High Temperature Stresses. Vol I, New York, Academic Press, 487p.

Levitt, J. (1972). Responses of Plants to Environmental Stress. New York, Academic Press, 697p.

Lewis, E.B. (1954). Gene-environment interaction. Heredity 8: 333-356.

Ludlow, M.M. and Muchow, R.C. (1990). A critical evaluation of traits for improving yields in water limited environments. Adv. Agron. 43: 107-153.

Matthews, R.B.; Reddy, D.M.; Rani, A.U.; Azam-Ali, S.N. and Peacock, J.M. (1990). Response of four sorghum lines to mid season drought I. Growth, water use and yied. Field Crops Res. 5: 279-296.

Mishra, B.; Singh, R.K. and Senadhira, D. (2008). Enhancing genetic resources and breeding for problem soils. http//www.idrc.ca November, 2008.

Morgan, J.M.; Hare, R.A. and Fletcher, R.A. (1986). Genetic variation in osmoregulation in bread and durum wheats and its relationship to grain yield in a range of field environments. Aust. J. Agric. Res. 37: 447-457.

Narayan, D. and Mishra, R.D. (1989). Drought resistance in varieties of wheat (*Triticum aestivum* L.) in relation to root growth and drought indices. Indian J. Agric. Sci. 89: 559-598.

Oretli, J.J. (1968). Extra cellular salt accumulation, a possible mechanism of salt injury in plants. Agrochimica 12: 461-469.

Pasternak, D. (1987). Salt tolerance and crop production: A comprehensie approach. Ann. Rev. Phytopathol. 25: 171-191.

Ray, J.D.; Yu, L.; McCouch, S.R.; Campoux, M.C.; Wang, G. and Nguyen, H.T. (1996). Mapping quantitative trait loci associated with root penetration ability in rice (*Oryza sativa* L.) Theor. Appl. Genet. 6: 627-636.

Schneider, K.A.; Brothers, M.E. and Kelly, J.D. (1997). Marker-Assisted selection to improve drought resistance in common bean. Crop Sci. 37: 51-60.

Sharma, J.P. and Mir, A.A. (1997). Crop Improvement Perspective in Cold Arid Zone-Ladakh. SKUAST, RARS, Leh, 79 p.

Singh, M.; Ceccarelli, S. and Hamblin, J. (1993). Estimation of heritability from varietal trials data. Theor. Appl. Genet. 86: 437-441.

Tal, M.L.; Rosental, L.; Abramowitz and Forti, M. (1979). Salt tolerance in *Simmondsia chinensia*: Water balance and accumulation of chloride, sodium and praline under low and high salinity. Ann. Bot. 43: 701-708.

Tangspremsri, T.; Fukai, S. and Fisher, K.S. (1995). Growth and yield of sorghum lines extracted from a population for differences in osmotic adjustment. Aust. J. Agric. Res. 46: 61-74.

Yadav, R.S.; Hash, C.T.; Bidinger, F.R. and Nguyen, H.T. (1999). QTL analysis and marker assisted breeding of traits associated with drought tolerance in pearl millet. In: Genetic Improvement of Rice for Water-limited Environments. (O Ito amnd J. O'Toole, eds) IRRI, Philippines. Pp. 211-226.

Yadav, V.K. and Sharma, J.P. (2008). Adaptability of bread wheat cultivars and breeding lines under cold arid conditions of Ladakh. Int. J. Plant Sci. 3: 428-431.

Yadav, V.K.; Sharma, J.P. and Roy, D. (2004). Population parameters and characters association in dwarf and tall populations of sunflower (*H. annuus* L.) under cold arid conditions of Ladakh. Envi. & Eco. 22: 470-472.

Yadav, V.K; Singh, H.V.; Singh, A.K. and Wani, M.A. (2003). Performance of desi and kabuli chickpea genotypes for yield and wilt resistance under cold arid condition of Ladakh. In: National Symposium on pulses for crop diversification and natural resource management. 20-22 December, 2003, Indian society of pulses research and development, IIPR, Kanpur, pp. 24.

□□□

10

Greenhouse Technology for Higher Himalayas

SANJAI K. DWIVEDI AND ELI PALZOR*
CEPTAM (DRDO), Metcalfe House, Delhi-110054
*Defence Institute of High Altitude Research (DRDO), C/o 56 APO

Protected cultivation using greenhouse technology tempts ones mind as it permits enormous modification in microclimate enabling the cultivation of crops in adverse climate, caring least for the outside environment. Vrikshayurveda, an epic of 11th Century AD by Surapala, state that any plant/tree could be grown anywhere provided king, treasury and destiny are favourable. This is an indication that agricultural experts of that era were aware of the protected cultivation methods. The idea of growing plants in environmentally controlled areas has also existed during Roman times. The Roman emperor Tiberius ate a cucumber-like vegetable daily. The Roman gardeners used artificial methods (similar to the greenhouse system) of growing to have it available for his table every day of the year. Cucumbers were planted in wheeled carts which were put in the sun daily, then taken inside to keep them warm at night. The first modern greenhouses were built in Italy in the 13th Century to house the exotic plants that explorers brought back from the tropics. They were originally called *Giardini Botanici* (botanical gardens). The concept of greenhouses soon spread to the Netherlands and then England. Jules Charles, a French botanist, is often credited with building the first practical modern greenhouse in Leiden, Holland to grow medicinal tropical plants. In the 19th Century the largest greenhouses were built. The conservatory at Kew Gardens in England is a prime example of the Victorian greenhouse. In Japan, the first greenhouse was built in 1880 by Samuel Cocking, a British merchant who exported herbs. In the 20th Century the geodesic dome was added to the many types of greenhouses (Anon., 2008). In the country like India with its vast and diverse agro-climatic conditions, protected cultivation technology in the last three decades has made good progress and proved as a boon for production of high value, low-volume crops with better productivity and quality.

1. Uses

Greenhouses are often used for growing flowers, vegetables, fruits, and other plants. Many vegetables and flowers are grown in greenhouses in late winter and early spring, and then transplanted outside as the weather warms. Started plants are usually available for gardeners in farmers' markets at transplanting time. The closed environment of a greenhouse has its own unique requirements, compared with outdoor production. Pests and diseases, and extremes of heat and humidity, have to be controlled, and irrigation is necessary to provide water. Significant inputs of heat and light may be required, particularly with winter production of warm-weather crops. Special greenhouse varieties of certain crops, like tomatoes, are generally used for commercial production. Greenhouses are increasingly important in the food supply of high latitude countries. One of the largest greenhouse complexes in the world is in Almeria, Spain where greenhouses cover almost 50,000 acres (200 km^2) and where almost 5% of Spain's salad vegetables are grown. Greenhouses protect crops from too much heat or cold, shield plants from dust storms and blizzards, and help to keep pests out. Light and temperature control allows greenhouses to turn inarable land into arable land. Greenhouses can feed starving nations where crops can't survive in the harsh deserts and Arctic climates. Hydroponics can be used in greenhouses as well to make the most use of the interior space. Biologist John Todd invented a greenhouse that turns sewage into water, through the natural processes of bacteria, plants and animals (Anon., 2008).

2. Greenhouse Technology

Greenhouse technology implies production of plants for economic use in covered structure that allows rapid harvesting of solar radiation and modification of agro-climatic conditions conducive for plant growth and development. The technology embraces infrastructure modelling, selection of plants for adaptation, production economics, agronomic management and commercial potential *etc*.

"A greenhouse is a framed or an inflated structure covered with a transparent or translucent material which permits at least partial control of plant environment and which are large enough to permit a person to carry out cultural operations" (Chandra and Panwar, 1987). While full advantage is taken of the available sunshine for crop production by way of selecting proper covering materials, the enclosure provides an opportunity to control the other environmental parameters. As a result greenhouse crop productivity is largely independent of outdoor environmental conditions. However, environmentally controlled greenhouses are not popular in hilly region simply because of acute shortage of electric power in the region. Hence, greenhouse technology is discussed with special emphasis on crop production in hilly region. Types and design of protected structures mainly depend on the availability of materials, purpose of protection, climate of the region and market for quality produce. Environmental conditions in general include light, temperature, air composition and nature of rooting medium. Depending upon the local climate and crop requirements, adequate environmental control is attempted rather than full control of full components to maximize the profit.

3. Principle

A greenhouse is generally covered with transparent material such as polyethylene, glass or polycarbonate. Depending upon the clading material and its transparency, major fractions of sunlight is absorbed by the crops and other objects. These objects in the greenhouse in turn emit long wave thermal radiation for which clading material has lower transparency with the result solar energy is trapped and resist the temperature inside the greenhouse (Nimje and Shyam, 1991). This is known as greenhouse effect. This rise in temperature in greenhouse is responsible for crop forcing in adverse climate. During winter months loss of trapped temperature should be minimised whereas, during summer months air temperature in greenhouse is to be brought down by providing cooling device or ventilation.

4. Agro-climate

There are several climatic factors that need attention but in hilly region following factors are important for winter cultivation:

4.1 Sunlight

The direction of the greenhouse should be such that it receives sunlight for maximum hours. In hilly region, greenhouse should face the South-East direction and its location should not be under shade. Sunlight is essential for augmentation of the process of photosynthesis. In addition, it provides requisite thermal environment for various other physiological processes including photosynthesis. Hilly regions receive clear sunny days in winter so sunlight as such may not be a limiting factor but its tapping and conservation could be a major concern for further investigation.

4.2 Temperature

Sunlight is the major rather sole source of temperature. Examples of providing external heat source other than sunlight are rare in hilly region. Temperature during night hours may drop down to below 0° C if heat obtained from trapped solar radiation is not properly conserved. Following steps are advantageous to maintain thermal environment of greenhouse:

- Blackening of internal surface of brick wall and soil bed surface
- Providing insulation by doubling brick wall
- Covering polyethylene or glass external surface with multi layer covering during night hours
- Mud plastering in case of stone walls
- Placing of black coloured stones in the greenhouses
- Keeping containers filled with water

4.3 Relative Humidity

The minimum limit of relative humidity for normal physiological functioning of the plants is 50%. Extreme xeric conditions hardly allow relative humidity to rise to this level in outer environment but inside greenhouse sufficient water application and restricted air circulation gears up and maintains relative humidity at higher level. It may be increased by keeping pan evaporation. Mechanical device which automatically controls relative humidity by periodical sprinkling of water is not used in hilly region.

4.4 CO_2 Enrichment or Maintenance

Carbon dioxide (CO_2) is the basic ingredient required for manufacturing of organic matter by process of photosynthesis. In cold arid region, CO_2 concentration is below normal (NAEDB, 1992) and in the closed structure like greenhosue it may drop down further because of its consumption by the plants. At relatively high sunlight intensities and temperature, photosynthetic rate is directly proportional to CO_2 concentration upto 2,400 ppm. Hence, supplementary application of CO_2 will boost plant productivity inside the greenhouse. Under hilly region situation, appropriate air circulation is better and economic means to prevent undesirable lowering of CO_2 while for conservation of CO_2 produced at night, a tight closed greenhouse is pre-requisite.

4.5 Soil

Soil serves as growth medium for plants. Sandy loam soil is the best choice. It should be supplemented with well rotten FYM, compost or biofertilizers. Inorganic fertilizers need to be used as per recommendations. Continuous cultivation inside the greenhouse may give rise to pathogens. To destroy pathogens pasteurization of soil is recommended (steam aeration). A temperature of 60°C for 30 min is good enough to accomplish the task. Chemical fumigation can also be used but steam is the best as it is not selective thus attacking all sort of pathogens.

5. Benefits of Greenhouse in Hilly Regions

Major benefits of greenhouse in hilly region (Singh, 1995b; Chandra, 1991) are given below:

- Cultivation of crops during winter months at subzero temperature when it is not possible to grow them in open fields
- Tropical vegetables like cucurbits, capsicum, brinjal, okra etc. are rare in hilly region but these crops can be grown in greenhouse
- Greenhouse provides an excellent opportunity to produce quality crops for export
- Raising nursery earlier and advancing the availability is also possible by the use of greenhouse
- Productivity in greenhouse is increased manifold in comparison to open field

- Conservation of valuable planting materials and its cultivation is also possible under greenhouse
- Greenhouse can also be used for growing of flower plants, and propagation of fruit plants like strawberries, grapes etc.
- Insect-pest and weed management is easier in greenhouse than open field
- Greenhouse conserve the moisture. Hence, frequency of watering to plants get reduced
- Greenhouse is ideally suited for hilly region farmers having small holdings
- Productivity per unit area and time can be increased by adopting suitable crop sequences

6. Infrastructure of Greenhouse / Glasshouse

6.1 Types of Glazing Materials

a. Glass

Glass has been the preferred covering material for greenhouse world wide becuase of its light transmissivity characteristics. Transmissivity of 40-50 years old glass differs a little from that of new glass. Temperature retension in glasshouse is pretty good. However, high installation cost is major limitation. Moreover, transportation of materials and frequent damage of glasses due to high wind velocity, glasshouse is discouraged in hilly region.

b. Polyethylene

Plastic polyethylene is the most widely used greenhouse film in hilly region. It is produced by mixing homopolymers of ethylene with or without an ultra-voilet (UV) inhibitor package but only UV stabilized polyethylene sheet is recommended for hilly region because non-UV inhibitor package added polyethylene will break down after 3-5 months due to photochemical reactions. Polyethylene film is tough, flexible and relatively inexpensive. It can withstand as low as -50°C, but 80-90°C temperature will melt it. Above 60°C it losses much of its strength and may stretch markedly. The life of UV-stablized polyethylene film is 1-3 years. However, regidex polyethylene sheet may be used for 5-6 years on trench.

c. Fibreglass Reinforced Panel (FRP)

These panels consist of fibre glass reinforced polyester. The panels have been very popular in areas of high light intensity such as southern USA but about 10-15% heat loss due to increased exposed area makes it unfit for hilly region region.

d. Polycarbonate Panels

The panels are available in double skinned sheets. SDP polycarbonate sheets are similar in physical dimension to the SDP acrylite but are stronger and have a lower light transmissivity. Its price is approximately 25% higher than SDP acrylite. Further, with its age under use, the transprency gets reduced due to yellowing.

6.2 Construction Materials

Besides glazing materials (polyethylene/glass) several other materials are also required for construction of a greenhouse. It is always better to prefer the use of locally available materials for the construction of a greenhouse. It not only reduces the installation cost but also helps to popularize the technology. The major construction material required for greenhouse is its frame structure which could be either of a metal or wood. Iron angle, GI pipes or alumminium frame are the most common metal used in greenhouses. Wooden frames can also be used which are locally available in the region. Such frames are cheaper than those of metal frames and also easy to use. Wooden frame should also be preferred in hilly regions since metal is a good conductor of heat thus its temperature gets reduced during winter months; during summer it gets hot and polyethylene sheet gets stuck to the metal which tears off the sheet. Sometimes supporting wires are also used in greenhouse to hold the polyethylene sheet. It is always advisable to avoid direct contact of metal with polyethylene sheet. If at all it is essential, gunny bags or cloth should be wrapped over the metal. Other precautions include use of smooth surfaced frame and removal of sharp edges of frame. Strong foundation of frame is essential in cold arid region to resist the strong winds. Stability is further ensured by constructing earthern brick wall at the back and on either side.

6.3 Types of Greenhouses in Hilly Region

Though hilly region is predominated by Ladakhi polyhouses however studies carried out by Field Research Laboratory (FRL) Leh recommend trench types of greenhouse for the region (Dwivedi and

Singh, 1999). Various other types of greenhouses are also seen in hilly region (Singh *et al.*, 1998a). The brief description of each type is given below:

a. Glasshouse

As its name indicate, glass is used as glazing material in this greenhouse. Perhaps this is the first and oldest among all types of greenhouse structures. Glass panels are fitted with the help of wooden or metal frame. It can be of any shape and size and it is pretty effective for winter cultivation but due to increase in day temperature in summer, it becomes unfit for cultivation during summer. High initial cost, difficulty in construction and frequent damage of glass panels by strong winds are other discouraging factors. In hilly region only a few such structures are found that too for research purpose. One oldest glasshouse constructed in 1964 is available with FRL, Leh.

b. Polyhouse

Ideal features of polyethylene have increased the use of polyhouse inplace of glasshouse throughout the globe. It has not only reduced the initial cost but also increased the popularity of greenhouse by simplifying the installation technology. Generally two types of polyhouses are there in hilly region.

c. Prefabricated Polyhouse

A prefabricared metal structure of convenient size is installed and polyethylene film is covered over it. This structure is not recommonded at all for the region because of poor temperature retension, low crop yield and high installation cost (Singh *et al.*, 1998b).

d. Ladakhi Polyhouse

The most common greenhouse in Ladakh region is *Ladakhi* polyhouse. It is similar to normal but the only difference is its surrounding mud brick wall in place of polyethylene sheets which not only cuts down the installation cost but also reduces the adverse effects of strong winds and also increase temperature retention, and ultimately net profit gets increased. This is generally lean type, have three sides made of mud bricks. The back wall is 7 ft in height, while front has no wall. The length is 32 ft with a width of 16 ft (Singh *et al.*, 2000a). The polyethylene is supported on wooden poles and side walls; the two side walls one descending towards the front. J&K State Deptt. of

Agriculture/Horticulture provides cash subsidy besides 32 x 16 ft free polyethylene sheet. This has brought boost to greenhouse revolution in Ladakh.

e. Trench (Underground Greenhouse)

This is a very simple, cheap and useful greenhouse structure for hilly region and thus has unlimited potential in the region. This may be of any convenient dimension. However, a trench of 30 x 10 x 3 ft size is ideal. In this pit type of structure, wooden poles are used to hold UV stabilized polyethylene film. The polyethylene is also covered by an additional woolen or cotton sheet polyethylene film during night to reduce the heat loss during extreme winter. The damage of blowing off the polyethylene film by strong winds is minimised by putting stones along the sides. Cultural practices and other operations are done by removing polyethylene sheet from top of the trench. The structure does not require much skill in its construction and management. Its cost is lowest among all other greenhouses and being an underground structure heat loss is minimal and temperature retention is high (Singh and Dhaulakhandi, 1998) and thus yields good crop. Strong winds do not affect polyethylene cover much and hence it is long lasting. This structure is therefore, being recommended as most suitable greenhouse for the region (Singh *et al.*, 1998a, 2000b).

f. Plastic Low Tunnel

It is a small semi sphereical structure frame made of metal, wood or plastic and covered with polyethylene or fiber reinforced plastics to create protected environment. Plastic low tunnels are flexible transparent coverings that are installed over rows of individual beds of transplanted vegetables to enhance the plant growth by warming the air around the plants (micro-climate) in open field during winters. These help in warming the soil and protect the crop against the hails and snow and advance the crop by 30-60 days than their normal season. This low-cost technology is very simple and highly profitable for off-season cultivation and catching the early market. Generally, galvanized iron arches are fixed manually 1.5-2.0 m apart to support the plastic tunnel. Width and height is kept about 45-60 cm to cover the plant as per the crop requirement. Transparent non-perforated plastic of 30-50 mm thickness is sufficient. It partly reflects infrared radiation to keep the temperature of low tunnel higher than outside. Small vents can be made at the side of tunnel to facilitate proper aeration in tunnel. These structures serve as mini greenhouse. This is a temporary structure which can easily be shifted to any place. However, major problem in tunnels is faced in watering, weeding and harvesting which involves removal of cover and again putting them back.

g. Double Wall Polyench

A structure designed by FRL, Leh, to harness the soil and solar heat for growing the vegetables especially during winter months. The polyench refers to a polyhouse erected over a trench where back and sidewalls are made up of mud bricks. The sidewalls are provided with double walls filled with insulating material available locally, i.e., sawdust for better retention of heat inside the polyench. The inner walls are painted black to absorb more solar radiation during morning hours. The structure is also like gable uneven span with modified roof having polythene sheet towards sun facing side and grass thatch support over wooden frame opposite to sun facing side with provision of ventilators in the roof. Since locally available material has been used in this greenhouse, the cost is reduced drastically compared to normal greenhouse with better efficiency of temperature retention than trench and polypouse. Considering the wear and tear occurred due to high wind velocity in the hilly regions, the polythene sheet has been replaced with FRL sheets to increase the life span of the structure. The structure has been accepted by the Ladakh Autonomous Hill Development Council (LAHDC) and State Govt. is providing subsidy of 50 per cent to the farmers for promoting the protected cultivation in Ladakh region.

h. Shade House/Net House

To reduce the adverse effect of scorching sun and heavy rain, shade houses are becoming popular for growing crops and nursery during summer season. Net houses are used for raising vegetables/ fruits/flowers/medicinal plants in high rainfall regions. Roof of the structure is covered with suitable cladding material. Slides are made of wire mesh of different gauges. Such structures are popular in North-eastern region of the country.

6.4 Current Status

Acute power crisis in hilly region has resulted development of solar greenhouses in the region. There is not a single greenhouse available in the region with controlled climate device. The most common greenhouse is polyhouse are constructed using locally available materials like mudbricks, wood and polyethyene. UV-stabilized cladding film is used for these Ladakhi polyhouses. Commercial greenhouses with climate controlled devices are very few in the country and the technology is in early stage of development. Although, a few commercial greenhouse units in India existed even during 60's and FRL (DRDO) at Leh attempted research on solar greenhouse vegetable production during early 60's, the impetus for greenhouse area expansion came in mid 80's with the development of low-cost greenhouse structures and manufacturing of UV-stabilised LDPE. Indian Pertochemical Corporation Ltd. (IPCL) boosted the greenhouse research and application by providing UV-stabilised cladding film in India. The low-cost greenhouses in Leh have permitted the expansion of growing season from four to eight months. Outstanding contribution of FRL is that almost every rural family in Leh valley possess a polyhouse and almost 70 per cent of total area under polyhouse in India is in Ladakh region. The greenhouses in India today are being increasingly used for off-season vegetables, export-oriented floriculture, medicinal and aromatic plants, and to some medicinal and aromatic plants, and to some extent, for raising timely nursery of horticultural crops including vegetables. Applications for cold water fisheries and drying of horticultural produce is also coming up. State Government (J&K) provides subsidy through various agecies like Agriculture/Horticulture Deptt. to tribal farmers for constuction of Ladakhi polyhouse. Several other agencies also promote the construction of the greenhouse in the region. This has resulted in a greenhouse revolution in Ladakh. Approx. 20,000 greenhouses are availabe in Leh district which is perhaps largest number in the country. 'Nang' the remotest village of Ladakh, is having 55 greenhouses one each with all 55 families.

7. Features of an Ideal Greenhouse for Hilly Region

Peculiarity of hilly region weather, i.e., very high day-night temperature fluctuations, strong winds and high light intensity coupled with UV radiation needs a different type of greenhouse for the region. Since, hilly regions are remotely located and isolated from main land hence all the required materials for greenhouse must be locally available and cheap in cost so that poor farmers can afford the expenditure. The greenhouse must be capable enough to trap the solar energy and produce the crops even at subzero temperatures (Tewari and Dhiman, 1986). It should also be strong enough to resist the snow load, strong cold waves and extremely low temperature. The structure should also be able to retain the temerature for longer period besides being simple and easily manageable so that even a layman can built and manage it without any difficulty. All the features may not be accommodated in one type of greenhouse (Singh *et al.*, 1998a).

8. Selection of Suitable Crops and Production Technology for Greenhouse

Selection of crops for production in greenhouse is the most important factor in success. The crop should not only be able to tolerate the low winter temperature but should also grow under the cover where very little and diffused sun light is available. Fast vertical growing and ratoon yielding crops should be prefered so that horizontal and vertical space is fully utilized, more number of cutting/harvesting can be taken. Root crops like radish, beetroot, carrot, turnip can also be grown (Singh *et al.*, 1997). A list of vegetables suitable for winter months, their recommended varieties, sowing time and planting distance is given in (Table 5), (Singh, 1995b). Among all, most important factor is sowing time. Even one day delay in sowing causes significant reduction in yield. Thus crop should be sown before the onset of winter which enables the plants to resist low temperature.

There are certain other managemental tips required to be adopted for greenhouse cultivation which are listed below:

i. Use of well rotten FYM @ 1.5-2.0 kg per m^2 atleast 30 days before sowing improves the growth and yield of crop. Use of chemical fertilizer is generally avoided. However, 10-12 g of DAP per m^2 before sowing and 5-7 g per m^2 urea enhance the growth of plants.

ii. Since every bit of soil under cover is highly precious, hence it should be utilized properly. Even ridges of beds and channels should be used for growing of the crops.

iii. Water soaking of seeds for 36-48 h before sowing results faster and higher germination percentage. Soon after sowing, light irrigation is beneficial.

iv. Frequent light irrigation has been recommended for winter cultivation. It not only enhances the crop growth but also improves low temperature tolerance of the plants.

v. Since greenhouse temperature falls down below 0° C during peak winter months, cold protection measures are essentially required for crops growing under the cover. For this, one additional covering of polyethylene just two ft above the plants (inside greenhouse) has given encourging results. This improves microclimate around the plants and hence crop growth increased. This covering is required especially during night in the month of December and January. Depending upon environmental temperature and sunlight, the polyethylene film may be removed during day time.

9. Future Thrust

Greenhouse technology development has made good progress in India during last two decades and steps necessary for promoting greenhouse cultivation of vegetable crops have been initiated. Energy efficient greenhouse cultivation continues to be an area of active research and development and this is sought to be achieved through precision equipment and protocols. While the gap between the demand and supply of most horticultural crops remains wide and country plans to double the production of horticultural crops by 2012, the protected cultivation technology holds the key to meet the targets. It clearly emerges that greenhouse technology has multipurpose application for sustainable development of hilly zone. Harsh winter which otherwise threatens the survival of human being in these remote hilly areas, popularization and further improvements in greenhouse management could provide a sign of relief in solitude of white snow cover. Thrust areas for future research are enumerated below:

- Conservation of more thermal radiation in winter
- Durable infrastructure modelling
- Development of successful commercial models
- Modernization of poly-greenhouse to permit agro-climatic manipulations of higher magnitude
- Identification of new crops/varieties suitable for cultivation inside the greenhouse
- Socio-economic impacts of poly-greenhouse technology in hilly region/cold arid zone
- Plastic films of suitable grades need to be made available in adequate quantity

Table 1: Performance of beet leaf (var. All Green) in greenhouse in Ladakh during peak winter (November-March)

Structure	Av. plant height (cm)	Av. yield (kg/m²/cutting)	Total yield (kg)
Glasshouse	21.8	1.1	3.3
Ladakhi polyhouse	22.0	1.8	5.4
Polyhouse	19.3	0.9	2.7
Trench	26.8	2.1	6.3
Tunnel	17.0	1.1	3.3
Unit polyhouse	21.4	1.5	4.5

Table 2: Performance of various leafy vegetables grown under polyhouse in Ladakh during peak winter (November-March)

Crop/variety	Av. plant height (cm)	Yield /m²(kg)
Beet leaf (All Green)	21.36	1.25
Beet leaf (Mongol)	34.44	2.50
Coriander (Local)	39.8	2.25
Fenugreek	39.92	2.50
Lettuce (Great Lake)	23.84	4.50
Veg. Mustard (ARU Black)	46.56	4.50

Table 3: Cost of different greenhouse in Ladakh per m² per year

Structure	Total cost (Rs.)	Life span (yr)	Area (m²)	Cost (m²/yr)
Glasshouse	1,15,200	10	40	288.00
Ladakhi polyhouse	5,240	3	50	34.94
Polyhouse	52,000	2	70	125.00
Trench	1,960	6	24	13.59
Tunnel	6,240	2	140	22.29
Unit polyhouse	5,300	2	70	37.86

Table 4: Cost of production of beet leaf (palak) in various greenhouse in Ladakh

Structure	Cost (Rs.) (m^2/year)	Produce (kg/m^2)	Cost @ 15/kg	Profit (Rs.)
Glasshouse	288	3.3	49.5	-238.5
Ladakhi polyhouse	35	5.4	81.0	46.0
Polyhouse	125	2.7	40.5	-84.5
Trench	14	6.3	94.5	80.5
Tunnel	22	2.2	33.0	11.0
Unit polyhouse	38	4.5	67.5	29.5

Table 5: List of vegetables, varieties, sowing time and planting distance for winter cultivation in Ladakh

Crop	Variety	Sowing time	Planting distance (cm)
Beet leaf	All Green, Mongol	First fortnight of Sep.	20
Beet root (leaf)	Detroit Dark Red	-do-	20
Celery	---	-do-	25
Chenopodium (Bathua)	---	First fortnight of Sep.	10
Coriander	RIL-195	End of Sep.	15
Dums	Local	End of August	20
Fenugreek	Kasuri	Second fortnight of Sep.	15
Karam sag	Khanyari	Second fortnight of August	20
Lettuce	Great Lake	End of Sep.	25
Mint	---	-do-	20
Parsely	Curld Leaf	-do-	25
Radish (leaf)	Pusa Himani/Local	-do-	20
Veg. Mustard	ARU Black	End of Sep.	20

Table 6: Yield of *Capsicum* and tomato under greenhouse at different locations in India

Location	Capsicum (t/ha)	Tomato (t/ha)
Pune	203.0	124.0
Coimbatore	148.0	186.0
Bangalore	147.0	152.0
Solan	79.0	95.0

Table 7: Comparative performance of solanceous crops in greenhouse

Crop	Fruit yield (kg/m²)		
	FRP	Trench	Open
Capsicum	1.82	0.83	0.54
Brinjal cv. PH-5	1.25	1.09	0.63
Chilli cv. BSS 344	1.35	1.24	0.13
Tomato cv. BSS 347	12.69	-	7.25

Table 8: Effect of greenhouse in achieving earliness in some crops

Crop	Days of first picking		Harvesting period (days)	
	FRP	Open	Polycarbonate	Open
Tomato	70	105	100	50
Capsicum	90	127	90	45
Brinjal	92	128	-	-

Table 9: Performance of *Capsicum* in various protected structures

Treatments	Days to 50% flowering	No. of fruits/plant	Av. fruit wt (g)	Fruit yield/plant (g)
Greenhouse	63.00	10.4	100.05	937.33
Double wall polyench	82.33	7.30	73.33	604.53
Polycarbonate	71.33	8.50	83.50	719.27
Polyench	74.33	10.4	67.69	642.13
Open	79.00	8.91	42.19	471.93

Table 10: Effect of various structures on *Capsicum* seed production

Treatment	Fruits/ plant	Av. fruit wt (g)	Fruit yield/ plant (g)	Seed: flesh ration	Seeds/ fruits	1000-seed wt (g)	Seed yield/ plant (g)
Trench	7.27	80.0	550.33	0.27	87.33	5.44	3.56
Polycarbonate	7.93	90.4	634.00	0.22	99.67	5.76	4.39
Polyench	4.50	90.2	384.67	0.19	84.67	5.58	2.12
Open	4.50	63.2	294.00	0.21	52.67	5.42	1.40

- Development of area specific suitable greenhouse designs
- Standardization of package of practices for growing various crops in greenhouse for specific regions
- Varieties suitable for greenhouse cultivation need to be made available to enhance the productivity
- Vocational training to the unemployed youth and greenhouse practitioners in remote areas to update their skills in this technology
- Financial assistance as low interest loans should be made available for poor farmers to adopt this technology
- Insurance of greenhouse structure and crop to protect the farmers against the loss caused by natural calamities
- Rigorous extension programme to disseminate the technology information to the poorest of the society.

References

Anonymous (2008). http://www.wikipedia.org/green house.

Chandra, P. (1991). The features of solar greenhouse for hill regions of India. Workshop on renewable energy option for Ladakh. SESI, H.P. Chapter, Forestry, Solan, May 3-4, 1991.

Chandra, P. and Panwar, J.S. (1987). Greenhouse technology and its scope in India. Paper presented in National symposium on use of plastics in Agriculture. Feb.-6, 1987 New Delhi. Proceedings pp. 62-66.

Dwivedi, S.K. and Singh, N. (1999). Roots of green revolution in Ladakh. Indian Farming 49(I): 20-21.

NAEDB, (1992). Action plan on cold deserts : An integrated approach for sustainable development. NAEDB, UHF, Solan - 173230 (H.P.)

Nimje, P.M. and Shyam, M. (1991). Greenhouse as an alternative technology for commercial vegetable crop production. Indian J. Agri. Sci. 61(3): 85 -89.

Singh, B. (1995b). Greenhouse development in Leh. Paper presented in Natiional workshop on commercialization of tissue culture and greenhouse technology, Banglore, India 5-6 May 1995.

Singh, B. and Dhaulakhandi, A.B. (1998). Application of solar greenhouse for vegetable production in cold desert. Renewable energy : Energy efficiency, Policy and the Environment. Proceedings of World Renewable Energy Congress, V, 20-25 September 1998, Florence, Itali 2311-2314.

Singh, B.; Jadhav, K.L. and Paljor, Eli (1997). In high altitude cold desert of Ladakh. Protected vegetable farming potentials. Indian Hort. 42(3): 16-17.

Singh, B.; Dwivedi, S.K. and Paljor, Eli (1998a). Studies on suitability of various structures for winter vegetable production at subzero temperatures. Paper presented in 25th IHC - Belgium, 2-7 August 1998. Abstract, pp. 290.

Singh, B.; Dwivedi, S.K.; Singh, N. and Paljor, Eli (1998b). Sustainable horticultural practices for cold arid areas. Paper presented in International symposium on sustainable agriculture in hill areas. 29-31 Oct 1998. H.P.K.V. - Palampur, Absctact pp. 35-36.

Singh, N.; Dwivedi, S.K. and Paljor, Eli (1999). Protected vegetable cultivation in Ladakh (In Hindi). Field Research Laboratory (DRDO)- Leh –Ladakh (J&K) India.

Singh, B.; Dwivedi, S.K. and Paljor, Eli (2000a). Studies on suitability of various structures for winter vegetable production at sub zero temperatures. Acta Hort. 517: 309-314.

Singh B.; Dwivedi, S.K. and Sharma, J.P. (2000b). Greenhouse technology and winter vegetable production in cold arid zone. In: Dynamics of Cold and Agriculture. Sharma, J.P. and Mir, A.A. eds. Kalyani Publishers, New Delhi. pp. 279-93.

Tewari, G.N. and Dhiman, N.K. (1986). Design and optimization of a winter greenhosue for Leh - type climate. Energy Covers Mgmt. 26(1): 71-78.

□□□

11

Weeds and Their Management

AJAY KUMAR PANDEY, DORJAY NAMGYAL*
H.V. SINGH AND TAHIR SALIM
Regional Agriculture Research Sub Station, Sher-e-Kashmir, University of Agriculture Sciences and Technology-Kashmir, Kargil-194103 (J&K)
*Regional Agriculture Research Station, Leh-194101

Weeds are well known to man and have been present ever since the cultivation of crop started from around 10,000 BC and they have been recognized as a problem from the beginning (Hay, 1974). Generally, the term weed is used to describe any plant that is unwanted and grows or spreads aggressively and which compete with crop to utilize land and nutrients consequently adversely affect the human welfare. Weeds infest large areas and cause economic and ecological damage. What is considered a weed in one area may not be a weed in another. These undesirable plants also flourish in cropped land as well as in non-cropped land such as roadsides, railway lines, airfield, water tank, waterways or any aquatic system. Therefore, all plants may become weeds depending upon particular situations. In cultivated field, weeds compete with beneficial and desired vegetation thereby they reduce the quantity and quality of produce on the other hand weeds are helpful in management of land and water resource depending upon the objective. However, their importance is more in agriculture where they cause losses by competing with crops. Out of total annual loss of agriculture produce from various pests, only weeds account for 45 per cent followed by insects (30 per cent), diseases 20 per cent and 5 per cent by other pest, (Rao, 1983).

Weeds cause enormous direct or indirect losses to the ecosystem by obstructing forest flora, cultivation of crops and pasture land while aquatic weeds hamper irrigation, increase pollution and cause inconvenience to man. In India, many of the noxious weeds are of foreign origin. Some are well known for lethal effects on man and his live stock. Variable amount of losses caused by weeds to the crops has been reported by different workers. Pandey (1993) estimated nearly 33 per cent loss in crop yield by weeds. Singh *et al.* (1993) reviewed that losses caused by weed in vegetable crops ranged from 6 to 82 per cent. Vere and Auld (1982) cited aggregate losses of thousands of million dollars. They estimated that weeds caused a reduction of crop yield 2.5 times greater in under developed countries (at 25%) than in countries with an intermediate development of agriculture.

Classification of Weeds

1. Based on Life Span

Based on life span weeds are classified into three broad groups

a. *Annual weeds* : Annual weeds complete their life cycle in one year or less and propagate by seeds. They are considered easy to control but being persistent due to abundance of seeds which germinate easily and emerged seedlings grow fast until condition become unfavorable.

b. *Biannual weeds*: Biannual weeds have life span of two years and germinate in spring or summer. During the first year vegetative growth takes place and in the second year they produce flower and seeds and die.

c. *Perennial weeds*: Perennial weeds live more than two years and some of them almost indefinitely. They prorogate by seeds and underground storage organs like rhizomes, stolons, bulbs, tuber etc. Perennial weeds are very aggressive and competitive weeds. On the basis of vegetative reproduce perennial weeds are classified as follows:

- *Simple perennial weeds*: Simple perennial weeds reproduce mostly by seeds. Vegetative reproduction occurs when roots and stem are cut mechanically. Each cut piece gives out roots and becomes a plant.
- *Bulbous perennial weeds*: Bulbous perennial weeds multiply through the underground parts like bulb and tuber as well as by seed.
- *Creeping perennial weeds*: This type of weeds spread by stolon, the lateral extension of the creeping, above ground horizontal stem rhizome root or seed.

Most of the weeds belong to the group of perennial weed followed by biannual. Biannuals comprise only a small percentage of weeds. In the USA, 45 per cent of the weeds are perennial, 34 per cent annual and only 7 per cent biannual (Jensen, 1977) while in India the ratio is 43, 40 and 6 per cent, respectively. Most of the weeds produce seeds.

2. On the Basis of Site of Predominance

a. *Obligate weeds*: Obligate weeds are such weeds that grow only in association with man and his agriculture. It is never found in wild form e.g. *Chenopodium, Anagallisels* etc.

b. *Facultative weeds*: Facultative weeds grow both as wild and in cultivated habitats e.g. *Argemone, Euphorbia hirta.*

3. On the Basis of Dependence of Host

a. Total root parasite: *Orobanche cernua*

b. Total stem parasite: *Cuscuta*

c. Semi root parasite: *Striga*

d. Semi stem parasite: *Loranthus*

4. On the Basis of Botanical or Morphological Character

a. *Grass* : Cylindrical and hollow stem having nodes and internodes and leaf emerge from node. This has fibrous root and belongs to Graminae family like *Echinochloa, Phalaris* etc. Most of the weeds belong to this group.

b. *Sedges* : Triangular stem, no nods, very large inter node and leaf at top and belongs to Cyperaceae family. It has no branch, usually three leaves at top e.g. *Cyprus rotundus.*

c. *Broad leaf*: Broad leaf, alternately arranged on stem, cylindrical stem, usually taproote e.g. all dicote weeds like *Chenopodium album*.

5. Other Weeds

a. *Relative weeds*

b. *Absolute weeds*: such weeds are known due to their characteristics like *Cyprus rotundus*.

c. *Rogue weeds*: The off type crop variety in the field of same crops variety is called rogue.

d. *Mimicry weeds*: Such weeds being similar to external morphology of crop e.g. *Phalaris* (wild oat) in wheat.

e. *Volunteer weeds*: grow from fallen seeds of previous crop in the same field.

f. *Noxious weeds*: Such weeds are declared expensively noxious and poisonous by the law e.g. *Parthenium, Cuscuta*.

Losses Caused by Weeds

Weeds cause severe loss by competing the crop in the following ways:

1. *Reduction of Crop Yield* : Weeds compete with crop plants for nutrients, moisture and light. The loss depends on intensity of weed competition which depends on type of weed competing ability of crop plants and climatic condition. Weeds utilize nutrients more efficiently than crop plant. In drought situation (Ladakh) they thrive better than crop plants. When left them undisturbed, some weeds can grow faster and taller than crop plants (EG) which inhibit the tillering and branching of crops. Depending on degree of competition, weeds reduce crop yield from 10 to 25 per cent. Parker and Fryer (1975) estimated that the world is losing 11.5 per cent of the total food production. Crop yield loss from weeds in highest in the tropics.

2. *Reduction in land value*: Heavy infestation of perennial weeds or aquatic weeds can make the land unsuitable or less suitable for the cultivation which results in loss of its economical value.

3. *Limited choice of crops*: Crops differ in their ability to compete with weeds. In many instances, presence of particular weed in the field inhibits the choice of crop to be grown. Very heavy weed infestation *render some economically inputs of crop.*

4. *Loss of quanlity*: The quanlity of crops including vegetable crops suffers in the presence of weeds. For example wild oat (*Avena fatua*) impures the quality of green vegetable. Contamination of other noxious weed seeds greatly reduce the value of crop seeds and grain. The weeds present in agriculture product always reduce their marketable value.

5. *R educe human efficiency*: Weeds reduce the human efficiency through physical discomfort due to allergic and poisonous nature. Weeds such as *Parthenium hyslerophorus, Rhus radicans* (poison ivy) and *Ambrosia* spp. (ragweed) caused itching, hay fever, and allergic problems in northern plain of India. Similarly, in Ladakh spiny weeds are major problem in vegetable crops during cultural practices. These spiny weeds resist the movement of farm worker which carry out different agricultural practices like fertilizer application, pest management practices etc.

6. *Increase the cost of pest control*: Besides the actual host crop, weeds serve as alternative host for survival of insect, disease and nematode. Insect such as aphid and thrips survive on weeds. Similarly, cabbage butterfly survives on a weed (*Lepidium* spp.) which is very common in Ladakh. Wild oats serve as alternative host of stem rust of wheat. The alternative host or weed maintain the inoculums for further attack on crop thereby they increase cost of production of the crop.

Crop Weeds Competition

Competition of weed with crop is a complicated process in which various factors are responsible which determine its extent. The effect of weed crop competition results in reduction of crop growth and yield due to utilization of nutrient, moisture and sunlight by weed. Competition between plant and weed is maximum when crop required resources are limited. Competition becomes more severe when crop and weed have same habitat and requirement of resources. A crop can easily compete with weed if they have vigorous growth compared to weed but advantage is lost in absence of advance and proper control practices. The extent of competition is determined by the weed species infestation of the area, density and duration of infestation. Example *A.fatua* (wild oat) was found more competitive with wheat than *Setaria viridis* (green faxtail). Increase in weeds population has direct effect on production. Alkamper (1976) found that yield was reduced substantially with an increase of weed in density of weed population.

In long term study Kolbe (1977) found that for every one per cent increase in yield, weed infestation should be eliminated to the extent of 3 per cent in winter barley, 2 per cent in winter wheat, 1 per cent in potato, 3 per cent in tomato and 8 per cent in apple. This study indicates that crops differ with each other in their competitive ability. Mani *et al.* (1968) reported that yield reduction in wheat ranged form 6-35 per cent, in maize 29-74 per cent, in millets 6-82 per cent, in peas 25-35.5 per cent and in carrot 70-78 per cent.

The weeds on crossing over the threshold level increase the competition for growth factors resulting in reduction of crop production. The weed and crop compete for such common growth factors when these are available in less amount. Normally weed competes with crop for nutrient, water, light etc. as described below:

1. *For nutrient*: One of the important characteristics of weed is to absorb more nutrients from soil compared to a crop. Analysis of weed and crop indicated that the concentration of nutrients was more in weed than the associated crop. This may be because of weed quality which completes its life cycle in 50-60 days while crop takes 120-130 days. Some weed like nut-sedge release a type of organic scented chemical in the soil which increases the bacteria and thus leaves very less amount of nitrogen in the soil for utilization of crop. Some of the weeds are also reported which directly absorb the nutrient from plant.

2. *For water*: Water is important constituent for which weeds compete with associate crop for utilization. This competition becomes very tough in un-irrigated dry land area. In dry land area like Ladakh, scarcity of soil water is the most common feature and the water uptake is comparatively more in weeds. The transpiration rate of weed is comparatively high and the amount of water required to produce unit amount of dry matter, called transpiration rate, is also high. It indicates that weed require more water than crop. In Ladakh condition where sandy soil is common, soil water is quickly absorbed by weeds and crop suffers adversely. The weed is capable to extract soil moisture even available at above 0.3 atmospheric meteric tension. Therefore, weed control is utmost important in Ladakh condition or any dry land area.

3. *For light*: The rapid biomass production of weeds cause shading and smothering effect on the associated crop. Most of the photosynthetically active radiation (PAR) is harvested by weeds which reduces the photosynthesis process in the crop.

4. *For carbon dioxcide* (CO_2): Most of the weeds are C_4 plants. These weeds deplete CO_2 to a very low concentration in comparison to crops. Heavy infestation of weeds utilize most of the CO_2 and create the competition between weed and the crop.

5. *Allelopathic effect*: A large number of weeds are known for their allelopathic effect to crop. Such weeds are *Agropyron repens, Cyperus rotundus, Euphorbia maculate, Ambrosia psitostachya* etc. These weeds release chemicals which have allelopathic effects on crops. The extract of *Ambrosia* rhizomes reduces germination and seeding growth of many crops like wheat, oat, tomato etc. Exudates of roots of wild oat at 2 and 4 leaf stage are toxic to wheat crop.

Characteristics of Weed

1. Most of the weeds spread vegetatively and are able to regenerate lost parts.
2. Compared to crop they are able to survive in adverse condition and compete with cultivated plant.
3. Weeds produce enormous seeds which mature simultaneously and remain viable for many years compared to crops seeds.
4. Weeds seed have some accessory appendages like hair, throne *etc.* which help in their spreading on wider area.
5. Some weeds have allelopathic effects on corps which reduce the crop growth.
6. Annual and biannual weeds produce large amount of seeds which ensure the high probability of dispersal and re-infestation.
7. Some weeds produce seeds after or between interval of normal disturbance associated with crop situation. Wild oat germinates at same time when wheat crop is sown and shatter their mature seeds before crop harvest.
8. Many weeds can produce large amount of seeds even after cut off after flowering and some of weeds produce seeds through apomixes, *i.e.*, without fertilization.

9. Some weeds seeds are very long lived, but they represent a small portion of total seed quantity/bank (Wilson, 1988).
10. Most of the perennial weeds multiply vegetatively which include rhizome, stolen, tubers, corm, root and stem besides producing seeds. These two methods of multiplication of weeds make more competitive and aggressive for cropping system. A seedling of a perennial weeds species growing from seed is not a perennial when it first emerges for the soil. Example *Cyperus esculentus* become perennial only 4 to 6 weeks after emergence.

Dispersal of Weed's Seeds

Dispersal of any kind of seed helps in spreading of plant for long areas where it does not occur. Dispersal of seeds may be occurred actively as well as passively. Some invasive leafy spurge seed pods build pressure and explode after ripening and seeds dispersed up to 15 feet away from the plant while some weed seeds dispersed passively by bird, air, water and small rodents. Most of the weeds seeds in Ladakh are dispersed passively for a long distance due to high wind velocity and run off of water as most of the land areas are sloppy. During the starting of summer season when snow starts melting and water goes down through small water channel, the seed of weeds grew on up lands come down by these channels. However, high wind velocity disperse the weeds seed which grew in lower land to up land. Birds also disperse the seeds both ways from low land area to up lands and vice versa.

Weeds of Ladakh

1. *Agropyron repens*: *Agropyron* is also known as Couchgrass, Twitchgrass, Quickgrass, Doggrass and locally known as Rampa. Quack grass is an aggressive perennial grass reproduces by seed and spread by shallow mass of long, slender, branching rhizomes. Rhizomes are usually yellowish-white, sharp-pointed and somewhat fleshy. They are able to penetrate hard soils or even tubers and roots of other plants. Leaf sheaths and the upper surface of leaf blades may be thinly covered with soft hair. Quackgrass grows in moist soils in cool temperate climates. It reduces productivity of crops and pasture. It is also a nuisance in lawns, ornamentals and home gardens and is believed to have allelopathic quality. Because of the ability of broken rhizome segments to grow and produce new plants, it is extremely difficult

to control mechanically. It has medicinal value which may be used in urinary infections such as cystitis, urethritis and prostatitis.

2. *Setaria* spp.: In Ladakh, this weed (Foxtail) is locally known as Yangjuk. The seedhead resembles to fox's tail so that name is fox tail and found throughout the Kargil and Leh district of Ladakh region, especially in fertile soil. Leaves are rolled in the bud, leaf sheaths are without hair, but the leaf blades have long silky hair on the upper surface near the leaf base. Identifying characteristic is foxtail-like seedhead that appears yellow when mature and leaves with long silky hair at the base only, which helps to distinguish this weed from both Giant Foxtail (*S.faberi*) and Green Foxtail (*S.viridis*). Giant foxtail has many short hairs on the upper surfaces of the leaf blades, unlike yellow foxtail which only has a few long, silky hair near the leaf base. Similarly, green foxtail has no hair on the leaf blades or at the leaf bases.

3. *Erigeron* spp.: This is a common weed which occure in almost all cultivated area. This is being prevalent in summer month or during the cropping season mostly in dry soil. The height of this plant is 12-24 inch. This is also used as medicine.

4. *Iris* spp.: Irises are a garden flower that can be grown in a variety of environment, and there are many kinds that grow well in cold arid region of Ladakh. They are very efficient in terms of water, and they grow very well in sunny climates. The word iris means rainbow because irises come in a variety of colors including orange, pink, brown, red, white, yellow, purple, blue and even black. However, blue colour is common in Ladakh which grows on bunds or in alfalfa crops and behaves like weed by competing with crop for nutrient, water and light. It survives well in drained soil where they will get a lot of sunlight.

5. *Mentha* spp.: The *Mentha* spp. is commonly known as wild mint and is classified within group Dicot of the Lamiaceae family. It is a perennial plant that means it lives or continues to survive more than two years. *M.longifolia* or wild mint is a fast-growing, perennial herb that has creeps along an underground rootstock. It can reach up to 1.5 m high in favourable conditions, but is usually between 0.3 to 0.5 m high and even shorter in cold arid region. This occurs in most of the areas of Leh and Kargil district

mostly margin side of cultivated as well uncultivated field. This is more common in those areas where moisture is high. The leaves are usually coarsely hairy and the edges sparsely toothed. The colour of the leaves varies from light and dark green to grey. The small flowers of *M.longifolia* are crowded into spikes at the tip of the stems. Varying in colour from white to mauve and survive throughout the cropping season (March to September) in Ladakh.

6.*Rumex* spp.: *Rumex* is very common in Ladakh which has been recorded in Nubra valley, Khaltsi blocks of Leh district and most of the areas of Kargil district. This is a perennial weed, 1–3 feet tall with little branchese, except where the flowers occur. Initially, it consists of a rosette of basal leaves about one feet across. Flowering stalk arises from this rosette during the season. The leaves are up to 6″ long and 1″ across (excluding their petioles). They are oblong-lanceolate, hairless, and dull green. Their margins are crisped and undulate up and down. Their petioles are up to 2" long, becoming broader at the base. The basal leaves have a similar appearance, except that they are somewhat larger in size, their petioles are longer, and their margins are less crisped or wavy. The inflorescence consists of a panicle of whorled racemes. The blooming period occurs during the warm weather, i.e., May-June in Ladakh. Each flower matures into a dry fruit that contains single seed. They are dispersed by wind or water. This plant spreads by reseeding itself.

7.*A maranthus* spp.: *Amaranthus* is a widely occurring annual broadleaf weed which is locally known as Homa in Ladakh region. It occurs in cropping and non cropping areas commonly in vegetable and crop fields. It is an erect summer season annual growing weed with the height of 40-80 cm and stem is round and ribbed. Leaves are smooth, pale green which blunt or shallow notch at rounded tip. Inflorescence is long, slender with a terminal raceme or axillary cluster. This weed grows successfully in soil having high organic matter and loamy texture. *Amaranthus* can be controlled by cultivation and hoeing. As seed germination is continuous in the season, herbicides which have longer persistence of activity are effective. Simazin, atrazin, trifluralin alfachlor etc. prevent establishment of *Amaranthus*. To control established weed, 2, 4-D, paraquat etc. are effective when applied on the foliage.

8. *Avena fatua* L.: Wild oat (*A.datua*) is an annual grassy weed which is very common in Leh and Kargil district of Ladakh. It is considered as one of the most competitive weeds in wheat, barley, buckwheat and other crops. The seeds shatter and fall to the ground before wheat or barley is harvested and persist in the soil. The seeds contaminate the crop seeds and get disseminated to new place. It is very much similar in appearance to cultivated oat. This annual grass has fibrous root system with flat leaves, broad base and acute apex. It grows on a wide range of light to heavy soil types and in both acidic and alkaline soil. Wild oats can be controlled by hand weeding; however, it is difficult to identify it at seedling stage. Application of trifluralin at preplanting stage gives very effective control. Triallate is particularly useful in wheat and barley.

9. *Convolvulus arvensis* L.: *C.arvensis*, (Field bindweed) locally known as Tiktikmo, is a climbing perennial weed which is very serious weed in tropical, subtropical and temperate regions. In India, it infests crops such as maize, potato, vegetable, tea and orchard. Stem of this weed is a twiner. It covers the host plant entirely due to its gregarious growth. Leaves are smooth, simple, alternate while flowers are funnel shaped and pink in colour. Flowering is occurs form June to September in Ladakh region. It propagates form deep underground roots and by seeds. It is extremely difficult to eradicate once. The seeds remain viable in the soil for long periods. Application of 2,4-D followed by cultivation is effective to manage this weed. However, some herbicides like MCPA and glyphosate effectively control when applied post. The seeds are disseminated by wind and man.

10. *Cuscuta* spp.: *Cuscuta* (Dodder) which is commonly known as Dodder, is very harmful parasitic weed in Ladakh which causes severe damage to alfalfa. *Cuscuta* spp. is an annual stem parasite and reproduce by seed. It emerges along with the germination of crop seedlings and parasitizes by attaching itself to the host plant and remains parasitic until harvest. It draws nutrient from the host for substance and provides a dense barrier that drastically retards growth and vigor of the host plant. The intensity of *Cuscuta* depends upon the capacity to rapid parasitization of the

host crop. For its management, non-chemical control methods may be useful in prevention or reducing *Cuscuta* infestation. Sowing of contaminated crop seed with *Cuscuta* seed is major means by which this parasitic weed spreads. Therefore, planting of pure crop seed could help in preventing its infestation and spread. After the appearance of *Cuscuta* in the field, farmers need to uproot the weed before they produce seed, followed by drying and burning.

11. *Panicum rapens* L.: Torpedo grass it is an aggressive perennial grass which propagates vegetatively through rhizomes. The rhizomes are knotty and swollen in appearance and send out erect culms from the rather distant nodes. It grows in different soil ranging from sandy to heavy clay. It occurs in moist and irrigated area of this region. Tillage can control this weed only partially.

12. *Chenopodium album* L.: *C.album*, commonly known as Bathua; in Ladakh it is known as Sogaani, is an annual, erect, herbaceous weed which grows to a height of 1-1.5 m. It is one of the most widely distributed broad leaf annual weeds in the world. Leaves are small, greenish or grayish-green, usually crowded into dense rounded clusters. It is distributed over a wide range of pH and soil types. The plants occur in groups and most of the seeds are deposited near the mother plant. They are prolific producers of the seed with each plant producing several thousands of seeds. The seeds have longer viability in the soil. It is sensitive to foliage applied herbicides such as 2,4-D, MCPA, paraquat, diclofop etc. as well as many soil applied herbicides including triazines and ureas. It occurs in wheat, barley, oat and in almost all vegetable crops.

13. *Bromus* spp.: *Bromus* spp. is causing serious problems in wheat and barley in most of the area of Ladakh. However, this weed has recorded in Turkey, Cyprus, Syria, Israel, Jordan, Egypt, Uzbekistan, Europe, Africa, Britain, North America, Australia, New Zealand, South Korea, Japan and India. However, the damage and intensity of this weed in Ladakh, varies from one area to another according to the agro-climatic conditions and cropping systems. This weed tolerates a wide range of climate and grows on acidic or alkaline, sandy or loamy soils especially once fertilized with nitrogen and phosphate. *Bromus* spp.

contaminants crop seeds and act as host for different serious diseases of crops. *Bromus* competes strongly with wheat. Seed production can range from 600 to over 3,000 per plant. Its life cycle ranged from May to September and at the end of the active period it produces seed. Seeds can remain viable in the soil surface for 2-3 years.

14. *Poa* spp.: This is a common weed of orchards, wheat, barley and vegetable fields. The weed starts emergence in March-April and flower and fruiting takes place from June to September depending upon weather condition. However, this weed causes higher infestation in cultivated field by creating competetion with crop. Stem of this weed is erect, slender and smooth with nodes and internodes in purplish colour. Leaves are narrow lanceolate 3-7 cm long mostly crowded at the base. This weed spreads fast. It produces seed in large number and seeds germinate readily on suitable substrate. It prefers moist situations but grows very well in open dry situation.

15. *Lepidium latifolium* L.: Broad-leaved Peppergrass *L.latifolium* which is commonly known as shangsho, is an extremely invasive perennial weed. It is also known as pepperweed, peppercress or perennial peppergrass. It is native of southern Europe and western Asia but is found in many parts of the world. It has been declared noxious in number of countries. It is 1 to 3 feet in height and leaves are lanceolate, bright green to gray-green, entire to toothed, basal leaves larger than upper leaves. Inflorescence is raceme and flowers are white in dense clusters near ends of branches. Cruciform flowers are small and white, borne of erect stems. A large number of seeds mature by late summer. Deep-seated rootstocks make this weed difficult to control. It grows in waste places, wet areas, ditches, roadsides, and crop land. *L. latifolium* can be found in a wide range of habitats. The weed occupies habitats such as the lower edge of coniferous forest 3,300 m in elevation to saline and alkaline deltas at sea level. For example, it is increasingly prevalent in saline soils and waste areas. *L. latifolium* may become an economic threat to alfalfa in Ladakh.

16. *Pennisetum* spp.: An occasional weed of dry location mostly occur in cultivated field and orchard. It is tufted grass with a creeping rootstock. The length of stem goes up to 70 cm, erect and generally

un-branched. Leaves are linear lanceolate, flat and glabrous. Inflorescence is dense spike like raceme. Spikelets, generally one on a stalk are densely arranged all over the rachis and stalk is small, surrounded at the base by numerous bristles.

17. *Sisymbrium* spp.: *Sisymbrium* is annual herbaceous weed and is distributed in Europe, Asia, Africa, India etc. In India, it is also reported to occur in Himalaya from Kashmir to Kumaon. In Ladakh region, it commonly occurs in wheat, barley and pea field besides vegetable fields.

18. *Portulaca* spp.: It is a cosmopolitan weeds which is very common in Kargil and Leh district. However, this occurs more frequently in orchard, cultivated, vegetable field and fallow land preferably moist and shady location. People of Kargil district use this weed as vegetable which posses sour test.

19. *Malva* spp.: It is biannual/perennial herbs with thick stoloniferous rhizome. It is very common in Leh, known as Sulcheli, occurs in vegetable field and fallow grasslands preferring day situation. It is distributed is most of the countries like Africa, Central Asia, Afghanistan, Pakistan etc.

20. *Trifolium* spp.: *Trifolium* is a perennial creeping weed commonly associated with crop. It propagates vegetatively as well as through seeds. It is considered as occasional weed of lawn and orchard growing along water channels. It is distributed in Europe, North Africa, Iran, Turkey, Asia, Afghanistan and India including Ladakh region. This weed looks like cultivated alfalfa and is very common in alfalfa crops besides on bund and grassy area where moisture and organic matter is more. It germinates in the month of March-April and propagates through seed as well as rhizome.

21. *Taraxacum* spp.: This weed is very common in cold arid region. It occurs in almost all cultivated and uncultivated areas of Ladakh having organic matter and moisture. This weed is also common in Kargil district. The roots of this weed are thick, tapering and deep in soil. The leaves of weeds are ablanceolated or linear, sessile, forming a rosette at the ground level. The flower is yellow. This produces large number of seeds which are disseminated by winds. It has simple tap root, generally, unbranched which goes upto 20 cm deep in soil.

22. *Artemisia* spp.: *Artemisia* is a perennial herbaceous plant. Stem of this weed is erect, slightly weedy at the base and up to 65 cm long. Leaves are of variable shapes radical, sessile and somewhat glabrous with spreading pinnatified 3-5 lobes. Flowering starts from last week of June. The seeds are disseminated mostly by wind. This weed reproduces vegetatively also. It grows mostly in uncultivated land but sometime it can be seen in cultivated area. It has medicinal properties.

23. *Sonchus* spp.: It is a cosmopolitan weed which most frequently occurs in orchards and some times in cultivated areas where sufficient moisture and organic matter is available. The stem is erect, weak, glandular towards the apices of the branches. Leaves are lanceolate, toothed margins, long unequal. On plucking of leaf, milky secretion oozes out from breaking part of leaf which is most identifying characteristics of this weed. Flowers are homogamous and bisexual. In Ladakh, seeds remain dormant for at least nine months and sprout in April. The seeds are mainly disseminated by wind and to some extent by water.

24. *Plantago* spp.: Plantago is a common weed of orchards, vegetable fields, lawns and grasslands which prefer moist situation. This weed is distributed in Europe, Siberia, Turkey, Iran, Afghanistan, Pakistan and India, and is very common in Ladakh region. Weeds have small rootstock and fibrous bunch of adventitious roots. Leaves are radical, broadly ovate or oblongovate distinctly stalked. The veins run parallel and convergent at the top. The seeds get disseminated by wind, animals and birds. This weed is very common in alfalfa crop in Ladakh.

25. *Urtica* spp.: *Urtica* spp. is a common weed of cultivated fields especially in fallow land. This weed is reported form Europe, North Africa Siberia, Afghanistan and India. This weed is common in Ladakh region. Stem is erect which may go up to 1 m with branch from the base and main stem is generally un-branched. The active period of this weed ranges from April to October in Ladakh region.

26. *Equisetum* spp.: It is a fast spreading weed of fields, fallow place and distributed in most of the country. This weed is very common in Ladakh region and occurs in those areas where moisture is high and organic matter is available. This weed was recorded in

Zanskar, Kargil, Drass, Btalik area of Kargil district. However, severity is less in Leh compared to Kargil district. Branches, arranged in whorls at the node, join longitudinally, striated. Leaves in the form of small scales at the nodes fuse to form a toothed sheath and generally numerous. This weed has been recorded in wheat and barley field mostly in irrigation channel. The root goes 2-5 feet deep in soil and very difficult to eradicate.

27. *Cirsium* spp.: This weed is reported from temperate Himalaya and also occurs in Ladakh region. A common weed of roadside and margin of the fields mostly occurs in moist condition. However, in this region this weed occurs in vegetable fields and severe infestation was recorded in Kargil district. The stem is erect up to 0.5 to 1.0 m. Leaves are up to 25 cm ling bearing spines. The seeds are disseminated by water and wind. This weed creates uncomfort in cultural practices due to spines on whole plants.

28. *Cichorium* spp.: This is a common weed which is noticed in apple and apricot orchard in Ladakh region. It also occurs at the margin of cultivated land where the moisture is high. Like other weeds which occur in Ladakh it also emerges in the month of April and completes its lifecycle up to September. Flowering and fruiting takes place in the month of June to August. The seeds are disseminated by human activity.

29. *Echinochloa colonum* L.: *Echinochloa* is distributed in warm countries of the world. This is a very common weed in almost all cultivated areas of Ladakh region. It occurs in wheat, barley, pea, lentil crops and in orchard. The seeds are disseminated by irrigation water and animals that feed upon it. The seed of this weed germinates in the month of May depending upon the weather condition. Farmers use it as fodder for cattle.

30. *Descaurainea* spp.: it is very common weed in cold arid region. It is widely distributed in Kargil district. It causes infestation in vegetable, alfalfa, wheat, barley and in fallow land. However, this weed occurs in irrigated land having organic matter. This weed causes severe loss to vegetable and alfalfa field by creating competition for nutrient, water, light and space.

31. *Neslia apiculata* Fisch. Mey & Ave-Lal: This is a common weed in Ladakh which is widealy distributed in Kargil district. Sometime

it is widely distributed in orchard, grassland and vegetable field including fallow land. Stem of weed is 30-50 cm tall erect and branched. Flowers are long, slender and glabrous and yellow in colour. Pod is 0.5 to 1.0 cm in diameter, slightly globular, little broader than long and hard. This weed propagates by seed which undergoes in dormancy during the winter season and germinates in the month of April-May and flowering takes place in the month of May-June.

32. *Borger*: A type of weed which is commonly known as Borger in balti language; the scientific name is not known. This weed is common in wheat, barley, pea and oat and causes sever infestation and creates competition with crops. Its flower is white in colour.

Besides these, some other plants are also known which behave like weed with some medicinal properties. Example, *Cicer microphyllum* (wild gram), *Echinopus cornigerus, Bera* spp., *Peganum harmala, Sassurea gossypiphora, Podophyllum hexandrum, Datura, Arctium lappa* etc. However, some of the weeds are yet to be identified.

Management of Weed

Weed control is a process of reducing weed population and minimizing their growth to an acceptable level. In weed management, primary objective is to maintain an environment that is detrimental to weeds possibly by employing both preventive and post-infestation control measures, using various methods either alone or in combination. However, weeds can be resistant to a particular control measure over a period of time and this demands manipulation of weed management practices. Biological control (using organisms such as insects or diseases to suppress weed population) is effective in reducing the spread of weed but this measure can not eradicate the weed. Manually pulling or digging of weed is effective in small infested area. Use of herbicide is effective in controlling weed and their spread especially when infestations are detected early. Land managers generally take an integrated approach, using a combination of these methods. A special emphasis is placed on early detection of infestations and prevention of new infestations. Freed (1980) observed that without herbicides the world could have been lacking many tones of grain and, the same varieties of wheat under the same climatic condition yielded 600-900 kg ha^{-1} more simply due to control of weed. Various methods of weed management are grouped under three broad groups, i.e., traditional, chemical and biological.

1. Preventive Measures

A preventive control measure is concerned with all the measures which prevent the introduction, establishment and/or spread of specified weed species in area which is not currently infested with that plant species. No weed control measure can be successful if adequate preventive measures are not applied to reduce the infestation of weed. However, success of preventive measures depends on efforts made to make it effective. Preventive methods of weed control include weed-free crop seed, weed-free manure, clean harvesting, cleaning of ploughing equipment and elimination of weed from irrigation channel and cultivated field.

a. Weed-Free Crop Seeds

Most of the crops seed are contaminated with weed seed, mostly at the time of harvesting. This is because of similar biology of weed to the crop when seed mature at the time of crop maturity. Some qualities like similarity of weed seed to crop the seed in size and shape are helpful in contamination. The presence of very small amount of weed seed may be sufficient to cause serious infestation. Fallowing points should be considered for producing uncontaminated seeds:

I. Cleaning of all farm machinery and produce to avoid production of uncontaminated seeds.

II. Separation of weed seed from crop seed on the basis of size, shape, specific gravity and colour before storage.

III. Cleaning of adhere weed seed and vegetative part form plough, harrow, drill etc. before using for cultural practices.

IV. Use of well decomposed farmyard manure and compost can reduce the risk of spreading the weed seed.

b. Seed Certification

Seed certification comes under Central or State Government which provides standard seeds which are free from contamination. In India, National Seed Corporation of Government of India and the Seed Corporation of various State Government take the responsibility for producing disease and weed-free crop seed and certify the same produce thereby help in supplying of genetically pure seeds and propagative materials. In Ladakh, where pesticides use is not common they should strictly enforce the rules and regulation to produce and supply absolute weed-free crop seeds. Alfalfa seed are very common which farmers exchange from each other which help in spreading of *Cuscuta* (parasitic weed) seed from one place to other. Therefore, farmers should grow their own seeds carefully which will prevent contamination. In Ladakh, a seed certification unit should be established to grow authentic seeds which will prevent contamination of seeds and will fulfill the requirement of area specific seeds.

c. Weed Law

Most of the farmers use unlabeled seed which are contaminated with weed seed and disease propagules. The weed law is important in reducing the spread of weed seed and increase the high quality of authentic seed. It is also helpful in protection of mislabeled contaminated seed. To enforce the weed law, prohibited area should be separated where the noxious weed occurs thereby to reduce the spreading of weed. For example, *Cuscuta* which is a parasitic weed of alfalfa and is very common in most of the areas of Kargil district but it is not present in Leh district. Therefore, the import should prohibit form Kargil to Leh to avoid the spreading of weed.

d. Quarantine Law

Quarantine laws enforce isolation of an area where problematic weed has been established, to prevent the movement of the weed into un-infested area. The normal quarantine laws restrict inter state and international movement of plant, seed and other materials.

2. Post-Emergence Management

This includes control or eradication of weed which are already infesting the crop. Post-emergence management depends upon the aim, i.e., control or eradication. Eradication involves killing of existing plants and destroying all reproductive parts. This measure applies when the weed species is very noxious and crop is very costly. Control measures include reduction or killing of plant growth. Various control measures of post emergence are as follows:

1. Eradication

Under this practice weed seed or vegetative part are killed or completely removed form a given area and weed plant will not appear unless reintroduced into that area. Though eradication is an idea of weed control measures but rarely achieved because of high cost and application is possible only in small area. This measure is applicable for valuable crops which are grown in small area.

2. Control

Under control we reduce the infestation of weed but do not eliminate. Control measures range from poor to excellent. Following measures come under this category:

A. Manual Methods

I. Hand Weeding

Hand weeding can be used as very effective control measure in Ladakh because the cultivated area is very small and weeds are used in animal husbandry where grass in used as fodder for cattle. Therefore, hand weeding can be done by physical pulling of weeds by hand or removal by implements like Khurpi. Long handle weeder, peg type weeder etc. are improved implements which will be useful for efficient management of weed in Ladakh condition.

II. Digging

Digging is very effective for perennial weed having very deep root system, for example, *Equisetum* is an important weed which occurs in most of the area of Kargil district and heavy infestation was recoded in Zanskar valley. Though digging is a labor intensive and slow process hence it is restricted to perennial weed where physical and chemical methods are not effective. This practice is more effective than hand pulling if the weeds have taproot and deep rooted vegetative propagules.

III. Sickling and Mowing

Sickling is also done by hand with the help of sickle to remove the top growth of weed to prevent the seed production. It is useful in sloppy areas where only tall weed growth is sickled leaving the root system to prevent soil erosion. Mowing is a machine operated process mostly done along the road side and on lawn. Therefore, mowing is effective for control of weed but due to small cultivated land and un-plain area mowing is not effective in Ladakh. Mowing is commonly used in meadow, pasture, along road side and waste place.

IV. Crop Management Practices

i. Use of fast growing varieties as better competitors which will give leaf canopy and cover the ground rapidly thereby reducing the weed growth.

ii. Prevention of weed seed production in cropped and uncropped areas before flower and fruit formation takes place.

iii. Some practices like proper placement of fertilizer, judicial and better irrigation, higher plant population in per unit area and crop rotation prevent the establishment of weed.

B. Mechanical Control

Under mechanical control we include tillage, hoeing, burning, flooding and mulching.

i. *Tillage*: Tillage removes weed from the soil resulting its death. It may weaken weed through injury of root and pruning, reducing its competitiveness or regenerative capacity. Weed control by tillage is achieved by burial of small annual weed. Disruption of relation between weed and soil leads to the check the supply of water and nutrients. Besides controlling the weed through tillage, it also improves the physical condition of the soil and turns crop residue under soil. Before the crop is sown or transplanted extensive tillage operation is used to promote the germination. The pre-planting tillage practices are used to promote germination of weeds through soil turnover and exposure of seeds to sunlight which can be effectively destroyed later. However, types of weed largely determine the effectiveness of weed control operation.

ii. *Hoeing* ; Hoeing is an effective means of weed control which is a common practice. Hoeing and hand pulling of weed is applicable in certain row crops such as in vegetable. It is effective for annual and biannual weed species. It can be more effective for the management of creeping perennial weeds like *Convolvulus, Mikania* etc. which have shallow roots.

iii. *Burning*: Burning is an economical and common practice for control of weeds. This method is used to kill accumulated weeds. It also destroys dry top of weeds, green weeds, buried weed seeds and other prerogative parts. The critical temperature above which cambium cell may be injured ranges from 45 to 55°C. Fire is used to ruin stubble which will reduce the weed infestation in next season. When burning is used for the control of selective annual broadleaf weeds and grasses in crop rows than it is know as flame cultivation. It is effective for control of *Cuscuta* in alfalfa.

iv. *Flooding*: Flooding is used for weed control by maintaining water at 15 to 30 cm deep for 3 to 8 weeks. It is a common practice in plain particularly in paddy crop. However, this practice is not suitable in Ladakh region due to unavailability of irrigation water and sandy soil having no water holding capacity.

v. *Mulching* : Mulching is very effective control measure for weed in Ladakh. Mulching restrict the photosynthesis and than inhibit crop growth. In Ladakh conditions where soil is sandy having very low water holding capacity mulching reduces the loss of water from soil through evaporation and suppresses the weed growth thereby this practice eliminates the need for inter-cultivation. It is very effective in vegetable cultivation which is done in small area. Mulching acts as barrier for weed emergence. There is two types of mulching first organic mulching which consists of plant byproducts such as crop residue, straw, sawdust, waste papers (news papers) small stone etc. Second is mulch with synthetic product which includes polyethylene (plastic). Mulching can be used in large area to avoid the transpiration of water from soil or to reduce the water requirement which is major problem in this region beside the weed problem. Polyethylene can be used for dual purposes, i.e., for control of weed as well as to conserve the soil moisture. For effective control of weed through organic mulching mulch should be 10-15 cm thick. It also enhances the soil temperature which enhances the plant growth. The black polyethylene is most effective. Though mulching is an expensive method for weed control but in Ladakh it is found very effective due to shortage of labour of high charges. This method can use in almost all vegetable crops and flowers, where herbicidal use is not possible because of small area. Weed in tomato, cabbage, cauliflower, knol-khol are common vegetable crops in Ladakh is controlled by proper use of black polyethylene as mulch.

C. Nutrient Based Approach

Plant nutrients promote crop growth but in crop-weed situation weeds are benefited more because of their greater ability to accumulate minerals and have maximum efficiency of utilization. Among the nutrient, nitrogen is important for which great competition takes place. Increased application of nitrogen helps the crop plant to grow better which suppress the weed growth but where the weed population is higher, the nutrient favours weed growth which suppress or becomes dominant over crop. The ability of weed species to utilize available nutrient better than crop species can also provide an advantage to the former while competing for water and light (Carlson and Hill, 1986; Liebmen and Robichaux, 1990). Liebmen and Robichaux (1990) reported that mixed cropping of barley and pea was more competitive against *Brassica hirta* (white mustard) when nitrogen was not applied, as compared to application of high dosage of nitrogen condition. This corresponded with reduction in photosynthetic capacity, seed production and shoot biomass in mustard. However, at high nitrogen rates, white mustard produced more leaves was taller plants. Similarly, time of fertilizer application is important aspects for weed control. Time of fertilizer application may be manipulated to tilt the crop weed competition for nutrient in favour of crop. In case of wheat, the requirement of nitrogen is greater at 3-leaf stage when tillering begins (Davidson, 1984). In contrast to *Lolium rigidum* (annual ryegrass) weed, production and tillering is continuous and stimulate by late nitrogen application. Therefore, early nitrogen application reduces the competitive effect of annual ryegrass on wheat yield. Pandey *et al.* (1971) found phosphorus uptake in a number of crops and weeds and found that the rate of phosphorus uptake varied with age and species. Purple nutsedge (*Cyperus rotundus*) exhibited rapid phosphorus uptake until plants were 24 day old, while the demand for phosphorus in other weeds and a number of crops including wheat, chick pea was higher after 60 day, during the flowering and fruiting stages.

D. Biological Control

In this method we utilize the natural enemies for the control of weeds. The aim is to reduce weed population below the level of economic injury (Anderson, 1977), i.e., to a level which will be non injurious to man. Some biotic agents bore into the plant, weaken its structure and ultimately consume and destroy the vital plant parts, is called direct action. In direct action, the biotic agent reserves the competitive ability of weed over the plant and enhances the favourable condition for plant pathogens. Important bioagents used for control of terrestrial plant are phytophagous (plant eating) insects. Besides insects, other bioagents like microorganism, parasitic plant, sheep and goat, green snails, birds etc. come under this category. Herbicides account for 55 per cent of global market of all pesticides used in crop production. Although, the organic herbicides are playing a major role in providing the food and fiber required by world population. However, injurious effect ensure about the safety for food products, their adverse impact on environment and weed resistance to herbicides force the biological control as an alternative weed management method to play a major role in agricultural production systems.

Biological approach for weed management involves various biological weed control agent such as pathogen, insect, nematodes etc. Insect kill the plant by utilizing the weed plants as food and some time through destruction of photosynthetic parts. The damage of weed depends on intensity of attack, condition of plant and environment. The number of insect to kill a weed plant is important aspect. An aggressive weed in favourable environment requires a large number of insects. The insect used as biological control for weed mostly belongs to order lepidoptera, hemiptera, coleoptera, diptera, hymenoptera and thysanoptera etc. The success of biological control of weed by insect mostly depends on host specificity if insect does not attack other hosts plants. However, biological control can be affected by some factors like choice of enemy agent, aggressiveness of the weed, climatic condition and competition of plant species.

In Ladakh, a number of insects are known to attack on weeds. A coleopteran insect is recorded to attack on *Bera* spp. which is very common in Ladakh region. This insect attacks on root of this weed and feeds inside which results in dying of plant. Similarly, larvae of a butterfly were found as voracious feeder of *Echinopus* spp. weed which is very common in alfalfa and in wide area of Kargil district. Larvae feed on leaf and cause defoliation of plant. Cabbage butterfly, which is a pest of cabbage, attacks on *Lepidium* spp. which a harmful weed in Ladakh. However, cabbage butterfly can not be used as biological control agent because it is major pest of cabbage, cauliflower, and mustard. A leaf minor was also recorded in Kargil district which mine the leaf of *Rumex* spp. which is a common weed of almost all field crops. Besides this a large number of insects are also known which attack on weed and reduce their infestation. The identification of these insect-pest is important to successful utilization for the control of weeds.

E. Chemical Control

Chemical control measures are important in weed management. The word herbicide is derived from Latin words *herba* (plant) and *caedere* (to kill). Herbicides constitute one of the most vital and essential input to maximize crop reduction level. Herbicides development came at a time when world agriculture was entering an era of increased mechanization and intensive cropping programmes to increase yields and reduce production costs.

Chemicals are reported to have been used in agriculture for several countries. However, introduction of Bordeaux mixture 1986 stimulated interest in chemical control. Many chemicals like sulphuric acid, iron sulphate, copper nitrate etc. were discovered between 1986 and 1970. The systemic control of weed took place with the introduction of nitrophonal as selective herbicide in 1935. the discovery of 2,4-D in early 1940 revolutionized the chemical method of weed control. Pokorny in 1941 reported synthesis of 2,4-D which was found effective as fungicides and insecticides while in 1942 Zimmerman and Hitchock found 2,4-D as growth regulator. Marth and Milchell of the USA, in 1944 reported selective weed control by 2,4-D. However, the earliest attempt in India was made in 1937 to control weeds by herbicides in Punjab for control of *Carthamus oxycantha* by using sodium arsenite. 2,4-D was first time used in India in 1946. Paraquat was first time used in India as chemical concept for weed control in tea plantation. Currently, herbicides constitute 55 per cent of world pesticide market.

Many pesticides are mostly used in formulated form instead of active ingredient. It makes their application easier and improves the effectiveness and cover wide area. Formulation is the process under which technical active ingredients are made easy to use, by mixing with liquid, dry diluents by addition of emulsifier, stabilizer and other formulation adjuvant to from a commercial product. Active ingredient may be solid, liquid or gas, being actual herbicides/ insecticides/ fungicides of formulation, is responsible for controlling the pest.

Stability of chemical active ingredient is important which influences the effectiveness of a formulation. Among the large number of factors, humidity and temperature mostly affect the effectivity of a formulation. Herbicides are available in different formulations.

1) *Solution*: A solution is physically homogeneous mixture of two or more substances. Such type of formulation is used to design for use in water carriers. In solution, a certain percentage of solute molecules are dissociated into ions.
2) *Wettable powder*: A wettable powder (WP) contains finely divided solid particles (size less than 3 microns) that can be readily suspended in water to form a suspension. Usually a wettable powder contains a diluents, a surface active agent and an auxiliary material. Wettable powder containing 50 to 80 per cent active ingredients are the most widely used.
3) *Emulsifiable concentrate*: An emulsifiable concentrate consists of a herbicides dissolved in an organic solvent, with sufficient emulsifier added to create an oil in water emulsion. An emulsion is a mixture in which one liquid is suspended in other liquid, like fat globules in milk. The emulsifying agent binds the two phases, i.e., the water phase and oil phase together.
4) *Granules*: Granules contain 1-20 per cent concentration of herbicides in particulate solid carriers suitable for directly field application. In the formulation the technical active ingredient is mixed or coated on inert carrier material. Many materials are available in granules form as carriers, example bentonite.
5) *Fumigants*: Fumigants are volatile chemical applied in confined spaces or in the soil to produce gas that will destroy weed seeds and act as soil sterilent.

Classification of Herbicide

Many herbicides widely used in the 1960s and 1970s have been phased out and replaced by newer and more potent herbicides. Some of the herbicides have been restricted, reduced even eliminated due to environmental and toxicological problems and due to availability of safer herbicides. However, herbicides are classified on the basis of following factors:

1. Method of Application

a. *Soil applied:* Herbicides that are applied at pre-planting and pre-emergence to weed, crop or both come under this group. It may be either selective or non-selective.

b. *Foliage applied*: Those herbicides which are applied at post emergence of the plant parts are included in this group. It may be either contact or translocate (systemic).

2. Translocation

1. *Systemic herbicides*: These herbicides move within the weed either through xylem or phloem and consequently affect the whole system called systemic herbicides. Most of the systemic herbicides are selective at recommended dose e.g. propanil, 2,4-D, atrazine etc.

2. *Contact herbicides*: Under this group comes those herbicides which kill the weed when it comes in contact e.g. paraquat, diquat.

3. Selectivity

a. *Selective herbicides*: Those herbicides which kill only certain weeds without a significant injury on associated crop or other desirable plant. It kills only target weeds. Example simazine, atrazine, 2, 4-D, butachlor etc.

b. *Nonselective herbicides*: These herbicides kill weed as wall as vegetation, i.e., weed and crop both when it comes in contact like diquat and paraquat. However, selectivity depends on dosage of herbicide which is used to control a particular weed.

4. Application

a. *Pre-planting applied*: Herbicides like fluchlorale and alachlor are used before planting for control of weed come under this groups.

b. *Pre-emergence herbicides*: Such herbicides which are used before emergence of weed but post-emergence of crops. Most of the selective herbicides are used in pre-emergence time.

c. *Post-emergence herbicides*: Such herbicides are applied after the emergence of weed and crop. Example 2,4-D, paraquat, etc.

5. Chemical Structure

a. *Inorganic herbicides*: e.g. arsenic, sodium arsenate, borax, copper sulphate, copper nitrate etc.

b. *Organic herbicides*: These are 16-17 herbicides groups like alipatics, amides, uracils, urea, triazines, phonols, bipyridilims etc.

Mode of Action of Herbicides

Herbicides after application on vegetation or weed affect their physiological and biochemical process in the following ways:

1. Herbicides affect mitochondrial activity by uncoupling the reaction responsible for ATP synthesis or by interfering with electron transfer and energy transfer.
2. Herbicides affect photosynthesis by inhibiting or blocking the Hill reaction and production of oxygen from water in photosynthesis II (ureas, triazines, uracils), inhibiting electron transportation in photosynthesis II (nitriles), removing electron from ferredoxin in photosynthesis I (bipyridilliums) and inhibiting the noncyclic photophosphorylation or ATP synthesis (urea, triazines).
3. Herbicides affect protein synthesis which takes place in three stages, *viz.*, initiation of the polypeptide chain, chain elongation and chain termination.
4. Herbicides affect the production and activity of enzyme which become active during seed germination. The production of hydrolytic enzyme requires the synthesis and presence of protein polyrhizomes and nucleic acids.

Herbicides affect any one or more of these events consequently affecting the germination of the seeds. For example amides, thiocarbamate etc. inhibit the metabolic process vital for seed germination.

Herbicides Dose Calculation

$$\text{Weight of commercial product required} = \frac{\text{Weight of the chemical (a.i.) to be applied}}{\text{Percentage expressed as decimal (a.i.)}}$$

Example: For a herbicides with 80% a.i. as diuron, to be applied @1 kg diruon/ha. Therefore,

$$\frac{1.0}{0.8} = 1.25 \text{ kg}$$

1.25 kg of the commercial product is required to apply 1.0 kg of the active ingredient of diuron.

Quantity of herbicide formulation required for a given area can be calculated by using the following formula

$$F = \frac{R \times 100}{C \times 10000} \times A$$

where,

F= formulated produce required in kg or l/ha

R= rate kg/ha to be sprayed

C= percent concentration (a.i.) in the formulated produce

A= area to be sprayed (m^2)

References

Alkamper, J. (1976). Influence of weed infestation on effect of fertilizer dressings. Pflanzenschtz Nachrichten Bayer 29: 191-235.

Anderson, W.P. (1977). Weed Science Principles. 2nd edition, West Publishing Company, USA. pp. 65-122.

Carlson, H.L. and Hill, J.E. (1986). Wild oat (*Avena fatua*) competition with spring wheat: effects of nitrogen fertilization. Weed Sci. 34: 29-33.

Davidson, S. (1984). Wheat and ryegrass competition for nitrogen. Rural Res. 122: 4-6.

Dryden, R.D. (1974). Losses of nutrients under varying agro-climatical condition-weed control. FAI-FAO Seminar 1974 on "Optimizing agricultural production under limited availability of fertilizers', New Delhi.

Freed, V.H. (1980). Weed science: the emergence of a vital technology. Weed Sci. 28: 621-625.

Hay, J.R. (1974). Gains to the grower from weed science. Weed Sci. 22: 439-490.

Jensen, L.L. (1971). Morphology and photoperiodic responses of yellow nutsedge. Weed Sci. 19: 210-219.

Kolbe, W. (1977). Long term studies on relations between weed cover and yield increase with chemical weed control (1967-1976). Pflanzenschtz Nachrichten Bayer 30: 121-137.

Liebmen, M. and Robichaux, R.H. (1990). Competition by barley and pea against mustard: effects on resources acquisition, photosynthesis, and yield. Ecosystems Environ. 31: 155-172.

Mani, V.S.; Gautam, K.C. and Chakraborty, T.K. (1968). Losses in crop yield in India due to weed growth. PAN (C) 14: 142-158.

Pandey, H.K. (1993). IWM for sustainable agriculture. Souvenier Int. Symp. Indian Weed Sci. Hissar, Nov. 18-20, pp. 37-39.

Pandey, H.N.; Misra, K.C. and Mukherjee, K.L. (1971). Phosphate uptake and its incorporation in some crop plants and their associated weeds. Ann. Bot. 35: 367-372.

Parker, C. and Fryer, J.D. (1975). Weed control problems causing major reduction in world food supplies. FAO. Plant Protection Bull. 23: 83-95.

Rao, V.S. (1983). Principles of weed sciences. Oxford Publishing Co., New Delhi.

Singh, K.; Pandita, M.L. and Thakral, K.K. (1993). Integrated weed management in vegetable corp. Proc. Int. Symp. Indian Soc. Weed Sci., Hissar, November, 18-20, pp. 365-368.

Vere, D.T. and Auld, V.A. (1982). The cost of weed protection. Weed Ecology 4: 29-42.

Wilson, R.G. (1988). Biology of weeds in the soil. In: Weed Management in Agroecosystems: Ecological Approaches. Altieria, M.A. and Liebman, M. (eds.). CRC Press, Boca Raton, FL. pp. 25-39.

□□□

12

Agricultural Insect-Pests and Their Management

AJAY KUMAR PANDEY
Regional Agriculture Research Sub Station, S.K. University of Agricultural Sciences & Technology-Kashmir, Kargil-194103 (J&K)
Presently at : GB Pant University of Agriculture & Technology
Hill Campus, Ranichauri, Tehri Garhwal (Uttarakhand)

The dry weather and severe winter of Ladakh are the major constraints in development of agriculture. The high dry temperature and increase evapo-transpiration and salinity is prevailing condition in Ladakh. However, the economy of Ladakh is dependent on agriculture, animal husbandry and horticulture including forest plantations. The cultivated land available to the farmers in the cold desert environment is much less than elsewhere in the country and therefore, wherever farmers find a cultivable stretch or just any small piece of land, whether in the desert, on mountain slopes or pastures, effort is spared to cultivate it. Most of the cultivable land lies along the course of streams and some along the celebrated rivers. Agriculture forms the most important constituent of the economy of this region. Cultivation of crop in this region is a challenging task but has been undertaken since ancient times. Ladakh, as a cold desert at very high altitude does not receive an abundant snow, rain is scant and when it comes it is immediately absorbed by the sandy soil. To maintain their existence in such harsh condition, people of Ladakh grow a large number of field crops like cereals (wheat and barley), pulses (chickpea, moong and lentil), oil seeds (mustard), vegetables (potato, onion, cabbage, pea, tomato, radish, turnip etc.) and forest trees (poplar and willow) besides fruit corps (apricot, apple, walnut etc). However, as in entire low land ecosystem, in high altitude (Ladakh), the insects are also the dominant animals belonging to orders *Coleoptera, Diptera, Lepidoptera, Orthoptera, Homoptera, Hemiptera* and *Thysanoptera*. Most of the insect-pests, which are found in this region, cause significant damage to cultivated as wall as non-cultivated crops.

The importance of insect is in their being pest of crops and animals, and as carrier of diseases. Although some of them are also useful as parasite and some produce honey, silk, lac etc., which are useful to human being. They also act as pollinator thereby increase yield. However, a large number of insects cause injury to economic plants by feeding on them. An insect becomes a pest when the population increases above a certain level of economic injury and its existence conflicts with man's welfare, convenience and profit. There are three categories of pest:

a. *Regular pests*: These are generally found in abundance during a crop season. Example, aphid and thrips are regular pest in Ladakh.

b. *Spor adic pests*: These assume pest status occasionally in certain years and include cutworm, gypsy moth on apricot etc.

c. *Potential pests*: These pest normally cause negligible damage but may become highly destructive resulting from some disturbance in the environment and quickly increase in their population, example army worm in wheat crop.

1. Population Dynamics in Cold Arid Region of Ladakh

It is obvious that a large number of insect population is present in order to cause significant damage to the crop. Existence of large number of insect-pest indicates that they have successfully adopted themselves to change conditions. A species under optimum conditions has a certain biotic potential for increasing in number which meets resistance from a number of environmental factors and the difference between the two is reflected by the actual increase in a given time. The level of population of a species determined by variation in the environment as a whole and the physical factor such as climate is primary factor. Pest population is regulated by their innate capacity to increase as influenced by various abiotic and biotic factors. The magnitude of increase and decrease in number is governed by the level of influence of various factors, *viz.*, the environmental factor and abiotic (temperature, humidity, light etc.) and biotic factors (food) and other factors modifying indirectly. Following factors of cold arid region of Ladakh influence the biology as well as population dynamics of insect pest:

1.1 Abiotic Factors

Climatic factors that have a dominating influence on survival, development and reproductive capacity of pest, are considered to be the most important cause of fluctuation of population.

a. Humidity

Moisture in the air, known as humidity, affects the insect population and their activity. Most of the terrestrial insects live in an environment which is dry. The amount of moisture in the soil as well as in the atmosphere has a profound effect on survival of many insect-pests. The moisture in soil influences the quality and quantity of nutrients in the host plants which ultimately influences the biology of insect like aphid feeding on mustard and other vegetable corps. For example, aphids multiply very fast at higher humidity condition not only in plain but also in Ladakh having low humidity percentage compared to plain. Infestation of willow scale was observed in relation to humidity which varies with the slight change in relation to a particular location in Ladakh. It was noticed that infestation of willow scale was more in those area where willow trees are grown near the river side having higher humidity compared to rest of the area. In Sankoo and Drass areas of Kargil district which are located near the river side and have more humidity, infestation of willow scale is more compared to rest of the area of Kargil. However, infestation of scale insect is more in Leh district compared to Kargil district because most of the plantations are located either on the river side or where the humidity is comparatively high. Similarly, cabbage aphid and cowpea aphid, which are major pests of cabbage and alfalfa, respectively, in Ladakh, multiply very speedily in high humidity with cloudy weather condition.

b. Temperature

Insects are known to be active within a certain rage of environmental temperature. Above and below the optimum range of temperature, insects may tolerate the temperature to some extent depending upon a particular species and temperature above or below of tolerance limits proves lethal to all the insects. Many insects avoid unfavorable cold conditions by concealing themselves in protective places such as the ground liter or in soil, where the temperature extremes are not as great as they are in the atmosphere. Extreme and fluctuating temperature in Ladakh is operating and affecting many insect populations. The thin atmospheric pressure and thin air layer exposed to direct sun rays, rapid increase in temperature while surrounding air and shade area remain cold. This great difference of temperature of exposed and unexposed bodies or area influences the biology of insects. Due to increase in temperature within certain limit metabolism process of insect-pest become fast which accelerate the development, increase reproductive activity and ultimately increase in population number. Below a certain temperature development does not occur. This is growth threshold or minimum threshold development. Fluctuation in temperature may have accelerating or retarding effects on growth of insect, synthesis of glycerol in stimulated by exposure to 0°C and continues to accumulate below freezing temperature which helps in avoiding the freezing of insect due to very low temperature. Low temperature influences the insect population like in codling moth. It has been observed that 5-9% hibernating larvae of codling moth die due to very low temperature during severe winter (December to February).

c. Altitude

Altitude is also an important aspect which regulates the population of insect. As it is well known that most of the cultivation of crops in cold arid region of Ladakh is performed at different altitude which varies from 4,000 to 10,000 feets. Slight variation in climatic condition and daily weather condition in Ladakh may fluctuate the population level of insect-pest even at same place with slight difference in altitude. Variation in altitude influences the temperature and humidity in relation to a particular place as it is well known that there is negative correlation between altitude and air temperature. With increasing altitude, the air temperature decreases and the climate becomes more severe and the growing season becomes shorter.

It can be very easily observed in Khaltsi block, Nubra valley of Leh district and some part of Kargil district which is located at low altitude where people harvest two crops in one season due to long cropping season (February to October) while in rest of the area of Leh and Kargil district there is single crop because of shorter cropping season. This variation in longevity of cropping season is only because of altitude difference which regulates the sowing time and harvesting time by regulating weather or climatic condition (temperature and humidity). Therefore, the area where two crops are grown there is higher pest infestation compared to single crop grown area. Thus, altitude influences not only the crop biology but also magnitude of pest attack. Altitude also affects the quality of agricultural produce. For example, fruit produced at relatively high attitude, usually, has bright colour because of increase of UV radiation. Increase and decrease of altitude influences the crop maturity also. For example, increase of 30 m in altitude delays fruit maturity by day. Thus altitude influences the temperature and humidity which, collectively, influences the insect population, biology, behavior and damage caused by them.

d. Insolation

The effect of light on insect life is different from that of temperature and moisture. There is no lethal effect of low or high light on insect. Light acts as a stimulus by enabling insect to regulate and synchronize their life cycle with changing season. Several insects such as moth and hopper are attracted to the light during dusk or night. The light photoperiod regulates mating, oviposition *etc.* It is also responsible for inducing diapause which is governed by photoperiod. Because of the unique topography of Ladakh, having high naked sky touching mountain and relatively very thin atmosphere, it receives more insolation compared to the rest of the country, which influences the crop biology over which pest biology depends. South-facing area in relation to sun gets more light and temperature compared to North-facing areas due to which early rise in temperature causes early bud burst and thus early production of blossoms which are prone to frost. Similar effects occur on other crops like cereals and vegetable which force to early sowing of crop. The early sowing of crop results in escaping from pest attack. For example, early sowing of mustard and wheat in Ladakh reduces the aphid and thrips infestation, respectively.

e. Wind Velocity

High wind velocity is one of the characteristic in cold arid region of Ladakh and it is also important particularly in relation to their strength and frequency. It helps in migration of insect-pest from one area to other. High wind velocity increases the transpiration from the plant which results in low water content in plant which insects do not like *e.g.* aphid on mustard.

f. Location of Field

The intensity of insect and pest attack largely dependends on the site where crop is grown. To achieve the success production of crop with lowest pest attack, the site condition must match the site requirement of species varieties of crops as closely as possible. In Ladakh, most of the crops are grown in patchy areas which are surrounded by hills. Such types of location of cultivated fields influence the insect and pest attack. If the crop or plantation site is closed at all sides by woods, houses, hill or mountain leads to lack of air movement which results in increase of incidence of pest. Open location has advantage that they offer low incidence of disease and pest due to better and faster drying.

1.2 Biotic Factors

a. Food Supply

The quality and quantity of food supply has direct effect on the population density which reflects in the increase or decrease of incidence of pest or disease in a year. It is well known that a crop is susceptible to attack of a pest at a particular stage of its development and pest can attack the crop destructively only during certain stages of its development. For example, cabbage aphid attacks on cabbage and cauliflower more severely then mustard at vegetative growth stage. However, in Ladakh, on mustard it attacks at reproductive growth phase unlike the vegetative growth phase in plain. Cabbage butterfly survives on a weed which is very common in Ladakh in absence of host plant. Cow pea aphid was found to attack on weed in absence of the host alfalfa. Therefore, supply of food in quality and quantity influences the pest attack and damage.

2. Crop Production Practices and Pest Attack in Ladakh

Most of the crop production practices influence the pest incidence and population buildup. In cold arid region of Ladakh, people grow the crop in traditional way in which there is no proper sowing time, sowing method, irrigation, and other such practices which ultimately influence the pest attack. Some of the important features are given below:

1. Irrigation

The irrigation facilities have contributed to an aggravation of pest problems in many crops. Sap-sucking insects like aphid and thrips, are sensitive to change of water levels in their host plant. In Ladakh, irrigation is dependent on melting of ice, and availability of such water is not certain for a particular time due to which people irrigate (by flooding) their field whenever water is available. Regarding the frequency of irrigation, sometimes they provide 2 to 3 irrigation in a week sometime not a single irrigation within week. The frequent irrigation influences the infestation and biology of most of the insects. The attack of cutworm, armyworm is less in frequently irrigated areas than those where irrigation frequency is low. Maximum fecundity of mustard aphid was recorded when the water level was maintained above the field capacity. Availability of high water or succulence of plant also influences the effectiveness of pesticides. The mustard aphid, *Lipaphis erysimi,* feeding on lower soil water regime was found to be more susceptible to insecticides than those from higher soil water regime. Normal sown mustard crop requires one spray for control of mustard aphid under no water stress condition. But, under moderate and high water stress condition, no spray was required for aphid control (Narang *et al.,* 1993).

2. Sanitation

Clean cultivation as a method for minimizing the incidence of insect-pest was advocated more than a century ago. Destroying or removing crop residues from the fields is one of the basic customs to eliminate the pest from over wintering sites and reduce the infestation. In Ladakh, removal of crop residue and infested fruit of apple is an effective method that minimizes the infestation of codling moth in subsequent season.

3. Sowing Time

Sowing time in most of the area of Ladakh depends upon favorable weather condition. However, actual time of cropping season in Ladakh is from February to October in double cropping area while May to September in single cropping area. It has been observed that early sowing of almost all crops in cold arid region of Ladakh escape from pest attack. In double cropping area, pest attack is less than single cropped area. However, sowing time depends upon the favorable weather condition and availability of snow melt water which melts when temperature goes relatively high.

4. Spacing

The major objective of spacing in crop plants is to obtain maximum high quality and quantity of yield per unit area and per unit time. But, spacing may also influence the population and infestation of many insect-pest by modifying the micro-environment of crop. In Ladakh, it has been observed that the cabbage aphid, *Bravicoryne brassicae*, on mustard was more in widely spaced mustard crop than in closer spacing. In dense mustard crop (high seed rate), infestation of *B. brassicae* was more at border side compared to inside. Similarly, crop density influences the infestation of jassid, gram pod borer and thrips. Therefore, dense crop avoids the severe infestation of aphid but it is not applicable for all the pests and their related host.

5. Inter Cropping

Deterrence of colonization is one of the most effective means of controlling insect-pest intra field diversity. The diversity of a crop system can increase by intercropping. For example, tomato intercropped with cabbage can reduce egg laying by diamondback moth which is a pest of cabbage and cauliflower in Ladakh.

6. Fertilizer

Fertilizers also influence the infestation and population built up of insect-pest. In Ladakh, it is very common that most of the farmers use urea to accomplish nitrogen requirement but not bother about phosphorus and potash including other nutrients while these nutrients influence not only the production but also pest attack. High plant nitrogen levels generally provide favourable condition to insect-pest populations due to greater survival, increased feeding rate and increased fecundity, resulting in faster population growth. Potash application enhances protein synthesis resulting in reduction of amino acid content of the plant sap, which makes the plant less suitable for feeding, development and multiplication of not only of aphid but most of the insect. Application of potash has also been reported to increase the efficacy of insecticides against mustard aphid. Similarly, application of potash either singly or in combination with phosphorus reduces the infestation of *Aphis craccivora* on Lucerne (alfalfa) (Dhaliwal and Arora, 1993). An increase dose of nitrogen increased the built up of mustard aphid while increased phosphorus (0 to 30 kg/ha^{-1}) resulted in decrease of aphid population on mustard (Bhat, 1979). Increase in nitrogen application from 30 to 150 kg/ha resulted in higher incidence of army worm in wheat while attack of *H. armigera* on tomato was minimum at the highest nitrogen level and maximum number of irrigation. Crop-wise account of major insect-pest is discussed in this chapter.

3. Insect-Pests of Fruit Crops

In India, temperate fruits like apple, pear, peach, plum, apricot almond, walnut and cherry are grown commercially in North Himalayan hilly regions comprising of the state of Jammu and Kashmir, Uttarakhand, Himachal Pradesh, Arunachal Predesh and to some extent other North-eastern state such as Sikkim, Meghalya, Mizoram, Nagaland and Manipur (Solanki, 1999; Pratap, 2003). In Ladakh, apple and apricot are the major horticultural crops which play a significant role in boosting the economy of the grower as well as of the region. However, the productivity of these fruits is very low compared to other states as well as countries like the USA, New Zealand and China. It may be due to poor soil depth, poor soil nutrition and unfavourable climatic condition. Apart from this, insect-pest and diseases are extremely important. Followings are the major insect-pest which are prevalent in Ladakh and limit the successful cultivation of fruit crops.

3.1 Codling moth (*Cydia pomonella* L.)

Codling moth is one of the very destructive pests introduced into Ladakh from North frontiers province of Pakistan and has been reported as very serious pest on deciduous fruits (Pruthi, 1938). It causes severe damage to apple and some extent of apricot. Its distribution is restricted to Ladakh region (Malik *et al.*, 1972; Wadi and Sethi, 1975). It causes more than 65% damage in this region. However, Zaki (1999) reported that codling moth caused 48 to 89 per cent damage in apple with an average of fruit infestation of 49.7 per cent and 42.5 per cent in Kargil and Leh districts, respectively. The infestation of the pest on fallen fruits was more severe (69.9%) than that of fruits still on trees (27.5%). Sometimes more than 90% infestation has been recorded in some of the apple growing areas of Ladakh. The adult female of codling moth is approximately 10 mm long and gray in colour. The male is similar to female but is slightly smaller. The wing is remarkably dark gray near the base. The adult holds their wings tent shaped over the body when at rest. The larva being cream in colour and later turs into light pink with brown head. The fully grown larva is 10-15 cm in length. The adult emergence takes place in the month of May-June depending on the climatic condition. The female moth after mating starts egg lying on fruits, leaves and twigs. The eggs being flat and white transparent in colour. The hatching of eggs takes place after 7-15 days depending upon temperature and humidity. The newly hatched larvae bore through the fruit surface generally from the blossom on calyx end and feed near the surface for some time before boring. After that, some larvae bore into deep and feed on pulp and developing seeds until fully grown. Early stage larva being creamy white in colour, which later becomes light pink. Matured larvae come out from exit hole of the fruit and spin a cocoon in loose bark, cracks or on debris and may or may not pupate. The codling moth over winters as full-grown larva within cocoon under loose bark, leaf litter, or any other sheltered place nearby. Pupation takes place during second fortnight of May, which ranges from 10 to 20 days depending upon temperature and other climatic condition. The pest completes one to two generations in a year depending upon the temperature and other weather parameters. In Kargil, warmer area of Ladakh region, two generations were observed while in Leh (cooler area) only one generation occurred. In a trial conducted by RARS, SKUAST-K, Leh, the codling moth was able to complete two generations in warmer area of Ladakh region like Batalik, Garkone, Darchik, Behma, Dah, Hanoo while in cooler area like Leh, and around (Ranbirpura, Sakara and Nimo) the pest was able to complete only one generation.

Damage

The infestation of this pest in Ladakh was 69.9% in fallen fruit while 27.5% in fruit still on tree during the year 1991. However, in a resent study conducted by SKUAST-K, Leh, it was observed that in Leh district alone the damage was 80%. This pest causes the characteristic damage to apple fruit in the following ways:

- A tunnel emanating from the distal end of the fruit and extend to the core.

- "Sting" small shallow holes with a small amount of dead tissue on the cavity wall. The stings caused by early instar larvae that have been poisoned and die shortly after puncturing the apple skin.

Codling moth larva feeds on the core characteristically frass (excrement) exciding from the point of entry. Both types of injury lower the value of fruit in market.

Management

As it is known, codling moth is a major pest of deciduous fruit throughout the world. At present, control of this pest is principally achieved by application of organic insecticides in different apple growing areas. However, application of broad spectrum insecticides for the control of codling moth may result in several side effects such as development of insect resistance to pesticides and destruction of useful natural enemies that may cause secondary insect problem in the apple ecosystem. Thus, careful consideration and initiation of multiple tactics will allow growers to control this pest without interfering with natural regulation of sporadic and secondary pest. Based on biology of moth following control measures should be adopted:

- Use of light trap is important step to monitor the emergence of adult codling moth through which exact time of application of control measures can be decide.
- More recently pheromone traps baited with female sex attractant is found effective to monitor the adult emergence. There are two ways in which pheromone traps can be used
  - Counting the number of males caught per week to assess the potential of population for causing economic injury,
  - Using the time of emergence of adult as biological reference for predicting the time of egg hatch through which correct time of pesticides application can be determined.
- Dead bark of old trees provides shelter to hibernate the larvae during dormancy period. So, scrapping of lose and dead bark from tree trunk and branches is effective to minimize further population buildup. In orchard having young trees with smooth stem control would be easier than in orchard having old trees.

- Burlapping of apple trunk (after harvesting) using gunny bags at the height of 2-3 feet with weekly collection and destruction of larva, trapped in the gunny bags, is found effective.
- At the time of harvesting, fruit damaged by codling moth should be collected and destroyed to reduce the codling moth population.
- Fallen fruits and leaves on the ground provide shelter to hibernate the larvae. So, the destruction of fallen fruit and debris is effective.
- Some times farmers leave the infested fruit on tree, in which codling moth larvae survive. So, complete harvesting of healthy as well as infested fruits should be done.
- Release of *Trichogramma embriophagum* parasitoids at the time of egg laying is an effective control measure to minimize the damage. The egg laying stage of codling moth coincides with peanut stage of apple. At this time female moths start egg laying.
- Two or three spray of insecticides like Chlorpyriphos 20EC (0.075%) or Endosulfan 35EC (0.085%) at 10-15 days interval from peanut stage onwards is found effective.

3.2 Wooly Apple Aphid (*Eriosoma lanigerrum* Hausman)

This pest is native of America and cosmopolitan in distribution. It attacks on pear, crabapple, almond and elms. This pest causes severe damage to apple in Kargil district. The identifying characteristic of this insect is presence of woolly filaments of wax on body which is secreted by them. It was first recorded in 1889 at Coriour (Tamil Nadu) in India damaging apple (Mishra, 1920), after that it has spread to all the apple growing areas of India. The pest is active throughout the year and attacks primarily the underground roots, trunks, branches, stems, twigs and fruit stalks. Upward and downward irrigation are accentuated during hottest and coldest seasons, respectively (Lal and Singh, 1947). Migration from roots to aerial parts takes place in May-June and towards root during October-November depending upon weather condition. Due to infestation of this pest, the affected trees show sickly appearance, loss of vigour, and fruit bearing is also adversely affected.

Adults are about 2 mm long and of chocolaty colour. Aerial colonies are usually located at pruning cuts or the base of sprouts. The pest over-winters either eggs stage or young nymph in colonies on the root of host (Pruthi and Batra, 1960). Crawlers can move to roots at any time. They are active mainly in June and July. Wooly Apple Aphid (WAA) may move from tree to tree whether by winged migrants or by crawlers. After hatching eggs, nymph settled down in batches and start sucking the plant sap. Nymph secrets waxy filaments over their bodies. They attack on wound, injured branches and cracks. Crawler dispersal between trees is favored by close tree spacing, and clean, smooth soil surfaces. High soil temperatures, weed cover, and distant spacing of trees inhibit spread between trees. Nymphal period is of about 11 days in June which gradually increases with decrease in temperature and becomes 93 days by December (Rahman, 1941). The reproduction of this pest generally is not affected up to 32°C but ceases when the temperature falls below 3°C. Excessive humidity retards the multiplication of this pest. Frequency of infestation in 2-25 years old orchards is correlated with orchard age. Orchards more than 25 years old may have 79-100% of trees infested. Severity of root infestation ranges from 24-50%, with individual trees as high as 84% infested.

Damage

The aphid sucks the sap from branches and twigs and from under ground twigs. The important characteristic symptom of this pest is formation of galls or knots on roots. Such galls can also be seen on aerial twigs or branches when they attack on above ground. The nursery grown for plantation being more conducive to the attack of this pest. Such nursery leads to heavy infestation of this pest, causes population buildup. Infestation of wooly apple aphid can reduces the growth of young apple trees and reduces the vitality of trees.

Management

- Use of resistant and tolerant rootstock like M-21, M-25 and MM series *Malus buccata* L. var. Himaliaca for vegetative propagations (Thakur and Dogra, 1980, Rai and Tripathi, 1984).
- Infested twigs and branches should be pruned and burnt to minimize the infestation.
- Healthy plants should be selected from nursery and treated with Chloropyriphos (0.05%) before plantation.
- Infested nursery is main source for spreading of this pest so nursery/plant should be obtained from those which are away from wooly aphid attack.

- The aphid population parasitized by a parasitoids *Aphelinus mali* and can reduce the population effectively. Therefore, release of *Aphelinus* @1000 adults per infested tree has been found effective. During vegetative growth stage, root should be treated through Dimethoate (0.05%) or Carbosulfan (0.05%) by mixing through basin. At this time the pest migrate from aerial part to root to pass the winter period.
- Several natural enemies [*e.g.* syrphid larvae, lady beetles and parasites such as *Aphlinus mali* (Haldemann)] provide biological control.

3.3 Green Peach Aphid (*Brachycaudus* spp.)

Green peach aphid is polyphagous pest which attacks on peach, apricot, almond and plum in order of preference (Pruthi and Batra, 1960). However, in Kargil and Leh district it attacks mostly on apricot which is the main fruit of this region. This pest occurs in almost all apricot growing areas of Ladakh. However, the damage was recorded more in Sanjak, Dha, Hanu and Btalik areas which are the major apricot producing areas. The nymph and adult are dark green in colour. The colour varies according to the host plant. The pest appears in May in Kargil district while in late June-July in Leh district.

Damage

Adult and nymph suck the sap from leaf which results in curling of leaf in which aphids survive and remain smaller. The attack of aphid leads to poor fruit setting, premature fruit fall.

Management

- Removal of alternative host from field is effective. Most of the aphid survive in green house/polyhouse during winter because there is no green vegetation in open field.
- Release of green lace wing, *Chrysoperla carnea,* @ 20 first instar larvae per tree has been found effective.
- Spraying of Imidachloropid (0.05-0.075%) at pink bud stage (7-10 days before flowering) leads to good control of aphid.
- Later on if the infestation in more then spraying of Dimethoate 30 EC or 200 ml Monocrotophos 40 EC in 250 l water per acre has been found effective.

3.4 Blossom Thrips (*Frankliniella* spp.)

Blossom thrips is another important pest which attacks apple and apricot in almost all fruit growing areas of Leh and Kargil district. It is minute (about 1 mm long) slender and soft bodied fragile insect with wings heavily fringed. It has asymmetrical rasping and sucking type of mouth part which the insect lacerate the plant tissue and suck the oozing cell sap. Apple and apricot blooms are usually attacked by the polyphagous thrips (*Frankliniella* spp.). Eggs are laid in flower buds before bud opens. On hatching the nymphs feed on petals and vital of the flowers by sucking the sap, the form causing destruction of flowers (Sharma and Bhalla, 1963). As a result, on fruits setting weak fruits are formed which drop prematurely. In cold arid region of Ladakh, thrips pass sever winter as dipausing pupae which is the source of further population buildup. Therefore, exposing diapauseing larvae by digging the bases of the trees during winter is effective to minimize the pest attack in coming season. However, this pest does not cause economical loss.

Damage

The infestation starts from blooming and population increases with advancement of flowering of apple and apricot. Nymph and adult feed on floral parts, leaves and bud by sucking of sap which leads to low fruit formation and reduction of yield.

Management

- Use of yellow trap was found effective for monitoring of thrips in field.
- Spraying of Imidachloropid 0.02% is very effective to control thrips infestation before opening of flower (15-20 day before flowering).

3.5 Chaffere Beetle (*Adoretus* spp.)

Beetle is polyphagous in nature and attacks a large number of fruits. It appears in the month of May-June and feeds on foliage and developing fruits at dusk. It lays eggs in the soil near the plant.

Damage

At night, only adult beetle causes damage by feeding on leaves which get skeletonized. The maximum activity of beetle is recorded at slightly higher temperature. Larvae after hatching from eggs feed on soil humus, roots of grasses and other vegetation.

Management

- Collection and destruction of adult through using light trap is effective method.
- Deep ploughing around the tree is effective which kills the immature stage of this pest.
- Spraying of Carbaryl 50 WP (0.01%) in May-June, when beetles are observed in orchard at dusk, is also beneficial.

3.6 Apricot Chalcid (*Eurytoma samsonovi* Vasiljev)

Eurytoma samsonovi Vasiljev is widely distributed in Afghanistan, Pakistan and India. This pest is reported to attack on peach, plum, prune and apricot. However, in Ladakh apricot is the major host of this pest.

Damage

Eggs are laid inside the tender fruits. On hatching, the grubs bore into the kernel and feed on the inner content and on developing seeds, which leads to fruit fall prematurely.

Management

- Collect and burn all infested fallen fruits.

3.7 Indian Gypsy Moth (*Lymantria obfuscate* Walker)

This pest has been reported from all over the world. In Ladakh, this pest attacks on apple, apricot, walnut including forest trees particularly poplar and willow. It is very harmful insect which defoliates the plants. However, the extent of defoliation varies according to host plant. In Ladakh, its preferred host is apricot and apple after poplar and willow. Female after mating lay eggs in cracks crevices, debris and below the bark and under roof. Damage of host plant was observed more near house hold area because they get shelter easily in these areas. The eggs can also be seen below the house roof and covered with yellow brown hairs. The hatching of eggs takes place in May-June which coincides with emergence of new leaf of host plant. The hatching of eggs also depends upon the temperature and humidity. Mature larvae pupate in soil or among the debris. The adult emerges in July-August. There is one generation in a year.

Damage

The damage stage is larvae which feeds on leaf at night and hide during day time at ground. In sever infestation they defoliate the tree leaving behind only veins of leaf.

Management

- Destroy the entire egg patches whenever it is seen. This will reduce the population of gypsy moth.
- Pruning and destruction of crop residue should be a regular practice to avoid further population buildup.
- Scrapping of dead bark of old tree in winter (January to March) is effective which exposes the eggs patch to adverse condition which leads to the killing of eggs.
- Bur lapping of tree trunk by loose gunny bag attract the mature larvae for pupation and adult for egg laying. Therefore, a regular opening of burlap to collect and destroy the larvae, pupae and egg mass will be effective in minimizing the further population multiplication and damage in coming season.
- Foliar spray with 200 ml Endosulfan 35 EC in 200 l water per acre was found effective.

3.8 Mite (*Tetranychus* spp.)

Mite (*Tetranychus* spp.) is found to attack fruit crops particularly apricot. Two species of mite, *viz.*, European mite (*Panonychus ulmi* Koch) and Spotted mite (*Tetranychus urticae*) are reported from Himanchal Predesh. The distribution and infestation in Ladakh varies according to altitude which influences climatic condition of that particular area. However, higher infestation of this pest was observed in Dha and Hnnu Villages Kaltsi block of Leh district.

Damage

They feed on green matter and suck sap which leads to brozing and leaf drop. The sever infestation occurs in June and July depending upon weather condition.

Management

- Remove the infested leaves and burn them.
- Use wettable Sulphur @ 100 g in 20 l of water, one spray when trees are dormant and mites are torpid with cold. One spray can be given after petal fall stage.

- Foliar spraying of Kanodane 20 EC @ 500 ml at petal fall and Nomite 18.5 EC at fruit development stage has been found effective.

4. Insect-Pests of Vegetable Crops

Vegetables are known as important supplier of vitamins, protein and minerals in adequate quantity. Among the vegetables, cauliflower, cabbage, capsicum, tomato, turnip, reddish, spinach etc. are important which are grown in cold arid region of Ladakh during main cropping season from March to October. However, these vegetable crops suffer heavy losses due to insect-pests attack.

4.1 Tomato Fruit Borer (*Helicoverpa armigera* Hubner)

Tomato is one of the most popular and widely grown vegetable in the world, ranking second in importance to potato in many countries. Because of its wider adaptability tomato is grown in almost all parts of India including cold arid region of Ladakh. High temperature and low humidity in Ladakh favours a good crop by avoiding disease. However, insects are one of the most important limiting factor in successful cultivation of tomato and pea. Tomato fruit borer is important pest which causes severe loss. This pest is polyphagous in nature and attacks on pea, soybean, safflower, chilies, gram, okra, tomato and many legumes and vegetables. However, in Ladakh most preferred hosts of this pest are pea and tomato. Moths are brown to yellowish in colour with slight strip and caterpillars are greenish. Immediately after hatching, the caterpillars crawl over the leaves and feed on vegetative parts. In Ladakh, fruiting starts from June-July depending upon temperature and altitude and pest attack starts from July and August depending upon weather/climatic condition. Female lays spherical yellowish eggs singly on tender parts. A single female may lay as many as 700 eggs in four days. The incubation period is 2-8 days depending upon climatic condition. The young larvae feed on foliage for sometime and later bore into the pods (pea) or fruit (tomato) and feed on developing grains/fruit. Characteristically, while feeding, the head is thrust inside leaving rest of the body outside. Cannibalism, among larvae, is common. The full grown larvae pupate in an earthen cell. The pupa is dark brown and has a shiny spine at the posterior end. The pupal period lasts 8-15 days in active season which may be prolonged in winter.

Damage

The larvae feed on leaves and fruiting bodies; later on, they bore into the pods or fruits in the case of tomato, completely eating away the internal contents. A single larva may destroy 20-35 pods before it reaches maturity owing to which yield loss in pea may go upto 20 per cent. This pest doses not cause sever loss to tomato but causes sever loss to pea in Nubera valley of Leh district and Zanskar area of Kargil district.

Management

- Collection and killing of caterpillars from infested fruits will reduce the infestation.
- Pheromone traps @ 5 g per ha was found effective to minimize the infestation. This will reduce the damage caused by this pest. The pheromone is easily available in the market.
- Alternate spray of Malathion and Carbaryl at 15-20 days interval was found better to control *Helicoverpa*.
- Spraying of Econeem (a neem based insecticides) was found to be the best means of controlling the pest (Talekar *et al.*, 1999).
- Various biocontrol agents have also been recommended for the control of this pest. *Eucelatoria bryani* was found very effective against *H. armigera* under field conditions, especially for controlling adult fruit worm (Sivaprakasam *et al.*, 1986).
- Spraying of NPV @ 250 LE + 0.1% Tinopal at early instar larvae was found effective. However, Ladakh is located at very higher altitude and intensity of UV rays is comparatively higher which degrade the nuclear polyhydrodsis virus. Therefore, NPV should be used at evening time.
- Endosulfan and Malathion 0.2% were found effective.

4.2 Cabbage Butterfly (*Pieris brassicae* L.)

In Ladakh, cabbage and cauliflower are the important vegetable crops and cabbage butterfly is highly harmful insect pest, which attacks severely on cabbage and cauliflower including knoll khal, turnip, radish and sarsoon and other cruciferous plants. Larvae of this pest cause damage to the crops. Freshly hatched larvae are pale white which turn greenish yellow later on. Short hair can also be seen on the caterpillar's body.

Young caterpillars of *P. brassicae* feed on leaf surface by scraping the green matter. Mature caterpillars are able to feed gregariously on whole leaves from margin to inward, leaving behind the main veins. Sometimes entire plants are eaten up. The pest attacks on the crop from June to October depending upon the length of cropping season. The female lays eggs on lower surface of leaves in groups (75-150 eggs) which are yellowish-orange in colour. After hatching caterpillars feed on leaves in mass scrapping the green matter and later they spread over whole plant. Pupation takes place on plant or debris. Two to four generations of this pest has been recorded in cold arid region of Ladakh.

Management

- In Ladakh, people grow cabbage, cauliflower, turnip, radish etc. on very small areas. In such situation, regular visit should be made to collect the mature larvae and destroy them, which is a very effective control measure.
- Growing of resistant variety is very effective to minimize the loss caused by cabbage butterfly. Red cabbage is found resistant to cabbage butterfly in Ladakh. However, it has low market acceptability compared to the white cabbage.
- Early planting can reduce the infestation of cabbage butterfly. However, in cold regions of Ladakh, sowing and transplanting mostly depends upon the availability of water (irrigation which comes from snow melt) and weather condition of cropping season.
- Spraying of Endosulfan @ 0.005% or Monocrotophos 36 EC 0.05% after 20-25 days transplanting was found effective against this pest.
- Among the natural enemies a parasitoid has also been observed which attacks on pupal stage of this pest. This indicates the application of bio-control agents in this region.

4.3 Cabbage Borer (*Hellula undalis* Fab)

This insect has a widespread distribution throughout the Middle-East, Asia and the Pacific. This pest was first identified in Italy. The cabbage and cauliflower are major host of this pest but it also attacks Chinese cabbage, Chinese broccoli, Chinese mustard, mustard cabbage and radish. It occurs in almost all cabbage growing areas of cold arid region of Ladakh.

Adults are grayish-brown moths. Forewings have wavy gray markings, a curved pale patch sub-terminally, and a kidney shaped mark at one-third length from the tip. Hind-wings are pale, with the tip being lightly colored. The wingspan ranges from 3/4 to 1/3 inch long. Adults live from 4 to 8 days, longevity increases with a decrease in temperature. The ovoid eggs are about 1/50 inch long, being slightly flattened upon the surface of deposit. Pearly white when laid, they become pinkish the next day and then turn brownish-red with the dark head of the larvae visible at one end just before hatching. Eggs are laid singly, or in groups or chains of 2 or 3 on the leaves of cabbage near the bud. Hatching of eggs takes place in 2 to 5 days depending on weather condition suction temperatures (Sivapragasam and Abdul Aziz, 1990). Caterpillars are greyish-yellow with five broad, irregular reddish-brown bands that extend the length of the body. Duration of larval stages varies with different hosts. On cabbage, larval development is completed in 16 to 19 days, but on cauliflower, it may require only 11-13 days. Pupation occurs in a silk cocoon. Early stage pupae are soft and pale yellowish-white in color with a bright red, dorsal blood. A few hours later, the pupae harden and become light brown. Cocoons composed of silk threads are generally spun between leaves or at the entrance of the feeding tunnel. Adults emerge in 8 days. Egg laying begins within 24 hours of adult emergence and continues for 3 to 10 days (Sivapragasam and Abdul Aziz, 1990).

Damage

This caterpillar is an important pest on head cabbage, Chinese cabbage, and mustard cabbage. Major damage occurs on young plants but caterpillars also feed on older plants. Damage is mostly severed between transplanting and the heading stage of cabbage even though the larvae are present in the field throughout the crop (Sivapragasam and Abdul Aziz, 1990). On cabbage seedlings, extensive damage can occur when the caterpillars feed on the growing points (apical meristem) and the developing leaves. Severe injury occurs when they tunnel into the main stem. Damage at this stage of plant growth results in stunting, and sometimes death of young plants. More often, however, damage to the growing point results in deformed plants and the formation of multiple growing points or heads. On older plants the larvae feed on leaves and by tunneling into leaf petioles. Because of their secretive nature, symptoms of imported cabbage webworm infestations are often not detected until plants appear stunted or deformed. However, silk webbing and frass are easily found when infested seedlings are examined closely.

Management

- Collection and mechanical destruction of caterpillar in early stage of attack can reduce the infestation.
- Screening of seedling beds and clean culture are helpful in minimizing damage caused by the cabbage webworm down. Plants should be 5 or 6 inches high and have a vigorous growth before being set out (Fullaway and Krauss, 1945). It is important that clean seedlings are used, if not it may be necessary to apply 1 to 2 insecticide applications if infestations are found.
- Several insecticides are efficacious in other parts of the world. carbaryl, permethrin, and trichlorphon are effective in controlling larvae of this pest. Spraying of Endosulfan 0.05% after 20-25 days transplanting is effective.

4.4 Cabbage aphid (*Brevicoryne brassicae* Linn.)

Cabbage aphid is most destructive pest in Ladakh which occurs in all the vegetable growing areas of Ladakh. It attacks on cabbage, cauliflower, radish, Karm sag, turnip etc. Adult and nymph suck the plant sap which leads to loss of vigor to the infested plants. On severe infestation curd or head formation does not take place. The active season of this pest is June to November. Severe damage leads to inconsumable cabbage head.

Management

- Use of yellow trap is very effective to find out the presence of the pest.
- Uproot the highly infested cabbage, cauliflower, turnip plants. This will delay or reduce the further population buildup.
- Early planting of cabbage, cauliflower and radish may escape aphid attack.
- Early infestation and colonies survive on leaves should be collected and destroyed. This will reduce the further population build up.
- Encourage the population of lady bird beetle which is predation of aphid.
- Spraying of Dimethoate 20 EC 0.25 l/ha or Endosulfan 0.05% is found effective.

4.5 Cut Worm (*Agrotis ipsilon* Hufn.)

This is polyphagous and most important destructive insect pest which occurs in all the vegetable growing areas of Ladakh. It attacks cabbage, cauliflower, capsicum, potato and so many other crops at seedling stage and maturing crop. Severe damage occurs at seedling stage. The caterpillars are grayish brown variously striped night moth. Caterpillars are 3.4 cm long gray or brown to almost black with marking.

Damage

The sever damage occurs not because of feed cutting the plant at ground level. The larvae hide during day time and feed and cut the seedling at night time. The female lay eggs in lower surface of leaves of host plant or in hole. Larvae feed on stem. In Ladakh, soil is light sandy which is very favourable for cut worm and dry season support the maximum damage by this pest.

Management

- As it is known that the larvae come out from hiding place at night so putting the heaps of grasses so that congregated larvae can be killed in the morning is effective.
- Collection and destruction of larvae during weeding/hoeing may reduce the infestation of cutworm.
- Following the crop rotation would be effective to avoid the damage.
- Destruction of the hibernation site by deep ploughing will suppress the population of this pest.
- Use of Quinalphos 5 g @ 20-30 kg/ha in the soil before transplanting found effective to minimize the pest infestation.

4.6 Diamond Back Moth (*Plutella xylostella* Linn.)

The diamond back moth is probably of European origin but is now found throughout the Americas, Europe, Australia, New Zealand, South-east Asia including India. This pest has also been recorded in cold arid region of Ladakh like Nubra valley, Neuma and Khaltis block of Leh district and almost all vegetable growing areas of Kargil district. Diamond back moth attacks only those plants which belong to the family Cruciferae. It attacks on all cruciferous vegetable crops including broccoli, Brussels sprouts, cabbage, Chinese cabbage, cauliflower, mustard, radish and turnip. These host plants are not equally preferred, however, cabbage is the most preferred host in Ladakh over which moth feeds and oviposit.

The adult moth is small, slenderical, grayish-brown in colour and with pronounced antennae. It is about 6 mm long and marked with a broad cream or light brown band along the back. The band is sometimes constricted to form one or more light-colored diamonds on the back, which is the basis for the common name of this insect. When viewed from the side, the tips of the wings can be seen to turn upward slightly. Adult males and females live about 12 and 16 days, respectively, and females deposit eggs for about 10 days. The moths are weak fliers, usually fly within 2 m of the ground.

Diamond back moth eggs are oval and flattened, 0.44 mm long and 0.26 mm wide. Eggs are yellow or pale green in colour and are deposited singly or in small groups of two to eight eggs in depressions on the surface of foliage. Females may deposit 250 to 300 eggs but average total egg production is probably 150 eggs. Depending on weather conditions, hatching of eggs takes place 5-7 days from the date of egg laying. Caterpillar has four instars which range between 3-7, 2-7, 2-8 and 2-10 days, respectively. The larvae are colourless in the first instar, but thereafter are green. The body bears relatively few hair, which are short in length, and most are marked by the presence of small white patches. larvae is leaf mining, although they are so small that the mines are difficult to notice. Throughout their development, larvae remain quite small and active. If disturbed, they often move backward, and spin down from the plant on a strand of silk. The second instar larvae emerge from their mines and thereafter start feeding on the lower surface of the leaf. Their chewing results in irregular patches of damage, and the upper leaf epidermis is often left intact. Pupation occurs in a loose silk cocoon, usually formed on the lower or outer leaves. The yellowish pupa is 7 to 9 mm in length. The duration of the cocoon averages about 8.5 days (range 5 to 15 days).

Damage

The first instar sometimes feeds in the spongy plant tissue beneath the leaf surface forming shallow mines that appear as numerous white marks. These mines are usually not longer than the length of the body. The larvae are surface feeders in all subsequent stages. These larvae feed on the lower leaf surface which results irregular patches in the leaves (Harcourt, 1957). All the leaf tissues are consumed except the veins. The last stage larva is a voracious feeder; it causes more injury than the first three larval instars.

Management

- Same as for cabbage butterfly (see page 252 of this chapter).

4.7 Cabbage Flea Beetle (*Phyllotreta* spp.)

The flea beetles attack almost all cruciferous plants and have been reported from Europe, North and South America, Australia, Japan and India (in West Bangal and Utter Pradesh). It causes sever damage to radish, turnip, cabbage, cauliflower and kno-khol.

The adults vary in colour from shiny black to black with a green sheen to black with yellow stripes on the elytra. The body is elongated, narrow in front, broad bestially and has round end. The grub is dirty white in colour with pale white head. This pest is active throughout the summer on different crops. During winter it hibernates in soil or among plant debris. The over-wintered adult beetle emerges in June-July depending on location. The female lays 50-80 creamy white eggs in the soil around the host plant. The larva is very active to feed on host roots. It pupates in the vicinity of infested host.

Damage

The larvae live in the soil and feed upon the roots of the host plants. The adults feed on the cotyledons and leaves of young plants making innumerable round holes. The stem, flowers and even pods may be attacked. The old eaten away leaves dry up and the young leaves remain unfit for consumption. The attacked plant of cabbage gives a special kind of decaying odour.

Management

- Deep ploughing in winter can suppress the pest population.
- 315 ml Dichlorvos 100 EC in 625 l of water is effective to suppress this pest.

4.8 Pea Pod Borer (*Etiella zinckenella* Treitschke)

This is major pest of pulses, attacks on pigeon pea, cowpea, lima bean and pea. However, in Ladakh this pest attacks on pea which is grown in most areas of Ladakh. Serious damage was observed in Zanskar valley and some surrounding areas of Leh.

The damage by this pest varies in different regions of Ladakh. The infestation of pest is higher in warmer belt of Ladakh than cooler areas. It causes 2 to 10 per cent damage to green pea by boring the pod. The moth is nocturnal. Female lays eggs singly on various parts of plant including developing pod. Larvae after emergence feed inside the pods and become fully grown in 10 to 25 days. Pupation takes place in soil. The actual damage is caused by the larvae which feed on floral parts, pods and on developing seeds. Owing to feeding on developing seeds the reduction in yield goes upto 5 per cent in Ladakh.

Damage

The injury is caused by the larvae which consume floral parts, newly formed pods and feed on developing seeds. After feeding on pod, larvae move to another pod. Owing to feeding on grains, the reduction in yield may be up to 20-45% as recorded in Zanskar valley.

Management

- Spraying of systematic insecticides is effective to control this pest.

4.9 Two-spotted Spider Mite (*Tetranychus urticae* Koch)

Two-spotted spider mite is polyphagous pest and causes damage to all the leafy vegetable like spinach, Karm sag etc. in cold arid region of Ladakh. This pest is very common in Nubra valley, Khaltsi, Nyuma area of Leh district and Khurbathan, Zanskar valley, Drass area of Kargil district. It is the most economically important mite species. It affects more than 150 economically important host plants, including all of the tree fruit, small fruits, vegetables, and ornamental crops. Adult females are oval, dome-shaped mites. Adults are pale straw-colored with two black spots (caused by the dark gut contents). Eggs are clear and spherical, deposited on leaves. Over-wintering females are bright orange without the dark spots.

Damage

Damage is caused both by nymph and adults. A large number of webs are formed on the leaves under which the pest sucks the cell sap. The affected leaves dry up and finally wither away giving an unhealthy appearance. Feeding results in diffuse stippling followed by bronzing. However, feeding of *Tetranychus* species has a greater impact on host leaf function. It reduces photosynthesis of affected plant which leads to loss of vigour of apricot. Infestations can reduce yield and return bloom.

Management

- The presence of mite and their density can be detected through hand lance during the season by counting mites per leaf.
- The infested leaf shows whitish-green appearance. Therefore, plucking and destruction of infested leaves may reduce the reduction to some extent.
- Spraying of Kalthan is found effective for the management of the pest.

4.10 Onion Thrips (*Thrips tabaci* Lind.)

Onion thrips, *Thrips tabaci* Lind., is a cosmopolitan pest found throughout India including most of the onion growing areas of the cold arid region of Ladakh. In Ladakh, it is a major threat for successful cultivation of onion. It also attacks on tomato, cabbage, capsicum, apple etc. Nearly all garden vegetable plants, some forage and grain crops and many weeds serve as hosts. They are generally congregated at the base of leaf or in the flower of onion. In cabbage and cauliflower, they feed on lower surface of leaves. The damage is done by both nymphs and adults.

Onion thrips, formerly thought to have rasping mouth parts, are now known to have asymmetrical appendages comprising a single mandibular stylet that is used to pierce plant tissue; maxillary stylets forming a feeding tube are used to remove plant liquids by cibarial pumping. The males are wingless however, the females have long narrow strap like wings. Females regularly reproduce without mating which results in rapid increase of populations, when conditions are favourable. Eggs are laid in onion leaves and hatch in 4 to 10 days, depending on temperatures. The nymph is selendrical in shape, brown in colour and looks like adult but wingless and slightly smaller than adult. This insect becomes active from June to mid October depending upon the weather conditions which influence the length of the cropping season. There are three to five overlapping generations per season.

Damage

This pest infests the crop throughout the cropping season. Attack of this pest results in spotted appearance of leaves and later on pale white patch gives slivery appearance due to drainage of sap. Due to drainage of the sap, leaf tips become fade and gradually dry up, it is the symptom of thrips attack on onion. Infestation leads to the production of low quality bulbs and no seed setting in the flowers. Damage to plant results from the destruction of epidermal tissue and removal of cell sap from individual mesophyll cells. Feeding by nymphs is generally concentrated on young tissues between the leaf sheaths and the stem, but extensive feeding by adults on more mature tissues is not unusual. Affected leaves display whitish or chlorotic blotches, as if they have been sandblasted. If feeding is severe, particularly under drought conditions, the tips of leaves become brown. Prolonged feeding of thrips may eventually kill seedlings, reduce bulb size and increase the incidence of leaf and bulb rots. It was observed that the population of thrips has a significant positive correlation with maximum temperature and negative correlation with rainfall/snowfall. This indicates that higher temperature might have favoured reproduction and faster multiplication.

Management

- Jet sprinkling of water through jet nozzles to prevent thrips multiplication.
- Commercial onion cultivars differ in their susceptibility to thrips attack, although research reports suggested that onion types with glossy, spreading foliage were less suitable for thrips development; the common types were upright, with waxy foliage with tight leaf interfaces. Because of their mobility, crop rotation or isolation from immigration sources have little effect on onion thrips infestations.
- Onion field should be maintained weed free for 4-6 weeks after planting by resorting hand hocing/hand weeding.
- As thrips attract to yellow colour, setting up of yellow sticky traps is often recommended as control method. Yellow sticky traps were used for monitoring thrips (Diruaviam and Uthamasamy, 1992) in different crops.
- At seedling seeding stage yellow sticky traps @ 5/ha may reduce the infestation.
- Early transplanting is found effective to minimize the infestation of thrips.
- The treatment of seeding root dip with 1% neem oil emulsion and soil application of neem cake @ 500 g/ha^{-1} proved to be effective to control the thrips (Rao *et al.*, 1999).

- Effective management of onion thrips relies primarily on foliar sprays of insecticides based on some treatment threshold, usually from two to four nymphs per leaf. Because nymphs feed between the leaf sheaths and the stem, they are difficult to contact with insecticide sprays, consequently high pressure. Spraying of Malathion 50 EC at the rate of 0.05% with waiting period of at least 12 days before harvesting is effective to control this pest.

4.11 Onion Maggot (*Hylemya antique* Meigen)

Onion maggot, *Hylemya* (=) *Delia antiqua* (Meigen), is the most important and consistent insect-pest of onion in cold arid region of Ladakh including Leh and Kargil district, particularly where onion is grown continuously on the same ground. Onion maggot attacks only members of the Alliaceae family, however, onion is preferred host, but it also infests garlic, leek, shallots and chives. Three distinct broods occur annually in most temperate regions. Larval offsprings from flies that emerge in the spring usually cause the most damage because the plants are most vulnerable during the seedling stage.

Adults are somewhat smaller and more slender than house flies, but closely resemble them in general appearance. Flies live for 2-4 weeks and are capable of migrating about a mile in search of suitable hosts. Mating occurs in a week to 10 days after emergence and female flies oviposit on the soil near the plants or occasionally on the young leaves or plant necks. Each female is capable of laying several hundred eggs. Neonate larvae move down through the soil and enter at the base of the onion plant by feeding on the tissues in the vicinity of the basal plate using their hooked mouth parts. More than one larva may infest single plant and if plants are killed before larvae complete their development, maggots immigrate to nearby intact plants in search of additional food. More than 20 seedlings may be killed by a single maggot. Onion maggot over-winters as a puparium in the soil. There are three larval instars.

Damage

The first sign of onion maggot injury is the wilting of foliage, after which it subsequently becomes flaccid and collapses. This damage is particularly noticeable when the plants are in the seedling stage. The maggots frequently destroy groups of plants that results in a patchy crop. Larger onions may survive infestation, but distorted growth accompanied by the rotting of tissue renders them unmarketable.

Considerable losses of bulbs at harvest have occurred because of maggot feeding during late-season when the bulbs are being field cured.

Management

- Deep ploughing after harvesting of onion (in winter) and 15 days before transplanting reduces the infestation of this pest.
- Crop rotation of two or more years or the isolation of onion fields from one another can significantly reduce maggot populations for several seasons.
- Destruction of debris from onion storages eliminates possible sources of onion maggot infestation.
- Seedling treatment with 1% neem oil was found effective.
- Hoeing and weeding of onion field may reduce the infestation of onion maggot.
- Mechanical injury to the growing plant should be avoided because damaged foliage attract ovipositing flies.
- The onion maggot has many potential natural enemies but none are known to be effective.
- Effective control of onion maggot relies primarily on the proper application of in-furrow insecticides. In sever infestation, spraying of Malathion 50 EC at the rate of 0.05% or application of Quinalphos 5 g in soil at the rate10 kg per ha before sowing is effective.

4.12 White Grub (*Adoretus* spp.)

It is the most destructive pest of cold arid region of (Ladakh) which attacks on the wide range of crops such as cereal (wheat and barley), vegetable (brinjal, cabbage, cauliflower, chilly and capsicum etc.) and fruit tree (apple and apricot). It has also been seen that some species of this pest attack on forest plantation. It is serious problem in almost all areas of Leh and Kargil district. Adult emergence takes place in the month of May-June.

The grubs are 'c' shaped and white coloured having broad head and prominent legs. The adult beetles are dull brown or black in colour. Larval development completes in 50-70 days depending upon weather condition. Pupation takes place in the soil. The grubs and beetles both damage the crops.

Damage

The grubs eat away nodules, fine roots and may girdle the main root thereby killing the plants. The damage becomes evident only when the entire plant dries up. Symptom can easily be see in crops like brinjal, cabbage, chilly and capsicum.

Management

- Pre-sowing soil treatment with Quinalphos 5 g at 30 kg/ha (1.5 kg per cannal) effectively protects the crop.
- Hand picking and destruction of adults need to be done at beetle emergence; last week of May to June.
- Hand picking and destruction of grub at the time of transportation of FYM to field. FYM is the major source of spreading of grub to field.
- Deep ploughing of the field is very necessary to expose the grub to adverse weather and birds may feed upon them.
- Application of Quinalphos 25 EC or Cholorpyriphos at 4 l per ha (200 ml per Kannal) with irrigation water needs to be done for three weeks is standing crop. The insecticide should be applied in the following way:

Required amount of pesticides be applied at the point from which the water enters into the plot. This can be done by putting the insecticides in an earthen pitcher or tin pot having a hole in bottom. A cotton wick should be placed in the hole for regulation of the drops. The drops should be calibrated by the movement of the cotton wick so that required amount is applied uniformly with irrigation.

4.13 Ear Wigs

It is very harmfull pest of almost all leafy vegetables particularly spinach and Karm sag grown in Ladakh. It hides during daytime and becomes active at night. At night it causes damage by feeding on leaf.

4.14 Jassid

Jassid is also noticed in Ladakh. It sucks the sap from the leaves causing curling of the leaves. This pest is not serious in Ladakh including Leh and Kargil district.

5. Insect-Pests of Cereal Crops

Wheat, barley, buckwheat and some small millet are common cereal crops which are grown in cold arid regions of Ladakh. However, barley is grown on a wide area in Ladakh which is considered as a poor man's diet. These crops are attacked by a large number of insect-pests among which thrips, aphid, white grub and nematode are major problem.

5.1 Wheat Aphid (*Macrosiphum miscanthi* Takahashi)

Wheat aphid, *Macrosiphum miscanthi,* attacks wheat, barley, oat etc. and widely distributed in India. Nymph and adult suck the sap from ears of wheat plant. The insects are green and appear on young leaves or ears in large number during cold and cloudy weather. The winged form appears only during summer (July–August) in Ladakh.

Like other aphid, wheat aphid breeds fast during the cold and cloudy weather and reaches to the highest population in July-August in Ladakh when grains are at milky stage. The females are capable of reproducing without mating.

Damage

The aphid sucks the sap from the ears and tender leaf and that decreases the yield of the crop. The damage is more severe when the climate is cold and cloudy which is common in Ladakh in July and August. A heavy manured, well irrigated and green and succulent crop provides conducive condition for the attack of this pest which results in greater damage.

Management

- Irrigation should be done when it is required. The source of irrigation in Ladakh is snow melt water and the farmers of Ladakh irrigate the field whenever water is available. Therefore, more irritation or water logging or high moisture level provides favorable condition for the attack of aphid.
- Yellow traps can be used to monitor the pest.
- Avoid the application of excess amount of fertilizer. In Ladakh, most of the farmers use urea to accomplish the requirement of nitrogen but are not worried about phosphorus and potash, which can make the plant healthy and resistant to aphid attack.

Therefore, required amount of phosphorus and potash should also be applied to minimize the infestation of aphid.

- Spraying of 375 ml Dimethoate 30 EC or Monocrotophos 36 EC or Formothion 25 EC in 500 l of water per hectare is found effective to control this pest.

5.2 Wheat Thrips

Thrips are also known to attack on wheat and barley in cold arid region of Ladakh. However, the infestation of thrips is recorded more on wheat compared to barley. This pest occurs in Nubra valley, Kaltsi and around Leh (Thiksay, Nimmu, Suspole villages) and Kargil district which are major wheat and barley grown areas of Ladakh. This pest attacks on ear (inflorescence) of wheat and sucks the sap. Adult females are 1.0 mm long and black in colour. Nymphs are similar in shape to the adult but apterous and mainly brilliant red, with antennae, head, legs and tip of abdomen black, can be easily seen in ear where grain formation takes place.

Adult appears in June–July on ear of wheat and barley. Thrips have unique type of mouth part through which they tear the surface, inject saliva and start sucking up the cell contents. The female deposits its eggs at the axil of spikelets or at the base of the lemma of ear which still is enclosed in its sheath. Eggs are elliptical, 0.3 mm long and lay in isolation or in clusters of 3 or 4 eggs. The embryonic development lasts 9-10 days. Nymphal development lasts 20-27 days. The young nymph feeds on the lemmae and then penetrates the flower which may become sterile due to the nymph's feeding. When the grain reach the milky-ripe stage, the nymph moves into the furrow of the grain and attacks the pericarp. Once it has completed its development, the nymph vacates the lemmae and falls to the ground. Adults emerge in April–May and fly in search of wheat. They show great resistance to cold.

Damage

This minute insect attacks on ear of wheat and barley. Nymph and adult confine to the ear and suck the sap from the developing grain. Feeding on the ovaries of tender wheat leads to distortions, degenerations of grains, sometimes aborting. This has considerable effect on yield as well as on the baking quality of flour. Bites on hard wheat cause the formation of flecks on the grains. Affected areas of the pericarp turn brown and cannot be removed during processing.

5.3 Army Worm (*Mythimna* separate Walker)

It is the pest of graminaceous crops all over the world. This pest is reported from Pakistan, Sri Lanka, Burma, Thiland, Java, the Philippines, Japan, Korea, China, Australia and New Zealand and India. In India, it is sporadic pest which attacks on wheat, rice, sugarcane, maize, millets, oat, barley, rye, pea, tobacco and various wild grasses. However, in Ladakh this pest causes damage to wheat and barley and found in Nubra valley and surrounding areas of Leh and Kargil district.

The adult moths are pale brown and freshly emerged larvae are active, dull white and later turn green. The female moths lay eggs in row or in clusters on fresh plants or on the soil. The eggs are round, light green when freshly laid and turn pale yellow later and finally black. The incubation period is ranged from 4-15 days depending on weather condition. After hatching of egg, larvae eat the leaves in the central whorl of the plant. Pupation takes place in the soil near the water channel.

Damage

The fully emerged larvae spin threads from which they spread themselves in the air and then with the help of air current reach on to another plant. They feed on the tender leaves in the central whorl of the plant. As they grow they feed on older leaves also. In sever infestation, whole leaves including mid ribs are consumed and field looks as if grazed by cattle. However, this pest is not serious in cold arid region of Ladakh.

Management

- The pest can be suppressed by collecting and destroying the caterpillar.
- Excess use of nitrogen in wheat or barley crop attracts armyworm and aphid. So, balance amount of fertilizer would be helpful to minimize the pest attack.
- The damage of armyworm has been found more in late sown crop. Therefore, early and timely sown crop will have less infestation of armyworm as well as other sucking pests like aphid and thrips.
- Spraying of one l of Quinalphos 25 EC in 500 l water can control the pest.

5.4 Wheat Seed Gall Nematode (*Anguina tritici*)

Among various pests, nematodes are of great importance which cause damage to a large number of crops in India. Among them, wheat seed gall nematode, *Anguina tritici* is considered as major nematode because of its common occurrence and more damaging nature. This pest is found in almost all wheat growing areas of Leh and Kargil district. Wheat is major host of this pest in cold arid region of Ladakh, however, it also attacks some other crops like barley, maize, rye in rest of the country.

Damage

The larvae is endoparasitic and are resistance to desiccation and remain viable inside the gall in anhydrobiatic form for several years. The nematode completes one generation in a year.

Management

- Winnowing the infested seed material in open air or in front of fan separates the cockles from the grain.
- *Sieving*: Complete removal of gall is not achieved by this method but some larvae or bold range galls are retained in the seive along with seeds.
- *Hot treatment:* putting the infested seed in hot water 54-56°C for 10-12 min. was found effective in destroying larvae of *A. tritici.*
- *Water flotation or sole sedimentation:* putting contaminated seed in drum containing water or salt solution is more effective practice of removal of cockle from the grain. The gall being lighter in weight float on the surface of water and can be removed with sieve and drainage. complete removal of cockle at above 10% concentration of salt solution, sediment is more effective then water floating.
- Crop rotation is effective method to minimize the damage.
- Sowing of non-host crop for about 1-2 years in infested area has been reported to reduce the nematode infestation to great extent.
- Infested plant should be removed and destroy.

6. Insect-Pests of Oil Seed Crops

In Ladakh, mustard is important oil seeds crop which is grown on very small area. The production of mustard is very low in Ladakh which can be increased through improved irrigation, fertilizer application, use of suitable variety and through pest control. Losses caused by the insect-pest of oil seeds reduce the yield to the great extent.

6.1 Mustard Aphid (*Lipaphis erysimi* Kalt.)

Aphid has a wide range of host and attacks on almost all cruciferous crops like cabbage, cauliflower, radish etc. In Ladakh, two aphid species, *viz., Lipaphis erysimi,* and *Bracsics brassicae* has been observed which attack on mustard; among them *B. brassicae* causes serious damage to mustard crop. Unlike plains where it attacks at vegetative growth phase of mustard, in Ladakh, it attacks at flowering stage of crop. Colony of aphid on mustard can easily be observed just around the flowers. The nymph and adult suck the sap from the succulent twigs, which results in yield reduction. In severe infestation, grains become shrunken or grain formation does not take place.

Management

- As like other aphid control method. (See page no. 40 of this book).

6.2 Cabbage Butterfly (*Pieris brassicae* L.)

Cabbage butterfly attacks on mustard. The larvae can be seen on leaves just below the inflorescence. They feed on leaf, some times soft twigs may also be eaten. However, severe damage has not been observed in Ladakh (for more detail see page no. 37-38 of this book).

6.3 Bug

This pest is very common in mustard growing area of Ladakh. It attacks at flowering stage of crop and sucks sap from succulent floral parts. Sucking of sap from the floral part reduces the yield of crop.

7. Insect-Pests of Pulse Crops

A large number of pulse crops like chick pea, pigeon pea, urd, moog, lentil, pea etc. are grown in India. It is one of the rich source of protein particularly in essential amino acid, Lysine. In India, the area under pulse crop is 24.4 million heaters and annual production is about 14.5 million tones. One of the major constraints for low yield of pulse is the extensive damage by insect pest. Some of the pulse crops like pea and gram are grown in cold arid region of Ladakh.

About 250 insects have been recorded feeding on pulse crop among which pod borer, stem borer, leaf miner, cutworm, jassids and aphids are the most important. However, in cold arid region of Ladakh main pulse crop is gram, lentil and pea which grown in different areas. The area of different pulse crops varies according to crop suitable to a particular area, for example, Zanskar valley of Kargil district and Nubra valley of Leh district are major pea producing areas where pea grows on about 35 and 10 ha, respectively. Some of the important pests that cause damage to this crop are pea pod borer (*Etiella zinckenella* Treitschke) and gram pod borer (*Helicoverpa armigera*). Besides these, thrips and aphid also attack these crops.

8. Insect-Pests of Fodder Crops

Fodder is basic requirement of the local people to feed the cattle, domestic animal due to their dependence on them for ploughing, meat, milk and as a means of transport. Fodder for animal may be obtained from the trees and grasses grown in the forest area. Alfalfa is one of the oldest cultivated fodder crops which is practically growing in almost all the parts of the world because of its adaptability to warm, temperate and cool subtropical regions. In cold arid region of Ladakh it is an important fodder crop, which is grown on a wide area in the month of April-May and harvested in August-September. It is the only fodder crop on which animal husbandry is dependent to feed cattle during winter. For economic point of view it is like a cash crop in Ladakh that provides good returns to the farmer. However, a large number of insect-pest attack on alfalfa which limit the successful production of fodder. Insect-pest that seriously damage alfalfa crop are as follows:

8.1 Alfalfa Weevil (*Hypea postica*)

Alfalfa weevil is the most destructive insect-pest that attacks alfalfa in Ladakh. Its outbreak can dramatically lower the yield and quality. The adult alfalfa weevil is 0.5 to 0.7 mm long and brown snout beetle. They chew small holes in the base of alfalfa stems and deposit their eggs. Larvae hatch from eggs and crawl to the tops of alfalfa, where they feed on developing leaf buds and leaves. Larvae are pale to bright green, with a black head and white stripe down the centre of the back. The first instar larvae are yellow-green and about 2 mm ($1/16^{th}$ inch) in length. They are hard to notice unless newly formed leaves are unfolded. Larvae grow to a length of about 8 mm at the fully grown fourth instar. This usually takes 3 or 4 weeks. Mature larvae create a cocoon and pupate inside. Adult beetles emerge in late or and early July, but usually do relatively little damage. Single generation has been recorded in a year. Clover leaf weevil is sometimes mistaken for alfalfa weevil. They grow much larger and have a light brown head. The white stripe has a pinkish edge. Clover leaf weevils rarely cause economic yield loss.

Damage

Larvae initially feed within the leaf buds and then move to the tips of the plant. Damage starts as pinholes and progresses due to larvae feeding between the leaf veins. Third and fourth instar larvae cause most of the damage. Heavy infestations can result in severely skeletonized leaves that gives the field a grayish white appearance. Loss of leaf tissue results in lower feed quality (digestibility and protein) as well as reduction in yield. Accumulation of carbohydrates in the roots is also delayed.

Management

- *Early Cutting:* peak larval activity usually coincides with the bud stage of alfalfa, which is also the optimum time to cut fodder. Therefore, early cutting of fodder reduces the damage. Cutting removes large numbers of these larvae from the field.
- Spraying of Endosulfan or Chlorpyriphos @ 0.05% is effective to minimize the damage of this pest. There should be sufficient gap between the pesticides spraying and crop cutting to avoid the presence of pesticide residue.

8.2 Alfalfa Caterpillar (*Colias eurytheme* Boisduval)

The alfalfa caterpillar is a representative of a group of butterflies known as yellows butterfly. The alfalfa caterpillar has at least two colour forms, one yellow and one nearly white. Intermediate between these colors also occur. Alfalfa caterpillars can exceed in number and sometimes become pest of the crop.

8.3 Cow Pea Aphid (*Ahis craccivoar* Koch.)

Aphid belongs to the order Homoptera and family Aphididae. It is polish black with black cornicles and found on various crops and very common throughout Ladakh region where alfalfa is grown. They have piercing sucking mouthparts that extract plant sap (from phloem). Aphid feeding can result in stunting, yellowing, leaf curling and other physical disorders. It causes severe damage to alfalfa crop. At initial stage it attacks on top of the twig but later on colony is established on the whole twig.

Management

- It has been observed that availability of water to the crop influences the population build up of aphid. Therefore, irrigation should be provided whenever necessary.
- Different species of coccineallids predate upon the aphid. So, the practices which favour to increase their population should be followed.
- Spraying of Dimethoate or Chlorpyriphos @ 0.07% will suppress the aphid population.

9. Insect-Pests of Forest Plantation

9.1 Gypsy Moth (*Lymentria* spp.)

The gypsy moth is the most important defoliating insect of poplar, willow, apple and apricot in cold arid region of Ladakh. Defoliation often occurs in populated areas, especially where houses are located in previously forested land. This pest occurs in most of the areas like Nimmu, Saspole, Khlaltsi, Saboo, Thiksay, Shalsar, Diskit, Hunder, Turtuk villages of Leh district and almost all areas of Kargil district.

Male moths usually emerge before females. Males are strong fliers and are usually most active within the forest canopy during the day. They fly in zigzag patterns and can be seen searching up and down tree trunks for female moths. The female moth has well-developed wings but can not fly. The female moth releases sex attractant from her abdomen that attracts the male for mating. Soon after mating, the female deposits her eggs in a single mass, usually in the same location where she pupated. The female gypsy moth deposits her eggs in a well-formed egg mass in the month of August-September depending on weather condition and dies shortly thereafter. An egg mass may contain from 75 to 1,000 eggs. It is dull white in colour when first laid but becomes black at the time of hatching. The eggs are covered with yellow hairy substance which protect the eggs from sever winter. Gypsy moth larvae usually begin emerging from individual egg mass in the month of May. Hatching of egg is determined by temperature and usually coincides with bud-break of poplar and willow. The egg mass hatches completely over a period of 3 to 5 days depending on the location of egg masses and their exposure to sunlight. Newly hatched larvae are buff colored but turn black after a day of emergence. They remain on or around the egg mass for several days if temperature is low. When conditions are favorable, larvae leave the egg mass and climb trees in response to overhead light. First instar larvae, which usually have two or three feeding periods during the day, chew small holes within the perimeter of the leaf. When larvae are not feeding, they rest on a mat of silk, which they make on the underside of a leaf. Male larvae molt (shed the outer skin) through five stages, females through six. Each larval stage lasts for 4 to 10 days, the time depends mainly on temperature during each developmental stage.

Advance stage larvae feed at the leaf margins and characteristically stay on the tree tops, migrating to the undersides of branches and twigs when not feeding. The behavior of mature larvae changes dramatically. They feed during the night and then descend the trees at dawn in search of protective locations where they rest for the whole day. At dusk they climb on the trees again to feed. The movement up and down the tree is apparently triggered by light intensity. It is during this stage of development that defoliation can be severe, enough to be noticed. Larvae usually complete their development in July-August. Males usually pupate before females because they pass through one less larval stage. Female pupae are usually larger than male pupae. However, in dense populations where available foliage is limited, female and male pupae may be of the same size.

Management

- For monitoring the gypsy moth, burlap bands or skirts traps should be placed 5 ft above the ground level of tree trunks which can detect the presence of gypsy moth larvae before noticeable defoliation occurs. Larvae climbing down the trees early in the morning can be located and crawl beneath the burlap where they can be observed or periodically removed and destroyed which will minimize the infestation.
- Bark scrapping is effective method to destroy the egg mass which pass hibernating stage below the dead bark during the winter.
- Removal of crop residues can be a effective practice to minimize the further population buildup.
- Collection and killing of larvae which hide at the base of tree trunk effective method which can reduce the infestation of poplar and willow.

- Ground beetle larvae and adults feed readily on gypsy moth larvae and pupae and are usually most evident when gypsy moths are numerous. These beetles are very active and readily climb trees in search of their prey.
- It is very common that people grow alfalfa for fodder purpose where willow and popular are grown. Alfalfa grown around the tree provides the shelter to the larvae to rest during daytime. Therefore, avoid growing alfalfa around the tree trunk that will reduce the infestation.

9.2 Willow Scale (*Chionaspis salicis* L.)

A large number of scale insect has been reported to attack the forest plantation including many host plants. Most of the scale insects infest the twigs and small branches of host plant. The scale insect is a major problem in Ladakh which severely attacks the willow. In Ladakh, willow scale insect, *Chionaspis salicis,* is one of the major insect which attacks on willow at the point of branching from main trunk. Initial infestation starts at the junction of branches and it extends up to 8 to 9 feets. During a survey it was found that more then 65 per cent willow trees growing in Choglamsar, Thicksay, Nimmu, Suspole and in Nubra valley of Leh district are infested by willow scale insect. However, in Kargil district its infestation is comparatively less, *i.e.,* 36 per cent. This insect causes loss in wood production by sucking the sap from branches and twigs of willow. In sever infestation, cracks can be seen on infested bark which lead to drying of willow tree.

Female scale insects have a protective covering of wax, which is secreted from her body. The eggs are laid under the protective scale covering. After hatching, young crawlers move for a short period of time and settled down on suitable site and start sucking the plant sap. Identifying characteristics of this insect is that the infested branch looks white in colour and newly infested branch has dark pink appearance which is due to emergence of large number of crawlers which are pink of colour.

Damage

The nymph and adult suck the sap from the tree which reduces the vitality of tree. In sever infestation tree becomes weak and finally dries up. Before dying the bark of the tree cracks. Due to the infestation of this pest, wood production reduces.

Management

Management of this pest through pesticides is very difficult due to presence of protective covering, and scattered plantation. Following practices may minimize the infestation of scale insect to greater extent:

- The first step in successful scale insect control is prevention. Carefully examine all nursery trees prior to actual planting. If scales are present, discard the trees or exchange them with "clean" trees.
- New plantation should be maintained away from older planting where scale has been a problem.
- Removal and destruction of infested branch form older tree will control the further infestation on healthy branches as well as tree.
- Natural enemies are of great importance. A coccinellide predator of scale insect, *Chilocrous bipustulatus*, is very prevalent in most of the areas of Leh district and is voracious feeder of scale insect (Pandey, 2004a). It should promote to minimize the infestation
- Spraying of insecticide (Chlorpyriphos 0.05%) should be done when crawler emerges from the egg. This stage coincides with the starting of emergence of willow leaves.

9.3 Tent Caterpillar (*Malacosoma* spp.)

Tent caterpillar, *Malacosoma* spp., has been reported from hilly areas of Himachal Pradesh and Uttar Pradesh on a number of fruits as well as on forest trees. Besides, almond, cherry, gooseberry, peach, walnut etc., it also attacks on apricot in Kargil district. The forest tent caterpillar causes serious damage to apricot in Lato village and other surrounding areas of Kargil district. However, this pest has been reported to attack most of the forest plantations like willow, poplar, oak etc. The caterpillar webs tent like nest at the forking of the twigs in which they hide during day time. Larvae feed inside the tent and expand the size as they grow. At night, they feed on leaves voraciously leaving behind only the midribs. Defoliation leads to low bearing of fruits. Larvae are quite hairy and have a conspicuous white line down the middle of their back. There is only one generation in a year and most of the year including the winter is spent in the egg stage. Eggs are 1.0 mm long shinning and elongated thimble shaped and laid round the braches of tree as broad bands. Eggs are well protected with a coating of dark brown glue sticky substance. Full grown caterpillars are 4 to 4.5 mm long blackish brown in colour having a broad grayish band dotted with black on the dorsum. Pupation takes place in silken cocoon on stem, between leaves or in debris lying on the ground. Adults are light brown to light red in colour.

Management

- Sanitation of plantation area may reduce the infestation to some extent.
- Since, tent caterpillar lays eggs in mass around the branches so destruction of all the eggs bend at the time of pruning can be effective to minimize the damage of this pest.
- Tent caterpillars form tent in which they feed and survive. So, removal of tent is very effective which minimizes the defoliation by the tent caterpillar.
- Most of the apricot trees are smaller in size and collection and destruction of larvae can be done to suppress the infestation.
- Spraying of dimethoate 0.1% is found effective to control this pest.

9.4 Willow Leaf Beetle (*Altica* spp.)

Willow leaf beetle is an important insect-pest of willow which defoliates the plants in Ladakh. This pest occurs in Khalsti and Nubra valley of Ladakh and has also been reported from Surru valley. Adults and grubs feed on leaves. Immature larvae initially scrape green matter leaving behind main and sub veins thereby causing severe damage to willow plantation. The adult being shiny blackish blue in colour, 0.3-0.5 mm long while grubs are of 5-8 mm in size and black in colour. Adult feeds on whole leaves while grub feeds on lower surface of leaves at mid or tip side. Severe damage was observed in Khaltsi, Saspole, Nimo areas of Leh district.

Management

- Collection and destruction of adult and grub is an effective method to minimize the pest population.
- This pest does not cause severe damage to willow every year. In case of severe infestation spraying of 0.05-0.08% Chlorpyriphos may reduce the damage.

9.5 Goat Moth or Trunk Borer (*Cossus cossus* L.)

Goat moth, *Cossus cosses* L., is common in Europe, North Africa and part of Asia. It feeds on poplar, willow and about 30 other species of hardwood forest and fruit trees. The caterpillars develop in the trunks of numerous fruit trees like apple, cherry, pear, plum, olive, and other deciduous trees such as sweet chestnut, elm (*Ulmus*), oak (*Quercus*), poplar (*Populus*), chestnut (*Aesculus*), lime (*Tilia*) and maple (*Acer*). This pest has also been recorded in most of the areas of Leh district attacking on poplar.

Adult moth is 7 to 8 cm wingspan, grayish, with a massive body covered with hair. The fore wings have dull appearance and the hind wings are hairy on their basal part. Adult moth flies at dusk and at night in the beginning of summer. The female, due to a powerful oviscapt, inserts her eggs in clusters in the bark crevices. Average fecundity is 500 eggs. Embryonic development of eggs lasts 12 to 15 days. Eggs are elliptical, reddish-brown with black longitudinal striae, 1.2 x 1.7 mm.

The caterpillar is about 9 to 10 cm in length. The young caterpillar is carmine pink and the colour is darker and more pronounced in older individuals. The ventral surface is light yellow and the head is black with powerful mandibles. Glands present in mouthparts secrete an extremely smelly substance, that smells like old leather or poor quality wine. The caterpillar lives in galleries which it bores under the bark, then in the wood of the host tree. It ejects reddish granular mass formed of sawdust and frass mixed with silk which accumulates at the foot of the tree.

At the end of its development, it builds a cocoon out of wood particles and pupates. Pupa is about 5 to 6 cm long with crowns of sharp spicules which enable it to crawl towards the opening of the gallery just before emergence. The pupal stage lasts for about one month. The adult emerges from pupae and the exuviae remains stuck in the bark. The moths appear from the end of June to mid-August. Its life cycle extends normally over two years but can be prolonged to up to three years in colder climates.

Damage

The presence of very smelly worm-dust is the symptom of the attack of this pest. There are usually more than a few caterpillars in the same trunk. The caterpillars penetrate deep into the wood and can even drill into the heartwood. Some of the tree species such as elm, are extremely resistant to the attack of this pest. Fruit trees, particularly apple and apricot, are more sensitive to this pest and die rapidly. In orchards, the goat-moth generally attacks old or wounded trees in which the sap moves poorly.

Management

- Treating the trunks with Chlorpyriphos 20 EC by injection into galleries can destroy young larvae.
- Larvae pupate outside the trunk just above the ground on trunk. Therefore, regular visit should be made to destroy the pupa.

9.6 Ermine Moth (*Yponomeuta rorellus* Hubner)

The ermine moth, *Yponomeuta rorellus* is an introduced pest of Eurasian origin. It was first found in North-western Washington in the summer of 1985. This pest is also noticed in Leh and Kargil district of cold arid region of Ladakh. The ermine moth feeds exclusively on willow and apple. However, it attacks on willow in Leh and Kargil district. Higher damage of this pest was observed in almost all areas of Kargil district while less infestation of this pest was recorded in Nimmu, Saspole and Khalsi area of Leh district. This pest webs tent on the top of twigs, and if tents and larvae are numerous, serious defoliation can occur. In some cases, complete defoliation has been observed in May or June (in 2003, 2004 and 2005). The infestation of this pest has also been observed on apple which is very hard to notice.

Adults are striking silvery white in colour and fore-wings have rows of small black spots. The wingspan of the moth approaches 2.5 to 3.0 cm. Females lay 10 to 80 eggs in cluster in overlapping rows. Freshly laid egg masses are light yellow in color which transform in bright red about 2 weeks. Larvae vary from gray to cream color with pronounced spots along the sides which become more conspicuous with advancement of larval stage. At maturity, larvae are about 2.5 to 3.0 cm long. Pupae are contained within grayish white cocoons that are suspended in clusters within the feeding web.

Damage

The ermine moths, at first, mine the leaves, but towards the end of bloom they begin to feed. The webs are extended to engulf more leaves and can be as large as a tennis ball. The tent caterpillar *Malacosoma* spp., which attacks on willow and apricot in Kargil district also webs the nest which differs from the ermine moth. Unlike tent caterpillars, ermine moths will make several tents during larval development, not one large community tent. Feeding in this fashion continues till June, when pupation takes place. Pupation is striking in this pest as it is a communal event. Pupating caterpillars arrange themselves in a tightly packed cluster. Adult moths begin emerging in late June, and females lay eggs from July to September.

Management

- At initial stage larvae web two or three leaves and feed inside. So, hand picking of the infested leaves to destroy the larvae may reduce the infestation. However, this practice is not possible on big trees.
- When tents are limited in number and easily accessible, they can be very easily destroyed. Months of May-June are best time for this practice.
- Spraying of Chlorpyriphos 20 EC 0.1% can reduce the infestation.

9.7 Gall Insect

Galls are the major problem in Ladakh. In Ladakh, various types of galls have been observed on different plants like willow, poplar and apricot. The gall can be formed by the fungal pathogens, insects or by mites. Gall production is believed to result from the reaction of cambium and other meristematic tissue to stimuli produced by the gall inducers to cause the abnormal growth. Size of galls may vary depending upon host or gall inducers. Some galls may be 2 inches in diameter while some are so small that they are rarely noticed. The surface of gall may be smooth, hairy, or spiny. Generally, galls seriously harm the plant. If the trees are unhealthy or under stressed condition it will be damaged to a greater extent. Following galls has been observed in Leh and Kargil district of Ladakh region.

*a. Willow Apple Gall (*Pontania *spp.)*

Gall formation takes place on the lower surface of the willow leaf. The visible effect of the insect is swelling of affected leaf, which forms gall. This pest attacks on young shoot in favorable season, *viz.*, May to June when new leaf emerges from the twigs. Attack of this gall inducer (*Pontania* spp.) may be continuous along the shoots consequently large number of galls appear on same twig or branch. Regarding the number of galls per leaf (1 to 3) has been observed. The galls are globular or oval in shape and greenish, yellowish-green or reddish in color. The diameter of gall ranges from 2 mm to 20 mm. In severe infection, it was observed that entire leaf of twigs takes part in gall formation because of which branches become stunted and growth of the tree become retarded.

Adult emergence takes place in month of May-June. After mating females lay eggs singly into the leaves with the help of saw like ovipositor. The development of sawfly gall is well advanced before the eggs hatched. The gall is probably result of a chemical substance which is released at the time of egg laying. The eggs are creamy white (transparent) in colour, which in later stage remain white. The egg hatching takes place after 10-16 days and larva comes out inside the gall cavity. The larva is creamy white in colour with light brown head. Each sawfly gall contains only one creamy white sawfly larva. The larva obtains its nutritional requirement from the tissues of gall and survives inside till the pupation. During the development inside the gall, larva deposit excretory materials from the fresh tissue of which larvae used to feed. Full grown larva makes an exit by chewing the wall of gall and drop down on ground for pupation. Pupation takes place in upper surface of soil below the debris. Pupal stage passes through hibernation (from November to April). The larval and pupal period varies according to climatic condition. Adult emergence occurs in month of May-June. Only one generation was observed in a year (Pandey, 2004b).

*b. Poplar Petiole Gall (*Pemphigus *spp.)*

Such type of gall is formed on genus *Populus balsemifera* and observed in Nubra valley and around the Leh and most of the areas of Kargil district. The gall is caused by aphid, *Pemphigus* spp., which survives inside the gall. The formation of gall takes place at the junction of the petiole and is produced by a swelling at the base of the blade and at the top of the petiole. However it does not cause sever damage to the poplar plantation.

c. Gall Caused by Mite

Few species of mite belonging to Tarsonemidae and Eriophyidae cause the gall formation on poplar and willow. Eriophyid mites are thought to possess chemicals in their salivary secretions that act as growth regulators. When the mites feed, these chemicals are injected into the plant. Leaves may become discoloured or plant growth patterns may be changed.

The female eriophyidae, possess greater ability to induce the gall. In Ladakh on poplar, the gall can be seen in Diskit, Changmar and Hunder area of Nubra valley while on willow it can be observed in Dha, Hanu, Skurbachang areas of Khaltsi block of Leh district and Baru, Mingi, Khurbathang and other surrounding areas of Kargil district. These galls develop from epidermal cells that are infected by growth regulators injected by the mites. Although, the adult mites are microscopic in sizes, the galls they produce can be identified by the naked eye. The female mite lays eggs on any soft part of a plant with exception of the roots. The gall starts to develop soon after the eggs have been deposited and the juvenile mites begin to feed at the end of development, adult mites emerge from the gall.

Management

- Generally, galls do not cause sever damage to the plant. However, fertilizer, irrigation, pruning and other approved practices need to be carried out to keep plants healthy.
- If practical, some galls that occur on the stem and branch may be pruned and destroyed.
- After formation of gall, it is impossible to eliminate the galls or pests with insecticides because they are enclosed and well protected inside the gall. Thus, insecticides control is usually not practicable because: Most of the insecticide injury in not serious; Correct timing of application is essential to control the adult pests before the gall is formed and there is considerable environmental contamination and expanses are involved, especially on large tree.

9.8 Poplar Blotch Miner

This is the most important pest of poplar in cold arid region of Ladakh. It causes severe damage to poplar grown in Leh and Kargil district. It occurs in almost all the areas of this region where poplar is grown. The larvae mine the leaves and produce the blotch symptom on the leaves. This pest is active from June to July. The female starts egg laying in the month of June–July and after hatching larvae come out and immediately mine the leaves and survive inside till the maturity. Single generation of this pest is recorded.

9.9 Willow Aphid

Willow aphid is major problem on willow in Leh and Kargil district. Nymph and adult suck the sap from lower side of leaf which results in drying of leaf. The population of aphid increases with slight increase of temperature than normal temperature and moisture. Identification of attack of aphid on willow is the presence of honeydew below the plant canopies where few predators like coccinellide present for feeding of honeydew.

9.10 Hopper

Plant hoppers have also been observed on willow tree which suck the sap. However, they do not cause severe damage to the plant or tree.

Insect-Pests Management

- Farmers awareness of the problem and of potential solutions is the most important factor. The farmer should have ability to identify pests and symptom of damage.
- Farmers should determine the results of the control practices on per hectare cost benefit ration in terms of output (yield).
- Farmers should have idea about the cost of cultivation.
- Farmer should have ability to identify almost all important pests which cause severe loss. This will be helpful in identification of insect that is causing damage to the crop.
- Farmer should apply fertilizer and irrigation to the crop in required amount because these factors influence the pest incidence to great extent.
- Cultural practices should be followed in such way that can minimize the damage caused by different insect-pest.
- Preventive measures should be applied from early stage of crop. Insecticidal treatment should provide before the severe infestation takes place. As in the case of aphid which attacks on almost all crops and causes severe damage. This pest has ability to multiply very fast.

10. Future Work Plan

- Most of the areas of Ladakh region has not yet been explored to identify the those insect-pests which are not reported till date.

- There is need for intensive study of biology of insect-pest in relation to climatic condition and to crop phenology. This will be helpful in application of different control practices.
- There need to carry out intensive survey to identify the effective natural enemies of different insect-pest so that in future the natural enemies can successfully be utilized in IPM program.
- There is a need to explore the effectiveness of cultural practices to minimize the infestation of the pest.
- There is a need to standardized quantity of different macro and micro nutrients and irrigatiòn frequency in relation to pest attack.

References

Bhat, N.S. (1979). The role of some amino acids in the host plant on fecundity and excretion of the mustard aphid, *Lipaphis erysimi* (Kaltenback). Ph.D. Dissertation, Punjab Agricultural University, Ludhiana.

Dhaliwal, G.S. and Arora, R. (1993). Changing status on insect pests and their management strategies. In: Gill, A.K.S.; Dhaliwal, G.S. and Hansra, B.S. Changing Scenario of Indian Agriculture. Commonwealth publishers, New Delhi, India, pp. 98-145.

Diruaviam, J. and Uthamasamy, S. (1992). A new sampling technique involving yellow sticky trap for monitoring of thrips infesting different crops. *J. Ent. Res.* 16: 78-81.

Fullaway, D.T. and Krauss, N.L.H. (1945). *Hellula undalis* (Fabr.). In: Common Insects of Hawaii. Tong Publishing Company, Honolulu, Hawaii, 228p.

Harcourt, D.G. (1957). Biology of the Diamondback Moth, *Plutella maculipennis* (Curt.) (Lepidoptera: Plutellidae), in Eastern Ontario. II. Life-History, Behavior, and Host Relationships. Can. Entomol. 89: 554-564.

Lal, K.B. and Singh, R.N. (1947). Seasonal history and field ecology of the wooly aphis in the Kumaun hills. Indian J. Agric. Sci. 17(4): 211-218.

Malik, R.A.; Punjabi, A.A. and Bhat, A.A. (1972). Survey study of insect and non insect pests in Kashmir (J&K), Horticulturist 3: 29-44.

Mishra, C.S. (1920). Index to Indian Fruit Pests. Rept. Proc. 3rd Ent. Mtg., Pusa (Bihar), February 1919, 2: 564-595, Calcutta.

Narang, D.D.; Josan, V.S. and Dilawari, V.K. (1993). Control of mustard aphid, *Lipaphis erysimi* (Kalt.) on *B. juncea* Coss. and *B. compestris* L., grown in field under different soil water regimes. J. Insect Sci. 6: 195-196.

Pandey, Ajay Kumar (2004a). First record of a predator, *Chilocrous bipustulatus* (L.) (Coleoptera: Coccinellidae) of willow scale insect, *Chionaspis salicis* (L.) in Ladakh, India. Insect Environment 10(2): 86.

Pandey, Ajay Kumar (2004b). First record of willow apple gall, *Pontanina* spp. (Hymenoptera: Tenthredinidae) in cold arid region of Ladakh, India. Insect Environment 10(1): 26-27.

Pratap, Tej (2003). Horticulture in the Hills: Issue and challenges. In: Apple farming and livelihoods in the Himalyas. Jindal, K.K.; Bawa, Ranjan and Pratap, T.E., Bishan Singh Mahindra Pal Singh Press, Dehradun, pp. 17-28.

Pruthi, H.S. (1938). The distribution, status and biology of codling moth (*Cydia pomonella* L.) in Baluchistan with notes on some other insects infesting apple. Indian J. Agric. Sci. 9: 499-547.

Pruthi, Hem Singh and Batra (1960). Important Fruit pests of North-West India. bull. No. 80, 113 p., Indian Council of Agricultural research, New Delhi.

Rai, K.M. and Tripathi, G.M. (1984). Resistance of various rootstocks to the attack of apple woolly aphid, *Eriosoma lanigerrum* (Hausmam). Prog. Hort. 16: 359-362.

Rao, Mallikharjuna, Rao, N.; Muralidhara, G. and Rao, L. Tirumala (1999). Efficacy of neem products and their combinations against chilli thrips *Scirtothrips dorsalis* Hood. Pestology. 23(3): 10-12.

Sharma, P.L., Bhalla, O.P. (1963). Flat headed borer, *Sphenoptera lafertei* Obb. and its control in Himachal Pradesh. Himachal Hort. 4 (1): 1-5.

Sivapragasam, A. and Abdul Aziz, A.M. (1990). Cabbage Webworm on Crucifers in Malaysia. pp. 75-80. In: Diamondback Moth and Other Crucifer Pests. Talekar, N.S. (Ed.) Asian Vegetable Research and Development Center, 603 p.

Sivaprakasam, N., Balasubramanian, G.; Jayaraman, V.; Narayanan, A. and Venkatesan, S. (1986). *Madras Agril. J.* 73: 614-617.

Solanki, A. (1999). Diversifying Mountain Horticulture. Agriculture Today 2(5): 51-53.

Talekar, N.S., Hal, T.B.H. and Chang, W.C. (1999). Insect Environment 5: 142.

Thakur, J.R. and Dogra, G.S. (1980). Woolly apple aphid, *Eriosoma lanigerrum*, research in India. Trop. Pest Management 26: 8-12.

Wadi S.R. and Sethi, G.R. (1975). Eradication of codling moth- a suggestion., J. Nuclear Agric. and Biol. 4: 18-19.

Zaki, F.A. (1999). Incidence and biology of codling moth, *Cydia pomonella* L. In Ladakh (Jammu and Kashmir). Applied Bilogical Research 1: 75-78.

□□□

13

Diseases of Field Crops and Their Management

H. V. SINGH
Regional Agricultural Research Sub Station, SKUAST-(K) Kargil - 194 103 (J&K)

Ladakh region comprises of two districts, Leh and Kargil, is located in the higher reaches of upper Himalaya and provides distinct and natural ecosystem in cold arid climate. The cropping period is short and extends from March to September. Main occupation of the people is agriculture and animal husbandry, mainly goat and sheep. Altitude is a major factor which determines length of cropping system and type of vegetation. Almost all the *rabi* crops grown in the plains of India are cultivated in the region. Cold climate is highly unpredictable and difficult to manipulate because of several constraints. Due to increasing population and lacking of the region for almost seven months in a year pressurize to enhance the agricultural productivity per unit area so as to meet the food, fiber, fuel and fodder requirement of the people. Several systemic approaches made by the Field Research Laboratory, Leh and RARS, Leh. It is possible to identify some of the high yielding varieties in the region (Sharma and Mir, 2000). Introduction of new high yielding varieties certainly improved the production but it also increased the biotic stresses like disease caused by fungi, bacteria and viruses when grown in large area.

During the biological evolution not only plants were created before the origin of man but many other forms of life also developed that were incapable of synthesizing their own food. Evolution of organisms (the microorganism) gradually brought about changes in their capability for obtaining food from substrates and many others of them developed the faculty to live on living organisms including plants. These parasitic forms, which attack our crops more than often, upset our crop production programme. The technology for agricultural revolution was more oriented towards production rather than parallel protection. Maximum yield per unit area of land can be achieved and sustained only if along with ensured fertilizers, irrigation and seed of high yielding varieties there is also a provision for protection of the crops against it enemies. And all technologies for production and protection should be such that the small farmers get equal benefits. High yielding varieties are not necessarily disease resistant. If they are sown the resistance will be against one or two diseases, while crop is liable to suffer from more diseases. The uniformity of genetic material in the field, high, plant population, divers vegetative growth as a result of inorganic fertilizers and frequent irrigation favour survival, reproduction and spread of pathogen causing disease. Therefore, the methods that we are using for increasing production are increasing disease incidence also. As a result danger from disease will continue to increase in proportion to the development in technology of production without protection (Singh, 1998).

Management practices against disease should be planned in such a way that they can be conveniently accommodated in normal cultural practices for the crop and as far as possible the operations recommended should be such that they are effective against more than one disease. Success of management practices can be ensured only if they are recommended to and followed by majority of growers of the crop in the area. If a neighboring grower does not follow disease management practices his crop may harbor the pathogen in enough quantity to re-infest the crop in which management practices had been applied. This reduces the efficacy of the recommended practices and increases the cost of plant protection. The most appropriate approach for disease management is integration of practices directed against the pathogen (pathogen management), in favour of the host (host management) and for modification of the environment in favour of the host and against the pathogen. Genetic and chemical immunization is host management while the prophylactic measures including cultural practices and use of chemicals for protection and disinfection are interchangeably the practices for the management of the pathogen and environment.

Problems in management of vegetable disease is that the most vegetable crops differ from grain crops, in grain crops the produce is harvested only once at the end and of the most green vegetable crops like tomato, okra, cucurbits, table peas, beans, etc. harvesting is spread over a long period. This restricts the use of chemicals because of health hazards to the consumer. Ladakh is not studied well particularly the diseases of field crops, only few scattered information is available in the form of reports and some experimental findings. This information is not of the practical use of farmers and other extension agencies. The compilation of all information is needed about the biotic stresses in most of the crops grown in the region (Table 1).

Table 1: Crops of cold arid region of ladakh

Class of crop	Name of the crop
Cereal	Wheat, barley
Millets	Kagni, fox millets
Pulse	Pea, chick pea, lentil, faba bean, French bean
Oilseed	Mustard, sun flower
Vegetable	Potato, tomato, cauliflower, cabbage, knoll khole, cucumber, beat palak, spinach, onion, karam, turnip, Swede radish, chilies and carrot
Fodder	Alfalfa, oats
Other	Buckwheat

1. Diseases of Cereal Crop

Among cereal crops, barley is grown in large area than wheat since long. Due to changes in food habits, area under wheat is steadily increasing. However, the climatic constraints restrict its cultivation only to certain altitude (above 11,800 ft. ASI). Wheat also has been in cultivation in cold arid region since long. Research efforts, initiated in the recent past at SKUAST, Leh station, have resulted in the development of several high yielding varieties, namely, Kailash, Mansarover and Singchen which have resistant potential to yellow rust a major constraint in wheat production and lodging and responsive to high doses of fertilizers. After introduction of new varieties wheat productivity became double in this region. Importance of wheat in the cold arid region is subjected to survival of animal, human largely dependent of wheat productivity. Though various efforts on wheat production technology, wheat productivity is low in the region due to inadequate research on various aspects of production and protection technology and yet reach to formers levels whatever has been evolved (Table 2).

Table 2: Diseases of cereal crops

Disease	Pathogen	Host
a. Rust		
Yellow or stripe rust	*Puccinia striiformis*	Wheat and barley
b. Smut		
Loose smut	*Ustilago tritici*	Wheat
Covered smut	*U. hordei*	Barley
c. Bunt	*Tellitia carries* *T. foetida*	Wheat
d. Foliar Blight		
Spot blotch	*Bipolaris sorokiniana*	Wheat
Spot blotch	*Alternaria triticina*	Wheat
Stripe disease	*Drehcslera graminea*	Barley
Net blotch	*D. teres*	Barley
e. Powdery Mildew	*Erysiphae graminis* f.s. *tritici*	Wheat
	E. graminis	Barley
f. Ear Cockle	*Anguina tritici*	Wheat and barley
Tundu	*A. tritici+* *Corynebacterium tritici*	Wheat and barley

Barley is a traditional crop of cold arid region and consumed by people of all categories. Its cultivation in Ladakh extends from 8,500 ft to 14,000 ft. ASL. Besides a lot of variation for adaptation and different plant characters not a single germplasm has any appreciable response for resistant to yellow rust, the most devastating disease of naked barley. Even local germplasms are susceptible to lodging. Barley improvement efforts initiated at SKUAST, Leh have resulted in the development of varieties possessing moderate resistance to yellow rust and lodging as well as responsiveness to modern agricultural practices. These varieties are Sindhu, SBL-4 and SBL-8 in naked barley and VLB-1 and SBL-5 in husked barley. Besides the food of people all levels it is a major source of fodder for domestic animals and today it emerges as back bone of cold arid agriculture. After introduction of new production technologies and new varieties the production of wheat and barley increased in this area, but it has not reached up to desired level due to several biotic stresses. Among biotic stresses disease like rust, powdery mildew and leaf bight are more important in the region (Table 2). The awareness of farmers about these diseases is more resulted in loss to these crops.

1.1 Yellow Rust or Stripe Rust

Rust may debilitate and kill young plants, but more often they reduce foliage, root growth and yield by reducing the rate of photosynthesis, increasing the rate of respiration and decreasing translocation of photosynthates from infected tissues, instead diverting materials into the infected tissues. The quantity of grain produced by the rusted plants may be reduced greatly and the grains produced may be of extremely poor quality since it may be devoid of starch and may consist mostly of cellulosic materials that are of low or no nutritional value to humans.

Area and Distribution

The disease occurs less or high degree in all wheat and barley growing areas, though barley and wheat are grown in most parts of the Ladakh. The area covered by the barley crop is more so the rust occurrence is more on barley crop. The severity of the rust is comparatively higher in the cooler and humid area such as Timbis, Kanoor, Salaskut and Sangra villages of Sankoo block, and villages of Drass block having more or less infection. The occurrence is less in hotter places like Batalik, Sanjak while the area of Leh district like Chanthong, Diskit, Turtuk, Nemo, Saspol, Khalse etc. having rust infection in general on an average of 15 to 85.

Among three rusts, yellow rust is only found on both wheat and barley crops and is a major constraint for low production in the region. This rust is confined in North hills and cooler parts of India. This disease occurs at four leaf stage of the crops but more severe at ear emergence. Under severe condition whole foliage is covered with yellow uredial pustules even stalks and glumes also.

Symptoms

The green colour of the leaves fades in large strips on which row of small uredosori appear. Each row consists of series of oval, lemon, yellow pustules, arranged end to end, and each distinct from that below and above. In severe attack these serial arrangement is lost and large patches become covered with crowded pustule. In severe attack plant shows poorly developed root system (Martin and Handrix, 1966).

Environment and Incidence of Rust

This rust is highly susceptible to high temperature, optimum temperature for uredospore germination is 9 to 13° C, minimum 0-2° C and maximum 23° C. Germ tube growth is best at 10-15° C while the infection process requires optimum temperature of 8-13° C. Leaf wetness is major factor that determines the rust infection. For germination of urediospore free water on leaf surface for nearly two hours is essential. The infection of host and subsequent development in the plant population is influenced by factors such as light intensity, host variety, host nutrition and age of plant. At the time of passage of infection hypha through stomata, light intensity plays an important role (Hart, 1929). Chief stimulus for opening of the stomata is direct sun light. Incubation period may be increased by a week or more during cloudy weather. Temperature and light intensity also modify the type of reaction produce by different races of the fungus on the host (Shukla, 1953).

Host nutrition may directly influence rust development in the crop. Prasada (1964) had stated nitrogen fertilization favour rust development specially when in excess of P and K. In Ladakh, mostly the wheat and barley crop is irrigated frequently so the moisture regimes in the field is always high and increase relative humidity which favour the disease. Farmers of the region in general apply higher doses of N in form of urea as such the severity of rust disease is always high. Local varieties of both crops are small succulent and short-duration crops which are more favored by rust pathogen. Use of seed rate is higher so the plant population is also high which favored the microclimate of the pathogen.

Pathogn

The urediospores are nearly rounded, binucleate and unicellular. Their size is variable, being 23–35 x 20–25 microns. The spore wall is clourless, minutely echinulate and may posses 6-16 germ pore. The pathogen enters through stomatal opening. The mycelium extends rapidly and form club shaped haustoria and draw nutrients from adjoining cells. The teliospores are dark brown, often flattened at the tip and have two cells. They measures 35-63 x 12–20 microns. They may occupy the sorus in one group or many groups separated by row of sterile paraphyses. The further development of parasite is not known.

Origin of Races

The pathogen has physiological specialization and nearly fifteen physiologic races are being reported from India. Introduction of resistant varieties exerts selection pressures and impels the pathogen to counteract resistance, being more dominant than others. The origin of these physiologic races is probably not much different from that of morphological species. Adoption of races to become capable of infecting new varieties of the host after being grown for some time on intermediate host is not considered a mechanism of evolution of new races. Hybridization is the most common method followed by mutation in most of cereal rust. However, hybridization required the sexual stage of pathogen. Mutation in the diploid mycelium on the host is the possible source of origin of physiologic races. Mutation at the rate of 1.6 per 1,00,000 to 2,00,000 urediospore in *P. striiformis* had been detected.

Survival

Hills are the major source of survival. The pathogen survives mostly on wild grosses *Bromus japonicus* found in plain also. The volunteer crops of the family graminae at hills are harbor of the pathogen uredial spores.

Management

The impact of environmental relations on the rusts, its rapid dissemination and the frequency of evolution of new physiologic races make the management of rust a very intricate problem. The combined effect of suitable weather, sufficient inoculum, cultivation of susceptible varieties over long areas and the presence of suitable virulent races result in full fledged epidemic. The rust intensity can be minimizing by integration of all the possible means of control measures. However, breeding for disease resistance remains the most effective and practical method of rust management. The rust severity is managed by following way:

1.C lean cultivation helps in reduction of rust severity as rust develops on grass around and within the field.

2.M ixed cropping is also helpful in managing the rust. Mixed cropping of different crops like pea and mustard reduced the disease severity. Since the susceptible surface area is reduced in mixed crops, number of pustules formed are much less. This minimizes the inoculum of secondary spread of the disease. Seeds of different varieties with similar maturity and other characteristics but having different genetic composition if grown together can also serve the same purpose as rust pathogen has several physiologic races.

3.Since, nitrogen tends to increase susceptibility while potash has the opposite effect. Balance dose of N, P and K reduces rust incidence.

4.S everal fungicides had been tested and found effective in minimizing rust incidence and severity. The commercial control of rust with chemicals had many limitations. Success of protectant, non systemic fungicides such as dithiocarbomates (Zeneb, Maneb etc.) depends upon number and timing of spray operations which have to coincide with the time of introduction on inculum in the field. Some of the newly developed fungicides overcome this problem. Aujla *et al.* (1993) have reported that single spray of triadimefon or propiconazol (Tilt) proved very effective against the leaf rust and two sprays gave 100 per cent control of the rust disease. Singh and Tiwari (1979) observed application of 250 ml ai/ha of Bayleton a traizole gave good control of stripe rust.

5.The Mexican dwarf wheat when originally introduced in India had shown a high degree of resistance to stem rust. Other rust resistant varieties are HD 2278, HW 741, WL 614 for stem rust and WL 359, C 306 for leaf and strip rust (Aujla *et al.* 1986; Mutkekar *et al.*, 1985). In barley the cultivars BH 11, DL 435, P 147 and karan 435 are highly tolerant to yellow rust. In Ladakh the varieties Manasorover, Kailash and Singchen of wheat and SBL 4, SBL 8 and VBL 1 of barley showed tolerance to yellow rust.

1.2 Smut

The plant smuts, caused by Basidiomycete of the order Ustilaginals, occurs through out the world. Smuts causes serious loss than caused by the rust. In some respects, the smuts of cereals were dreaded by farmers even more than the rusts because many smuts attack the grain kernels themselves and replace the kernel contents with the black, dusty spore masses that resembles smut. Thus, the reduction in yield is conspicuous and direct, and the quality of the yield is drastically reduced by the presence of the black smut spore on the surface of healthy kernels.

A. Loose Smut

Area and Distribution

The loose smut occurs in the crops, wheat and barley but more common in wheat than barley. All the wheat and barley growing areas of Ladakh showed this disease symptom. However, humid areas are having high degree of infection. In severe condition the disease incidence goes upto 20 per cent while an average 5 -10 per cent disease is reported.

Symptoms

The destructive nature of the disease lies in the fact that every head of affected plant is converted into a black mass of spores and no grains are formed. In most of the varieties under cultivation the growth of the plants and its general appearance is not affected in any way until ear have appeared. The sori are formed in the spicklets. In early stage they are covered by a thin silvery membrane which raptures while ear is emerging out of the sheath. The loose spore mass is shed or blown away by wind leaving behind the bare rachis. Early symptoms of disease, before flowering can be detected some times by the discoloration of leaves (Beniwal *et al.*, 1990).

Pathogen

The spores are olivaceous brown, lighter on one side, spherical or occasionally oval, and measure 5-9 microns in diameter. Germination occurs in moisture by a germ tube which soon dies unless the spores have fallen on the feathery stigma of healthy flower. On germination over stigma the spores send their simple or branched germ tube in to the style. A month later all parts of the embryo except the root contain the hyphae and there is a very copious mycelium in the scutellum. Batts (1955) reported that direct penetration of the ovary caused by infective hyphae from England. In the ripe grains the hyphae of the fungus are dormant, thick-walled, oily and irregularly swollen. The species *Ustilago tritici* has many physiologic races.

Survival

The pathogen survives in the infected seed as dormant mycelium during storage. Fezer (1959) had reported that the mycelium could survive for as long as 11 years in barley seed.

Environment and Disease Development

The smut spore can retain viability for 5-6 months in open. Gera and Vashistha (1968) observed that air-dried spore can remain viable for 12 months at 0° to 10° C. However, this longevity of smut spore is not very important for disease cycle. There is direct correlation between embryo infection, colonization of seedling tissues and final expression of smut in the crop. Temperature more than 29° C during plant growth especially between internode elongation and boot emergence completely prevented spore formation in smut infected plant. If the temperature during plant growth has been kept at 19° C there is heavy incidence of smut (Dean, 1969). Optimum time for infection of flowers is when the flowers are in full bloom and the ovary is just commencing to develop after fertilization. High humidity of 60 to 85 per cent is essential. The smut spores germinate best at 28 to 20° C. Singh and Agrawal (1991) have reported that the flower of cultivars with loose spicklets are more susceptible than those with compact speckles. Low soil moisture encourages the development of the disease within the barley plants.

Management

Loose smut being an internally seed borne disease, can effectively be managed by eradication of the inoculum from the seed. Before the advent of systemic fungicides, hot water treatment on the loose smut mycelium in different varieties of wheat has also been reported. Systemic fungicides made the control of loose smut and most other cereal smuts, easy and highly reliable. Bavistin (carbendazim) has been extensively used for control of this disease. A dose of 2-2.5 g fungicide per kg seed gives almost complete control of the disease. The dose can be reduced according to load of infection in the seed lot. In Ladakh, varieties Singchen of wheat and SBL 4 of barley showed moderate resistant to loose smut.

B. Covered Smut

Area and Distribution

Covered smut is most common on barley and occurs in all the barley growing area of Ladakh. It is more common than loose smut.

The incidence of disease is varied from 5-20 per cent. The hotter areas are more severely affected. The disease incidence is comparatively higher in double-cropped areas than mono-cropped.

Symptoms

The smut dose not become evident in the field until the ears have formed. The affected ears emerge about the same time as the healthy ears, but remain shorter and are usually retained within the sheath for longer time before appearing or some time fail to emerge at all. Every ear in the disease stool and every grain in a disease ear is affected. The black spore mass of this smut remains covered by more or less firmly adhering membranes of the grain and the basal part of the glumes. As the crop approaches maturity the smutted ears become very conspicuous. The smut sori are mostly broken when the grains are threshed and then the spores get mixed with and stick to the healthy seeds.

Pathogen

The smut spores are round to elliptical, 6-9 or some times up to 11 microns in diameter and brown in colour. They appear black in mass; the colour on one side of the spore is lighter than other side. The wall is smooth. On germination, spores give out a typical 4 celled pro-mycelium. Sporidia are borne on the pro-mycelium near the septa and at the apex. Inter-sporidial anastomoses are common and bring the dikaryotic phase of the mycelium. The dikaryotic sporidia germinate to form infection thread.

Survival

The pathogen survives in the form of spore on the infested seeds.

Environment and Disease Development

Primary infection occurs on the very young seedlings when sporidia germinated on the seed coat of germinating seed. The germ tube penetrates through the young coleoptiles into the embryonic growing point below the soil. No infection occurs when primary shoot of the plant has grown out above the soil surface. Spores are produced by the cells of the hyphae and these replace the kernel. Optimum temperature for spore germination is 20° C.

Management

Since the fungus is externally seed borne, effective and economic control can be achieved by seed treatment with protectant or systemic fungicides. Bavistin at rate of 2 g per kg seed is effective in controlling disease. Resistant varieties of barley also help in managing the disease. In Leh and Kargil variety SBL 4 found resistant to this disease. While both Leh and Kargil local varieties are very much susceptible. Shallow sowing of seeds also helps in reducing disease incidence.

1.3 Bunt

Bunts of wheat are of two types, hill bunt and Karnal bunt, but hill bunt frequently occurs in cooler parts of Leh district like Nubra valley and sporadically in Kargil district. It is more prominent in temperate region of Jammu and Kashmir, Himanchal Pradesh, Uttar Pradesh and Punjab District. The pathogen causes considerable loss. Hill bunt is more common than Karnal bunt in Ladakh.

Area and Distribution

The disease is more common in Nubra valley of Leh. Some parts of Kargil districts are also affected but the severity is quit low.

Symptoms

Like other grain smuts, the disease can be seen only on the ear heads. Infected plants ripen usually more rapidly than the healthy ones and ears become more open and dark green in colour. The interior part of the grain is filled with the grey spore mass, which emits foul fishy smell due to the presence of a chemical trimethylamine. It is this smell of rotten fish that gives the name stinking smut to the disease.

Pathogen

The disease is caused by individually or jointly by both the species. The teliospores of *Tilletia caries* possess a rough wall, while *T. foetida* have smooth wall. In *T. caries,* the spores are irregular in shape and measure from 15 -21 microns while in *T. foetida* the spores are smooth and measure from 16-25 microns. The spores are borne singly, large in size and are found at the end of fertile swollen hyphae. The spores germinate by means of a short non-septets pro-mycelium, which bears apical whorl of hyaline elongated primary sporidia, less than 12 microns long. The uni-nucleate primary sporidia, fuse in pairs forming characteristic H-shaped bodies, which are dikaryotic. These germinate, producing dikaryotic hyphae, which bears sickle shaped hyaline secondary sporidia. These further germinate and produce infection hypha that penetrates host tissue intercellularly after coming in contact with cuticle of host seedling.

Survival

The pathogen, survives in the form of teliospores, is either carried on the seed or is present in the soil. However, survival in soil is not common in the region. The spores are sticky in nature and easily adhere to the seed during threshing and remain dormant on dry seed surface for long period.

Environment and Disease Development

Factors likes temperature, moisture, soil type, soil fertility, host variety, spore load and physiologic races of the pathogen, along with depth of seed sowing and seed rate have marked effect on disease development. In general, low soil temperature and high moisture are conducive for infection. The optimum soil temperature for spore germination is 18-20° C.

Management

Shallow sowing shortens the period of host susceptibility thus reducing disease. Use of disease-free seeds and resistant varieties; seed treatment with Carbendazim at rate of 2 g per kg gives very effective control of the disease. Avoid excessive irrigation particularly during flowering of wheat crop.

1.4 Leaf Spot and Leaf Blight

Leaf spot and leaf blight of cereals is caused by many species of genus Cochliobolus, *Pyrenophora* and *Alternaria*. The disease severity varied from crop to crop. In Ladakh, two pathogens are associated to leaf blight of wheat *Bipolaris sorokiniana* teliomarph *Cochliobolus sativus* and *Alternaria triticina*. However, pathogen, *A. triticina* is not reported from any other country. Wheat crop is more affected than barley.

Area and Distribution

The disease occurs in all the parts of wheat and barley growing area of Ladakh with low and high degree of disease severity. The maximum severity appeared in the hot and moist area of the region. In the Kargil district, the barley-wheat crop grown in villages of Sankoo block are heavily infected with leaf blight. The extent of damage depends on the wheat variety and the stage at which infection occurs (Chennulu *et al.*, 1967).

Symptom

The disease first appears as small oval, discolored lesions which are irregularly scattered on the leaves. The spot becomes irregular in shape with increase in size and appears brown to grey in colour. Several lesions coalesce to cover large area and cause death of the entire leaf. The lowest leaves, especially on plants along irrigation channels get the infection first. Similar symptoms appear on leaf shẹaths of ears and glumes.

Pathogens

Conidia of *A. triticina* are brown singly or chains on 2-4. they are smooth, irregularly oval, with both ends rounded, gradually tapering in to beak. The conidia germinate at 5-35° C. The conidia of *Bipolaris sorokiniana* is sub hyaline to light yellow in colour, slightly tapering, 1-7 septets. In culture media, the colour of the conidia varies from grey to olive or black. Conidia are brown on conidiophores. Germ tubes are produced by both end cells.

Survival

Both the pathogens survive as conidia on infected plants debris. When floral infection occurs, the mycelium gets established in the epidermis of peri corps and remain viable in dry seeds for an indefinite period. This seed born inoculum is the major source of prennation of the pathogen. The pathogen is soil as well as seed borne (Kumar and Arya, 1973).

Environment and Disease Development

The optimum temperature for disease development is around 25° C. When the average maximum temperature coupled with high atmospheric humidity and plant has reached the boot leaf stage disease, development of disease is maximum (Prabhu Prasada, 1966). Maximum period of 48 h of saturated atmosphere is prerequisite for successful infection.

Management

Seed treatment with 0.2% Thiram or Cereson controls seed borne iinoculum (Rout *et al.*, 1983). Foliar spray of Dithan M-45 2.5% at 10-15 days interval effectively control secondary spread of disease in field.

1.5 Stripe Disease of Barley

Area and Distribution

The disease occurs in the cool and moist area of Ladakh. Under severe condition the loss may be as high as 70-72%. In the badly affected cases the entire plant is blighted. The disease may occur on barley plant from seedling to maturity stage.

Symptom

The symptoms of the disease are conspicuous from late tillering stage until the crop is mature. The disease starts as small yellow spots on old leaf blades and the sheath. These spot elongate to stripes. These yellow stripes soon turn brown. In case of sever-infection spikes fail to emerge.

Pathogen

The conidia are sub-hyaline to yellowish-brown, thin-walled, straight cylindrical to slightly tapering having rounded ends 1-7 septets and emerge through the stomata in thick clusters. Conidia germinate from both ends and sometimes by the middle cells.

Survival

The pathogen survives in the seeds as seed born inoculum. Mycelium also perpetuates in soil on stubbles of the diseased crop.

Environment and disease spread

Sufficient moisture to wet the conidia is necessary for floral infection. Irrigation also favours the disease (Mathur and Bhatnagar, 1991). A cool, moist and fertile soil is favorable for stripe development in seedlings and developing plants.

Management

Field sanitation is recommended to reduce the soil borne inoculum; seed treatment with seed dressing fungicides at the rate of 2 g/ha seed is effective. Flier spray with 0.2% Dithan M-45 checks the secondary spread. Local varieties of Leh and Kargil showed the resistance to stripe disease but low yield.

1.6 Powdery Mildew

Area and distribution

This disease is more common in the cooler wet low laying fields. The disease occurs on wheat, barley, oats and some small millet.

However, wheat is more susceptible. The other grasses are also infected with this disease. The wheat crop grown in the villages of Sankoo block and few villages of Drass block of district Kargil and Khaltse area of Leh showed the symptoms of this disease.

Symptoms

The symptoms are much alike on all the cereals and generally limited to the upper surface of leaf. The fungus develops numerous superficial colonies. The mycelium formed flocculent growth, at first white in colour and thereafter changing to grey.

Pathogen

Conidiophores arise from hemispherical swellings on the mycelium web at right angle to the leaf surface. Each conidiophore forms a chain of 10-20 conidia. The conidia are hyaline, elliptical, uni-nucleate and 25-30 x 8-10 micron in size. In later stage, cleistothecia are globose, 160-192 x 120-130 microns black and partly immerged in the mycelial weft.

Survival

The pathogen survives as cleistothecia. It can survive for 13 years at low temperatures (Masemon and Powders, 1957).

Environment and Disease Spread

Conidia germinates at the temperature 5-9° C while as ascospores germinate best at 16-20° C. Over-managing specially nitrogen increase the incidence of the disease. The fungus is an obligate parasite and its development depends on growth of the host. Thus, the infection is favored by all those conditions which favour the vital process of the host.

Management

Best protection against the powdery mildew is a fungicidal cover of the foliage. Elasol 5 kg/ha and Karathan 2 kg/ha are effective against powdery mildew. Pathak and Joshi (1972) found complete control of the disease with the systmic fungicides. Banlet (0.1%), even up to 30 days after spraying. Some varieties are found resistant to diastase.

2. Disease of Millets

In general, the crops of small millets, kangani, foxtail millets are grown as second crop on doubled-cropped area of the region. These crops mostly grow for fodder and harvested before maturity of the crops. The disease (Table 3) occurs on these crops are not of much important and does not cause serious losses. The symptom and the control measures are similar to those followed for other cereals crops.

Table 3: Disease of small millets

Disease	Pathogen	Host
Kernel smut	*Ustilago crameri*	Seteria species (Kagni and Foxtail)
Head smut	*Sphacelotheca destruence*	*Penicum* species and *Echinochlova* species

Disease of Pulses

Pulse crops like pea, chickpea, lentil, faba bean and French bean are cultivated in the Ladakh. Among them pea is most common and grown as main as well as second crop generally in the double-cropped areas. It is a good source of protein, provides good fodder for animal also improves the nutrient status of soil, particularly nitrogen (Table 4).

Table 4: Diseases of Pulse crops

Disease	Pathogen	Host
Powdery mildew	*Erysiphae polygoni*	Chickpea, pea, lentil
Wilt	*Fusarium oxysporum* f. sp. *ciceri*	Chickpea
	F. oxysporum f. sp. *pisi*	Pea
Root rot	*Rhizoctonia solani*	Pea, chickpea

2.1 Disease of Pea

A. Powdery Mildew

The disease usually appears late in the season reaching its maximum intensity when the pods are farming. The loss is proportional to the disease intensity and varies considerably depending on the stage at which the disease occurs. Reduction in pod weight was about 24-27% if 100 per cent plants were infected (Munjal *et al.*, 1963).

Area and Distribution

The disease is more common in the hot and dry areas of the region, though it occurs in all of the pea growing areas of Ladakh.

Symptoms

The powdery mildew first appears on the leaves and then on other green parts of the plants; its symptoms characterized as formation of white floury patches on both the side of the leaf as well as on tendrils, pods and stems. When the disease advanced large area on the arial parts of the host may be covered with these white floury patches. There is increase transpiration and respiration and reduced photosynthesis.

Pathogen

Powdery mildew disease of pea caused by *Erysiphae polygoni*. The mycelium of *E. plygoni* is generally fine, persistent and rarely thick (Table 4). The conidiophores arise vertically from the superficial hyphae on the host surface. They are septet and their cells are not much different from conidia in shape. Each conidiophore bears several conidia in chain. The conidia are hyaline, elliptical, barrel-shape, unicellular and measure 25-35 x 13-16 microns. The ripe conida fall on quickly and are disseminated by wind. In late stage, several bodies (clieistothecia) may be formed as short, black, minute, scattered in the mycilial web. They are measuring about 90 microns in diameter; usually 2-8 asci are formed in each cleistothecium. The formation of cleistothesia on pea is not common.

Survival

The pathogen survives in the forms of conidia and spreads from one crop season to another. When crops in plain are not grown the pathogen survives on crops grown on hills.

Environment and Disease Spread

In general, three closely interrelated principal environmental factors, namely, temperature, humidity and light intensity a significant role in the development and spread of powdery mildew. The pea powdery mildew conidia germinate quickly in absence of water. The conidia can germinate at any temperature from 10-30° C but the maximum germination occurs at 20-24° C. Rain damages conidiophores and therefore, there is less production of conidia.

Management

Fungicidal management is the best in controlling powdery mildew of pea. Application of Karathene (2 kg/ha) is found effective in reducing the disease. Some fungi such as *Ampleomyces, Cladosporim, Tilleteopsis* and *Verticillum* have been found to parasitize the powdery mildew on host surface.

2.2 Wilt

The wilt disease of pea occurs thoughout the word where chickpea crop is grown. This disease occurs in other crops also like pea and lentil, but more destructive in chickpea.

Area and Distribution

It occurs in all pea growing areas of Ladakh. The area of Zanskar is severely affected by this disease. More than 50 per cent disease occurrence has been observed in the pea field.

Symptoms

The plants are attacked at all stages of growth. If the seedlings are attacked, further growth of plant ceases. Lower leaves become shriveled and wither away. When disease is advanced, the plants dry. Transverse sections of roots showed browning and presence of fungus mycelium in the vascular tissue.

Pathogen

The mycelium is septet, branched and intercellular conidia are hyaline and mostly 1-2 celled. The pathogen produced both macro and micro conidia. The macro conidia are fusi-form falcate and curved inwards at both ends. The pathogen forms chlamydospores.

Survival

The pathogen survives in the diseased crop debris and as chlamydospores in soil.

Environment and Disease Incidence

The disease is favoured by high soil temperature and low to moderate soil moisture. High nitrogen application increases the disease incidence. Farmers of this region sow their pea crops in upland soils which are deficient in irrigation facilities. In Kargil area, the disease occurs in late cropping seasons.

Management

The pathogen survives in soil the crop rotation is useful to manage this disease. Mixed cropping is useful for reducing the incidence.

Balance fertilizer, particularly incorporation of potash reduces the disease incidence and severity.

3. Disease of Oil Seed Crpos

Only two crops of oil seeds, rape seed and mustard, and sunflower are grown in the region. Oil seeds crop is comparatively less because of non-availability of fodder from this crops (fodder is essential for sustaining animal husbandry during long winter). Although several efforts made through agricultural department and SUKAST Leh, but the area is increasing steadily. Few fungal diseases occur on rape seed and mustard as well as sunflower but they are not of economic importance because of limited area and small fields (Table 5).

Table 5: Disease of oil seed crops

Disease	Pathogen	Host
Alternaria leaf blight	*Alternaria brassicae* *A. brasicicola*	Rapeseed, mustered
Septoria blight	*Septoria helianthi*	Sunflower
White rust	*Albugo candida*	Rapeseed and mustard
Downy mildew	*Peronospora parasitica*	Rapeseed and mustard
Powdery mildew	*Eryshiphae polygoni*	Rapeseed and mustard

4. Disease of Vegetable Crops

Ladakh the high altitude cold desert has short agriculture season (April to September). Production of maximum edible biomass with short growing season is possible through vegetable cultivation. In Ladakh, few vegetables were grown in early time mostly tuber and bulb such as turnip, radish, carrot, onion and potato and few leafy vegetables only in home gardens for own consumption. Several systemic approaches were made by FRL; and it is possible now-a-days to grow most of the transplanted vegetable. Over a period of three decades FRL has been able to identify a large number of vegetables and their verities suitable of the region.

Vegetable beings perishable commodity their shortage can not be met by imports. Available area under grain production can not be diverted for vegetable production as this region also needs grain and fodder to sustain the large animal population. This emphasizes the need for increasing the production and availability from the existing area under these crops and ensuring that the produce remains unspoiled before it reaches the consumer. Disease in the standing crops from the seedling stage till harvest and spoilage caused by microorganism during transit, storage and marketing are the major constraints in increased production and availability (Table 6).

Table 6: Disease of vegetable crops

Disease	Pathogen	Host
Late blight	*Phytophthora infestans*	Potato and tomato
Early blight	*Alternaria solani*	Potato, tomato, cauliflower, cabbage, knole khole, spinach, radish, turnip etc.
Wilt	*Fusarium oxysporum* f. sp. *lycopersicae* *F. species*	Tomato and chilies
Damping off	*Pythium* spp. *Rhizoctonia solani*	Tomato, cauliflower, brinjal cabbage, knole khole, spinach, radish, turnip etc.
Downy mildew of onion	*Peronospora destractor*	onion
Downy mildew of crucifers	*P. parasitica*	All crucifers
Downy mildew of cucurbits	*Pseudoperonospora cubensis*	All cucurbits
Powdery mildew	*Erishyphae cichoracearum* *Spharotheca fuliginia*	All vegetables
Fruit rot	*Pythium* species, *Fusarium* species, *Sclerotinia* species and *Phytophthora* species	All (cucurbits) vegetables
Viral disease		All vegetables
Blossom end or buckeye rot	*Phytophthora nicotianae* var. *parasitica*	Tomato
Sun scaled	Sun rays injury	Tomato
Common scab	*Streptomyces scabies*	Potato

4.1 Blight

Sudden necrosis of leaf, looks like burring symptoms in few days characterized as blight symptoms.

4.1.1 Late Blight

The disease occurs in both tomato and potato crops and causes severe loses to these crops; under severe condition whole plant dies and unable to produce any crops.

Area and Distribution

The disease appears in all the potato and tomato growing areas of Ladakh. The magnitude of disease varies from place to place in

Kargil district the cool and moist areas are more infected then dry areas. Similarly, in Leh district, the crops grown in Nubra valley get more infection.

Symptoms

At any time during plant growth but mostly at blooming time late blight appears on the foliage where it causes brown dead spot or extended dead areas more frequently until the leaves are killed. The dead spot appears at the tip or margin of the leaves and spreads downward or inward, the rate of spread is depending on weather. Generally, the lower leaves first show the disease but all portion are affected and there are numerous infections in severe cases. Primary lesion appeares on petioles, stems as well as on the leaves. The pathogen also infects tuber in potato crop in the field.

Pathogen

The mycelium of the fungus is endophytic consisting of hyaline, much branched, coenocytic hyphae which are intercellular with single or double club shaped haustoria. The sporangia are multinucleate (7 to 30 nuclei), thin walled, hyaline, oval or pear shaped with a distinct papilla at the apex and measure 22 to 32 x 16 to 24 microns. The sporangium may germinate directly as germ tubes but most commonly contents of the sporangium cleave to form number of zoospores which emerge through the papilla and swim away. The zoospores are biflagellate.

Survival

The fungus survives as dormant mycelium in tubers left in the field, or mycelium survives in seed tubers stored at low temperatures; are the main source of primary inoculums.

Environment and Disease Development

High humidity coupled with suitable temperature are the principal factors for germination of sporangia and further development of diseases. The optimum temperature for the growth of mycelium is 16-18° C, Rainfall at the time of sporulation of the fungus determines the level of tuber infection. For development of epidemic, night temperature below dew point for at least 4 h, temperature of 10° C along with cloudy day coupled with rainfall during next 24 h of at least 0.1 mm, are the pre-requisite factors.

Management

The disease can effectively be checked by timely spraying of fungicides. Field sanitation such as distraction of crop debris including removal of left-over infected potato tubers is helpful in managing disease. Seed tuber selection and treatment is also important; two spray of redomil MZ @ 0.255 at interval of 15 days in standing crops starting as soon as first visible symptoms appear in the field. These spray should be followed by weekly spray of 0.2% Mancozeb starting 15 after from second spray of Redomil MZ; two to three sprays of Mancozeb will be sufficient.

4.1.2 Early Blight

Like late blight this disease occurs on potato, tomato, turnip, radish, cabbage, cauliflower and other crucifer vegetables. The foliage is severely affected by this disease.

Area and Distribution

The disease is recorded from whole area of Ladakh. Despite late blight which is more severe in cool and wet area, this disease is free from any inhibition caused by weather conditions and occurs in cool as well as warm areas. Whole area of Ladakh where these crops are grown are reported to get infection of early blight.

Symptoms

This disease first becomes visible as small, isolated scattered pale brown spots on the leaflets. These spots are often covered with a deep grayish blue growth of fungus. Lowest leaves are attacked first and the disease progresses upwards. In necrotic tissue, concentric ridges develop to produce a target board effect, the most characteristic symptoms of disease. Stems may also be attacked especially in susceptible varieties.

Pathogen

The mycelium consists of septet, branched, light brown hyphae which becomes darker with age. Conidiophores emerge through the stomata from the dead centers of the spots; they are short, 50 to 90 microns long and dark coloured. Conidia are measured 120-296 x 12-20 microns. They are beaked, muriform, dark coloured and formed single. Both transverse and longitudinal septa are present in the each conidium.

Survival

The mycelium of the fungus can remain viable in dry infected leaves for a year or more. Conidia have also been found to remain viabile for 17 months at room temperature. Mycelia and conidia can thus survive in soil on diseased plant debris to cause primary infection in the next crop. Contamination of tuber with conidia in potato or mycelium is another source of primary inoculum. Perennial weed host also serves are perpetual source of inoculum.

Environment and Disease Development

The optimum temperature for conidial germination is 20-30° C. Moisture is essential for spore germination. Early blight may become serious when the season begins with abundant moisture or frequent rain followed by warm and dry weather which are unfavorable for the host and helps in rapid disease development. Weak or ill-nourished plants are more susçeptible to the disease than the plants with good vigour.

Management

Crop rotation and field sanitation are essential for effective check of disease. Timely spraying of fungicides is at present the best method of protection against the disease. Four to five spray of Dithen M 45 @ 0.2% has been found effective and economical. The number of sprys depends on diseases severity.

4.1.3 Wilt

There are two vascular wilt diseases of tomato, one is caused by species of *Fusarium* and the other is caused by *Verticellium alboatrum*. In temperate region, first is more common. Several farma species of *F. oxysporum* causing wilt diseases in other vegetables are also found.

Symptoms

Two of the earliest symptoms on young plants are clearing of venlets and druping of petiols. Later, the entire plants wilt. In field, the disease appears at any times if conditions are fevourable. Lower leaves show yellowing and die and symptoms continue to appear on successive younger leaves. The symptoms appear on the entire plants or on only few branches. Browning of vascular system can also be seen in a cross section of lower stem where disease advances.

Pathogen

The fungus is similar to species of *Fusarium* causing vascular wilt of pigeon pea etc. the mycelium septet hyaline at first becoming cream coloured with age. It produces macro-conidia, micro-conidia and chlamydospores. The macro-conidia are relatively less abundant than micro-conidia. They are fusiform, hyaline, mostly 2-3 septet and measures 25-33 x 3.5-5.5 microns. Micro-conidia are single celled, hyaline, ovoid to elipsoidal and measure 2.5-4.0 x 6.0-25 microns. Chlamydospores are formed in older mycelium.

Survival

The pathogen survives in soil as chlamydospores or some times as saprophytic growing mycelium in infected crop debris. Seeds also serve as source of primary inoculum.

Environment and Disease Development

Optimum temperature for disease development is 28° C. Soil moisture is conducive to best vegetative growth of the host is optimum for the disease development. Disease is more severe at low pH (bellow 6.4).

Management

Always use healthy seeds. Seed treatment with Bavisten @ 2.5 g per kg seed is effective. Resistant varieties also help in managing the disease. Long crop rotation is also helpful in reducing soil borne inoculum.

4.1.4 Damping Off

Damping off is a seedling disease common to most of the vegetables grown from transplanting or even when directly planted. The disease is responsible for poor germination and stand of seedling in nursery beds and often the infected seedling carry the pathogen to main field where transplanting is done.

Area and Distribution

All the vegetable growing areas of Ladakh showed the damping off symptoms at seedling stage. However, areas having high soil moisture and relatively high soil temperature favoured the disease development.

Symptoms

Damping off of tomato and other vegetables including chilies, brinjal, cabbage, cauliflower etc. occurs in two phases. Pre-emergence damping off and post-emergence damping off. In first one, the young seedling are killed before they emerge through soil surface. In fact, the seeds may rot or seedlings may be killed. Since this happens under soil surface the disease is often not recognized by the farmers, who attribute the failure of emergence to poor quality of the seeds. In the second phase (post-emergence), the mortality of seedlings is conspicuous. Infection usually occurs at the ground level or through roots, before stem hardened sufficiently. The affected tissues appear soft and water soaked. As the disease advanced the stem becomes constricted at the base and plant collapsed.

Pathogen

The fungus responsible for damping off includes species of *Pythium, Phytophthora, Glomerella, Phoma, Fusarium, Rhizoctonia, Sclerotinia, Sclerotiorum* and *Botritis*. Among these fungi the species of *Pythium* are more common in the region. The mycelium of the *Pythium* is cylindrical, coenocetic and much branched. Asexual reproduction is mostly by means of zoospores. The zoospores are kidney shaped, biflagellate. In the sexual reproduction the oogonium arises terminally less often intercalary. It is ovoid, spherical and smooth walled measuring 15-26 microns.

Survival

Most plant parasitic species of *Pythium* survives in soil through the colonized host residue carrying oospore and sporangium.

Environment and Disease Development

High soil moisture and relatively high soil temperature favour rapid development of damping off. In presence of high soil moisture there is rapid dissemination of these zoospores which attach to the seeds and germinating seedlings. The disease is most severe in ill-aerated, ill-drained soil. Temperatures between 27-31° C are ideal for disease development. High moisture, poor aeration and close planting lengthens the juvenile stage of the plants and pre-dispose them to damping off.

Management

Management of disease involves cultural practices such as thin planting, use of light soil in the nursery beds, light but frequent watering of the nursery, use of well decomposed manure, heavy doses of nitrogen fertilizer, chemical and physical disinfectant of nursery soil and use of chemical seed protectants. The seed protactent commonly used are Captan and Thiram @ 2.5 g per kg seed. Besides this, if the disease appears in the nursery beds then the application of 0.2-0.5 per cent solution of these fungicides as soil drench is quite effective. A number of bacteria such as *Bacillus, Pseudomonas, Streptomyces;* fungi such as *Trichoderma* and *Penicellium;* and some amoaebae have been reported as antagonistic (antibiosis or parasitism) to this fungi in soil (Singh and Singh, 1984) in other places but these organism have not been tested in the region so far.

4.1.5 Downy Mildew

Downy mildew is an important disease of all cultivated cruciferous and cucurbits vegetables and bulb crops like onion. The degree of infection varied with the species of vegetable.

Area and Distribution

The disease occurs in cooler and moist areas of the region. Only sporadic appearance is recorded and the damage is mostly noticed on cucurbits vegetables.

Symptoms

The first symptom on leaves resemble those of mosaic motling. The pale green areas are separated by islands of darker green on the lower side of these spots; purplish downy mildew growth appears in moist weather. The entire leaf quickly dies.

Pathogen

The mycelium is coenocytic and intercellular, small, ovate haustoria which some times develop finger like branches. The conidiophores arise singly or in group through stomata and conidia borne on conidiophore. The size and shape of conidia varies with species of pathogen associated to host vegetables. In the later stage a sexual spore, oospore, is formed.

Survival

The pathogen survives as oospore in soil. Seeds also carry the oospores (Vishunawat and Kolte, 1993). Bains and Jhooty (1976a) reported that the pathogen causing downy mildew in cucurbits survives as active mycelium on self-sown crops in shelter places or through oospore.

Environment and Disease Development

The development of sporangiophores and sorangia is affected by light and temperature. Dispersal of sporangia starts as soon as they are mature and detached. Low relative humidity favours dispersal of sporangia. Free moisture on leaf surface is essential for germination of sporangia. Low nutrition level of host plants pre-dispose them to infection.

Management

Some of the cultural practices that help in reducing losses from disease are early sowing, destruction of weed hosts, proper nutrition and reducing the amount of moisture among the vines. The fungicides recommended against this disease are Dithane M-45 0.1–0.3 per cent. Redomil MZ 0.1 per cent is quite effective in managing the disease. The number of sprays depends on the disease severity.

4.1.6 Powdery Mildew

This disease is more prevalent in this region and it occurs in all the vegetable crops grown in all the areas of Ladakh. Sometimes it becomes very serious disease of cucurbits.

Area and Distribution

All the vegetable crops particularly cucurbits are severely affected. The severity of disease goes up to 100 per cent in which whole plants are covered with the disease. No part of the region is free from this disease.

Symptoms

First symptoms are tiny, white to dirty gray spots on leaves and stems. Most extensive development of mildew occurs on the aboxial surface of the leaf. As the spots enlarge the surface becomes powdery. Black pin point bodies, representing the ascigenous stage of the fungus, appear rarely in late season. The effect of sever infection may be premature defoliation of the plant. The fruits remain under sized and often deformed.

Pathogen

Two species are associated to this disease: *Erysiphae cichoroceorum* and *Sphaerotheca fuliginea*. In both species, the mycelium is septate, hyaline superficial with uni-nucleate cells. The superficial growth of *E. cichrocearumi* is white, while that of *S. fuliginea* is rusty brown. The conidia of *E. cichrocearumi* are single-celled, hyaline, barrel-shaped and in long chains. They measures 30-64 x 13-32 microns. The mycelium of *S. fuliginea* is hyaline, occasionally brown when old. Conidia are formed in long chains. They often show fibrosin bodies and are ellipsoidal to barrel-shaped, 25-37 x 14-25 microns in size.

Survival

There are several possible ways in which the two fungi can live from one season to the next. The cleistothesia perennate from one crop season to the next. The main source of primary inoculum seems to be existed in wild or cultivated cucurbits in other locality, from where the conidia are blown to the new crops.

Environment and Disease Development

The fungi are greatly influenced by age of host plants and air humidity and temperature. Sixteen to 23 days old leaves are highly susceptible. The optimum temperature for conidial germination is 26-28° C.

Management

The cucurbitaceous weed should not be allowed to grow near cultivated fields of these vegetables. The disease crop debris should be burnt. Regular spraying of fungicides can effectively control the disease. Calixin 0.1 per cent was found effective to control the disease.

4.1.7 Fruit Rot of Cucurbits

Area and Distribution

This disease is reported in all the vegetable growing areas of both the districts of Ladakh, but more prevalent in wet areas having high relative humidity. It is not only a field disease but market and transit disease also. In poly house grown cucurbit vegetables this is most common.

Symptoms

This disease appears as a luxuriat wooly mycelial growth on the affected fruits which appear as wrapped in cotton. The disease starts with those fruits which come in contact of soil and cottony growth of mycelium developes under humid condition.

Pathogen

Several species of fungi are associated to this disease. *Pythium butleri* and *P. aphanidermatum* are the main cause of cucurbit fruit rot. All the characters of pathogen resemble as described in *Pythium.*

Survival

Pathogen survives as oospores in soil.

Environment and Disease Development

These fungi are soil inhabitants, living in pre-colonized tissues of the host or mostly as oospores. The important pre-disposing factors are abundant moisture and high temperature. A slight mechanical injury to fruit skin due to contact with soil particles facilities infection. If the fruit can be kept away from soil surface, chances of infection are reduced.

Management

It is not advisable to apply fungicides to prevent fruit rot. Chemical disinfectant of the soil on a field scale is uneconomical. Change in methods of cultivation of cucurbits so that the fruits do not come in contact with soil and also hygienic transport and storage are the methods to avoid the fruit rot.

4.1.8 Viral Disease of Vegetables

A variety of mosaic symptoms occur on different members of the cucurbitaceae. In Ladakh, these mosaic diseases are common on almost all cucurbit vegetables and often cause heavy losses. There are several viruses involve in causing mosaic type symptoms in cucurbits. The common mosaic is caused by the cucumber mosaic virus. In addition, cucumber green mottle mosaic virus is also common. Some of these viruses attack, in addition to cucurbits, a large number of other host plants including, spinach, bean, carrot, chilli, egg plant, tomato and potato etc.

Symptoms

The nature of symptoms depends on the host and the strains of viruses associated with the disease. If infection is early, the symptoms are aggressive and losses are high. When plants are attacked soon after emergence, the cotyledons are yellow and seedlings show symptoms of wilt. In older plants, the symptoms appear first on young upper leaves. There are alternate green and yellow patches on the leaf. These spots are irregular in shape and enlarge rapidly ultimately covering the entire leaf. The diseased leaves are mottled, deformed, small and some times curled downwards. Veins and veinlets also turn yellow. Young fruits are rough mottled and deformed. Often these fruits are white and much smaller in size.

Management

In absence of resistant varieties following measures can reduce severity of cucurbits mosaic viruses: destruction of diseased hosts, including weeds and use of seeds from virus-free crops minimize the disease. Use of insecticides such as Matasystox 0.1 per cent to suppress the insect vactors, because the viral disease is mostly transmitted through insect-vectors is advocated.

4.1.9 Blossom End or Buckeye Rot of Tomato

Area and Distribution

The humid areas of Ladakh are severely affected with this disease. The warmer areas with rainfall also showed this disease. The tomato crop grown in all the areas of Ladakh more or less showed this symptoms.

Symptoms

Spots on tomato fruits are marked with pale brown, concentric rings resembling slightly the markings on a buckeye nut. These spots may be small or they may cover a major portion of the fruit surface. There is no rotting of the skin but internal flash may be discolored up to the core. When green fruits are attacked they show brownish circular spots at the blossom end, shrink and are mummified. Leaves may be attacked but leaf blight is not a common symptom.

Pathogen

The hyphae of the fungus are tough, up to 9 microns wide, irregular in width but without marked hyphal swelling. Sporangiophores are thinner than vegetative hyphae, irregularly sympodially branched. These sporangia are papillate, occasionally have more than one apex. Their maximum size is 50-40 microns. Oogonia are usually produced in single culture though often very sparsely or not until after some weeks. The fungus also produces chlamydospores abundantly. These are less than 25 to up to 60 microns.

Survival

The pathogen survives mainly as oospore and chlamydospores in soil. The tomato seedlings get infection in nursery beds carry the infection to the main field without showing any symptom.

Environment and Disease Development

Infection occures only when the host cells garged with water. Fruits in contact with soil, which has been saturated with water for about 12 h, readily catch infection. The maximum disease incidence occurs at temperature of 20 to 25° C with average relative humidity of more than 60 per cent. High moisture is greatly responsible for the disease incidence and development.

Management

Crop rotation is helpful in reducing disease. Good soil drainage, selection of light soil, seed treatment and spraying in nursery beds as well as main fields are important for managing buckeye rot. Seed treatment with capton at rate of 2 g per kg seed is effective in reducing disease incidence. Three spraying of Mancozeb 0.25 per cent in standing crop gave good control of the disease. Staking of plants to prevent contact between leaves or fruits and soil, removal of lower leaves and fruits help in prevention of the disease and its spread.

4.1.10 Sunscald of Tomato

This is non-parasitic disease of tomato and occurs in all the tomato growing areas of Ladakh. Tomato fruits exposed to direct sun light and dry wind often suffer from sun scald, when foliage become sparse because of defoliation of leaves by any leaf spot or blight disease. Green fruits scald rapidly on bright sunny days. Blotches of dry skin appear on the exposed surface. The dry skin is at first white and then becomes brown. If fungi and bacteria enter this portion there may be fruit rot. To avoid this condition proper shading of plants should be arranged.

4.1.11 Common Scab of Potato

Common scab of potato is of world wide occurrence. In Ladakh, all the potato growing areas harbour the pathogen associated to this disease. In many fields, the percentage of scab bearing tubers are seen up to 40 per cent. The greatest loss from this disease is the reduction in market values of tubers due to rough and blemished skin. When scab is deep seated there is considerable loss in peeling of tubers.

Area and Distribution

All the potato growing areas of Ladakh are severely affected to the disease. Sankoo, Drass, Paschkum, Zhanskar areas of Kargil district and almost all the areas of Leh, where potato crop is grown, showed the symptoms of the disease. More than 50 per cent disease incidence have been observed from severely infected field.

Symptoms

Two types of symptoms are formed on the tubers, the shallow and the deep lesion of scab. In shallow scab, the affected tubers show superficial roughened areas, some times raised above but often sunken slightly bellow the plane of the healthy skin. The lesions consist of corking tissues which arise from abnormal proliferation of the cells of the tuber peridium due to attack of the pathogen. The lesions vary widely in size and shape, only sometimes darker than the healthy skin. In deep pitted scab the lesions are 1-3 mm or more in depth and are darker than the lesions in shallow scab. They also are corkey and may join together.

Pathogen

The pathogen is an actinomycetes, forms filaments, often assuming fungal appearance. The filaments are 0.5 to 0.2 microns in diameter. They form a well-developed branched mycelium that dose not fragment readily. The aerial mycelium forms chains of 3 to many spores at maturity. The spores are cylindrical, 0.8-1.0 x 1.2-1.5 microns in size; these spores germinate by germ tubes.

Survival

All the actinomycetes are basically soil inhabitants. The *Streptomyces scabis* persists in soil indefinitely as resting spore.

Environment and Disease Development

The young tubers during formation of internodes and before suberization of lenticels are most susceptible to infection by scab organism. Infection may occur through newly formed lenticels, stomata and wounds and directly through cuticle when cuticle is thin. The pathogen is susceptible to acidity. With the pH range of 5.2 and 8.0 the disease severity increases with rising pH. The disease can develop at a temperature range of 11-35°C. Optimum temperature for infection is slightly below 20°C and for lesions development slightly above 20°C.

Management

Tubers with deep lesions should not be used for seed. Keeping the field well irrigated from the fifth week after planting up to the ninth week can reduce infection. Seed treatment with 0.01 per cent streptocycline, reduces the disease incidence and severity.

5. Disease of Fodder Crops

Only two fodder crops are grown in the region, alfalfa and oat. In alfalfa, powdery mildew occurs late in season mostly in the second cutting and any chemical is not recommended for control of this disease because this crop is directly used as fodder for the animals. Similarly, oat is also grown for fodder and before maturity of grain this crops is harvested, though some grain and leaf diseases occur but these are not economically important.

Disease	Pathogen	Host
Powdery mildew	*Erysiphae polygoni*	Alfalfa
Loose smut	*Ustilago avenea*	Oats
Covered smut	*U. kolleri*	Oats

6. Disease of other Crops

Buck wheat is another crop grown as second crop in limited area. There is no disease reported which causes any significant loss to this crop.

References

Aujala, S.S.; Sharma, Y.R. and Nanda, G.S. (1986). Relative tolerance of wheat varieties to stripe and leaf rusts. Indian Phytopath. 39: 46-49.

Aujala, S.S.; Basandrai, A.K. and Ratton, G.S. (1993). Chemical control of leaf rust of wheat. Indian J. Mycol. Plant Pathol. 23: 209-213.

Bains, S.S. and Jhooty, J.S. (1976a). Overwinterring of *Pseudoperonospora cubensis* causing downy mildew of muskmelon. Indian Phytopath. 29: 213-214.

Batts, C.C.V. (1995). Infection of wheat by loose smut, *Ustilag tritici.* Nature (London) 175: 467-68.

Beniwal, M.S.; Gupta, A. and Karwarsa, S.S. (1990). Effect of loose smut on the growth and morphology of wheat cultivars. Indian J. Mycol. Plant Pathol. 20: 41-43.

Chennulu, V.V.; Singh, A. and Joshi, L.M. (1967). Estimation of loses due to *Alternaria* blight of wheat caused by *A. triticina.* Int. Symp. Plant Pathology New Delhi pp. 4.

Dean, W.M. (1969). Effect of temperature on loose smut of wheat. Ann. Appl. Biol. 64: 75-83.

Fezer, K.D. (1959). Survival of loose smut in Barley seed. Plant Dis. Rep. 43: 1287.

Hart, H. (1929). Relation of stomatal behaviour to stem rust resistance in wheat. J. Agr. Res. 39: 924-948.

Kumar, V.R. and Arya, H.C. (1973). Certain aspect of perpetuation and reoccurrence of leaf blight of wheat in Rajashan. Indian J. Mycol. Plant Pathol. 3: 93 -94.

Martin, N.E. and Hendrix, J.W. (1966). Influence of stripe rust on root development in wheat. Phytopath. 56: 149.

Mathur, A.K. and Bhatnagar, G.C. (1991). Barley stripe: Its prevalence and loss in grain yield. Indian J. Mycol. Plant Pathol. 21: 280-282.

Moseman, J.G. and Powders, H.R. (1957). Function and longevity of cliestothecia of *Eryshiphae gramimis* F. sp. *hordei*. Phytopath. 47: 53-56.

Munjal, R.L.; Chennulu, V.V. and Hora, T.S. (1963). Assesment of losses due to powdery mildew (*Eryshiphae polygoni)* on pea. Indian Phytopath. 16: 260-267.

Mutkekar, M.L.; Bhangale, G.T.; Patil, J.V. and Kalekar, A.R. (1985). Source of resistance to stem rust of wheat. Cereal Rust Bull. 13: 192.

Prasada, R. (1964). Host parasite relationships in rusts. Indian Phytopath. Sc. Bull. 2: 57-72.

Rout, J.G.; Guldhe, S.M. and Wangikar, P.D. (1983). Seed borne infection of *Alternaria tricina* in wheat and its control. Phytopath. 36: 274-277.

Sharma, J.P. (2002). Crop Production Technology for Cold Arid Region. Kalyani Publishers, New Delhi. 236 p.

Sharma, J.P. and Mir, A.A. (2000). Dynamics of Cold Arid Agriculture. Kalyani Publishers, 338 p.

Shukla, T.N. (1953). Factors affecting cereal rust reaction. II Variability due to light. Indian Phytopath. 7: 43-52.

Shukla, T.N. (1955). Factors affecting cereal rust reaction. I variability caused by temperature. Indian Phytopath. 6: 67-79.

Singh, A. and Agarwal, V. K. (1991). Floral morphological characters in relation to incidence of loose smut of wheat. Phytopath. 44: 514-516.

Singh, A. and Tiwari, A.N. (1979). Control of yellow rust of wheat by single spray of fungicides. Indian J. Mycol. Plant Pathol. 9: 264-266.

Singh, D.V. and Singh, V.K. (1971). Control of leaf blight of wheat with fungicides. Indian Phytopath. 24: 694 -697.

Singh, R.S. (1985). Disease of Vegetable Crops. Oxford & IBH Publishing Co. Pvt. Ltd. New Delhi, 406 p.

Singh, R.S. (1998). Plant Disease. Oxford & IBH Publishing Co. Pvt. Ltd. New Delhi, 416 p.

Singh, R.S. and Singh, U.S. (1984). *Pythium* in host and soil. Pp. 227-271. In: Progress in Microbial Ecology, Mukharji, K.H.; Agnihotri, V.P. and Singh, R.P. eds. Print House, Lucknow.

Vishunawat, K. and Kolte, S.J. (1993). *Brassica* seed infection with *Peronospora parasitica* and its transmission through seed. Indian J. Mycol. Plant Pathol. 23: 247-249.

□□□

14

Poultry Farming: A Potential Venture

BAL GANGADHAR ROY, DEEPOK GOGOI AND MANZUR AHMED
Defence Institute of High Altitude Research (DIHAR), C/o 56 APO

The term "Poultry" now includes a number of avian species such as chicken, duck, turkey, geese, guinea-fowl, quail, ostrich and emus domesticated for economic purpose. However, chicken accounts for more than 90% of the total poultry population of the country. Chicken and Ducks are reared for egg and meat production; turkey, guinea fowl, geese, ostriches and emus etc. are maintained only for meat production. The growth of poultry industry in India has been phenomenal in last two decade and has reached a stage of self-sufficiency and sophistication. Today, Indian poultry industry ranks 4th in egg production. However, the picture is very grim in cold arid part of Himalayas. Here, still it is a backyard venture with very less number of birds. The non-existence of any commercial hatchery, problem in transportation of day-old-chicks (DOC) and unavailability of poultry feed acts as the major obstacle in growth and popularization of poultry farming in this region.

Due to sustained efforts of State Gov., animal husbandry department, N.G.O'S and few agro-animal research organizations, the awareness level and importance of poultry farming among farmers has increased. The presence of Army, increased arrival of tourist, changed life style and food habits create huge demand of poultry meat and eggs but the local availability is very less and transportation becomes difficult in winter. Field agricultural activities ceases in winter so farmers can go for indoor occupation like poultry farming. In spite of various constrains, it indicates big scope for poultry industry in this region. Few farmers have come up with small scale broiler farms in cities like Leh and Kargil. At backyard level most of the farmers opt for 15-20 egg laying or dual purpose birds. Poultry population in Ladakh district is reported to be 8,267 (Bhagat, 2003) to 0.12 lacs (Phuntsog, 2003). For Kargil district it is 2,61,171 in the year 1999-2000 (Gupta, 2003).

Constrains

To popularize poultry farming in cold-arid region it is imperative to understand the constrains and difficulties of poultry production. Following are a few major hurdles:

a) *Religious sentiments*: Major population of this region is Buddhist community. They are vegetarian and avoid killing of animals. Their interest in poultry is limited to keep egg laying birds.

b) *Extreme cold and hypoxic environment*: A-day-old chick of any species of poultry is physiologically prone to adverse effects of cold stress because of its poor homoeothermic limit. The temperature range for optimal production is 17-23°C (thermo-neutral zone) only. Any variation of temperature above or below this range affects the growth and production parameters of birds. In this region, the temperature fluctuates from +35°C in summer (day temperature) to -25°C in winter (night temperature) which exerts huge stress on homoeothermic control system of birds as well as huge expenditure on maintaining the optimum temperature inside the house in extreme winter. The altitude varies from 5,000 ft to 12,000 ft. It directly affects the hatchability percentage of chicks because of reduced partial pressure of oxygen in air. This single factor is mainly responsible for non-existence of any hatchery in this region and in turn non-popularization of poultry farming. Cold stress coupled with hypoxic and dry environment (RH =35-45%) reduces the optimum production and livability potential of poultry birds.

c) *Un-availability of day-old chicks and poultry feed:* Due to non-existence of hatchery and poultry feed shops in nearby place, it becomes difficult to procure DOC and ready made poultry feed in winter from other parts of the country; when road route is blocked and the region remains cut-off for at least 6 month. Difficulty of transportation in high altitude mountains and high transportation charges add to the burden of farmers. Local availability of poultry feed ingredient is minimal and there is hardly any source to get them at regular interval at local level, along with non-availability of suitably developed poultry breeds that are adapted well to the region with optimum production potential.

Germplasm

Chicken Breeds: Important representative breeds of chicken are listed in Table 1 with their identifying characters. The varieties used for modern breeding are:

a) *Single comb white leghorn*: Most extensively used for commercial white shelled egg production.

b) *Single comb RIR*: for brown shelled egg production by crossing its male line with Barred Plymouth Rock female line.

c) *New Hampshire*: for meat production as female line.

d) *White Plymouth Rock*: used for broiler production.

e) *Cornish* : highly valued broiler parent male line.

f) *Barred Plymouth Rock*: for auto sexing and as female brown egg producing line.

Class: Denotes groups of breeds which have been developed in certain regions or geographic areas.

Breed: Refers to a group of fowls with a common ancestry and having similarity in shape, growth, temperament, shell colour of egg and breed true to their type.

Variety: It is a sub-division of breed. This term refers to fowls of same breed differing in plumage colour and comb type.

Strain / line: Refers to a group of fowls identified by a given name and produced by a breeder for a particular purpose. Strains in a breed/ variety may look alike, but differ in their productive performance.

Considering the climatic condition and constrains, selection of appropriate germplasm for broiler and layer becomes very important. Germplasm should be able to withstand harsh climatic condition and produce optimally. Research findings of FRL (DRDO), Leh, DARL (DRDO), Pithaoragarh and J&K state A.H. department suggests commercial egg laying strains of WLH and RIR are good enough for egg purpose. These two breeds survive well with good production performance even in backyard of farmers. F1 strains of slow growing commercial broilers are suitable for backyard level of farming. In poultry houses with better temperature control, fast growing venky broilers have been found more economical by virtue of their rapid body weight gain and lesser FCR. But in general, cockerels of slow growing strains (Chabro, CPBF, Chandigarh) and dual purpose Indian varieties like Dhanraja, Vanraja will be more appropriate because of their hardy nature and better survivability at farmer's level. Buragohain *et al.* (2006) reported average body weight of Vanraja chicks at day-old and at 6th week of age as 37.97 ± 0.60 g and 810.13 ± 29.60, respectively at the altitude of 5,200 ft, above MSL. The average FCR was 2.64 ± 0.04 kg at 6th week with 22.63% total mortality during this period. This shows that the growth of broilers at high altitude is lesser compared to plains and they consume more feed/unit body weight gain. To encourage poultry farming at backyard level there is need to introduce dual purpose broody Deshi birds. Till the time hatchery is established or facility for DOC is made available, at least farmers can perpetuate backyard poultry keeping by getting chicks hatched from Deshi brooding birds even in winter.

Other species of poultry which can be reared in this region (as reported by research findings of FRL) are Khaki Campbell (layer), Super Vigova (broiler) as Duck breed, Japanese quail, Guinea fowls and Turkey. Wild varieties of Duck (*Anas platyrhyanchos*), Quail (*Coturnix coturnix*) and Goose (*Anser anser*) have been reported to thrive in wild areas of Ladakh region. These hardy birds show good tolerance to harsh climate (Table 1).

Table 1: Important characters of representative breeds of chicken

Class	Type of comb	Colour ear lobe	Skin	Shank colour	Feather	Colour of egg	Used for (Variety)
American							
Plymouth Rock	Single	Red	Yellow	Yellow	No	Brown	Broiler (White) Autosexing (Barred)
Wyandotte	Rose	Red	Yellow	Yellow	No	Brown	Dual
Rhode Island red	Single & Rose	Red	Yellow	Yellow	No	Brown	Dual
New Hampshire	Single	Red	Yellow	Yellow	No	Brown	Dual
English Breeds							
Austrolorp	Single	Red	White	Slate	No	Brown	Layer
Cornish	Pea	Red	Yellow	Yellow	No	Brown	Broiler (Male line)
Dorking	Single & Rose	Red	White	White	No	White	Broiler
Orrington	Single	Red	White	White	No	Brown	Dual
Sussex	Single	Red	White	White	No	Brown	Layer
Mediterranean Breeds							
Leghorn	Single & Rose	White	Yellow	Yellow	No	White	Layer (White)
Minorca	Single	White	White	Dark slate	No	White	
Ancona	Single & Rose	White	Yellow	Yellow	No	White	
Andalusian	Single	White	White	Slaty blue	No	White	
Asiatic Breeds							
Brahma	Pea	Red	Yellow	Yellow	Yes	Brown	Brahmaputra valley, Broody, poor Layer
Cochin	Single	Red	Yellow	Yellow	Yes	Brown	Shangai Fowl, China Broody, poor Layer
Langshan	Single	Red	White	Bluish black	Yes	Brown	China, Broody, poor Layer
Indian Breeds							
Aseel	Pea	Red	Yellow	Yellow	No	Brown	Fighting, Meat, Good Mother
Chittagong	Pea	Red	Yellow	Yellow	No	Yellow	Dual
Kadaknath (Kalamasi)	Pea	Purple	Black	Slate	No	Brown	Meat, high protein
Bursa	Pea	Red	Yellow	Yellow	No	Brown	Poor Layer

Housing System

The housing of poultry birds is difficult in cold arid climate especially in winter. To maintain the optimum temperature of 25 to 35°C inside brooder and 15-20°C inside grower/layer house, special care, accessararies and some modifications in design of houses are required. The systems of housing prevalent in this region are semi-intensive to backyard. The types of houses used by farmers are given below:

1. Over ground mud walled polyshed: Made of double wall mud bricks whose inner 1/3 portion from floor is cemented. Roof is made up of bamboo, wooden, grass and locally available materials. 2/3 part of one side wall facing the sun is formed of polythene sheet supported by wooden logs at the centre of which is glass window. This whole portion is again covered with another white polythene sheet to trap sun light and thus maximum heat inside house. Inside, half part of wall is further covered by blanket and plywood at two different faces of wall to preserve the inside heat. False roofing has been tried with a thick woolen sheet just at half the height of house. At the centre of the house, there is local bukhari using wood as fuel, fitted with chimney and wrapped by wire mesh to prevent the birds coming close to bukhari. In peak winter when out side temperature falls to -4°C at day time, the inside temperature at birds level was noted to be +8°C without bukhari in Nubra valley of Ladakh.

2. Semi-under ground mud/stone walled house: To protect the birds from extreme cold, different types of underground and semi-underground sheds are used by poultry farmers those are made of mud, mud brick, stone and concrete. Some of these main houses are provided with poly houses as extension to expose the birds to direct sun light during day time. These semi-underground sheds are constructed of double wall mud or stone up to ground level and after that mud bricks are mostly used to construct the open part of wall as well as the roof. The roof is also double layered with mud bricks and inside part of ceiling is supported with wooden logs and locally available packing materials. False roofing is made by covering the inner aspect of roof with polythene or card board or plywood. One or two windows are

made at the interface of wall with ground. It is either fitted with wire mesh or adjustable wooden or glass panel opening outside or covered by curtain or polythene. Floor, corners and lower one third part of wall is cemented for easy cleaning and protection from rodents. Mud bricks are reported to be better insulator than stone. These houses are of low cost and affordable to farmers. Another type of semi-underground house has been tried by FRL. Walls and floor are made of concrete bricks and asbestos sheets have been used for roofing. At the centre of the roof, polycarbonate sheet has been provided to trap solar radiations. Double door is there to prevent incoming of air draft. In both type of houses, one opening is left open in roof for the passage of chimney of bukhari.

3. Trombay wall shed : This is scientific adaptation of local poultry house for farmers. This has been designed by FRL to cater the needs of farmers. It is made of double wall mud bricks and plastered with mud from all sides. This over ground shed is provided with two windows on both wall and floor is cemented. Lower one third of wall is covered with galvanized aluminum sheet for easy cleaning. Roof is same like farmers shed. Extra door has been provided at right angle of the first one. Some modification in this structure has also been tried by providing large double layered glass windows on whole face of wall facing sun. Outer side of wall is painted black and poly carbonate sheet at the centre of roof. Many small windows at the lower part of wall or at the bird's height have been provided to meet the ventilation requirement. Window doors open outside, partially or fully as per requirements. All these modifications augment the solar heat trapping capacity and improved ventilation of house. This is easy to construct with locally available materials and it costs less. Experimental trials for growth, Feed Conservation Ratio (FCR) and mortality of broilers in this house as conducted by FRL, show satisfactory results even in peak winter.

4. Solar poultry house : This house has been designed by FRL to provide maximum comfort, maximum protection from cold, better ventilation and hygienic condition to birds. This is highly costly but most efficient in maintaining optimum temperature and ventilation inside house.

Low-Cost Ideal Poultry House in Cold Arid Region

1. House can be semi-underground to over ground provided ventilation is optimum.
2. It should be double walled with some space for air.
3. Mud brick walls are better insulator and cheap.
4. Roof should be double layered with space for poly carbonate sheet to trap solar radiation in day time.
5. Outer surface of walls facing sun should be painted black.
6. Big glass walls or glass windows should be provided to trap heat in day.
7. Additional covering of black and white polythene over walls/ windows in day and night time will increase the heat retention capacity of shed.
8. Double layered curtains made of gunny bags can used as double door to prevent the incoming of chilled air as draft while opening and closing of door.
9. False ceiling is must.
10. House should not be very long, very wide, and very high. Preferably, smaller houses are better for maintaining temperature and humidity.
11. Ventilators should be so constructed to prevent direct incoming of wind.
12. Packaging materials that can be used to fill walls and roofs are wheat straw/chopped.
13. Lucerne hay/any cheap dried stacks of herbs/plants mixed in mud or any other insulator.
14. Inside, one third part of walls should be cemented/covered with tin/aluminum sheet and corners rounded.
15. In night, curtains can be used to cover glass windows/poly carbonate sheet to avoid chilling or plywood made folded shutters can be tried to cover windows/poly carbonate sheets.
16. The house should face South or South-east to get maximum sunlight.

Feed

Poultry birds are monogastric and their feed consist of mostly grains. Food grain production is limited in this region. Farmers are provided subsidized human ration through Public Distribution System (PDS). But for animal ration there is no such system and food grain/ poultry feed has to be purchased. Poultry farmers face three major hurdles in feeding their birds:

1. Limited Availability / Un-availability

Wheat, Barley, Pea and Mustard are the few major local food grains, which are produced by farmers in limited quantity. Very little amount of surplus grains are available with the farmer to spare for poultry. They import either poultry feed ingredients like maize, soyabean and wheat bran or purchase ready made feed from other parts of country, when road route remains open, i.e., summer only. Transportation and storage of feed from distant place exerts extra economic pressure.

2. No Feed Formulation / Improper Feed Formulation

Balanced ready made feed is not always procured by farmers. To save money, they purchase few ingredients and supplement that with locally available feeds. This unbalanced feed decreases the weight gain in birds and increases feed conversion ratio. Ready-made starter feed is given to chicks and maize-based-locally-prepared, poor quality feed is provided to birds above 4th week of age. In winter, at the time of scarcity of feed, farmers make their own poultry feed with locally available vegetable, cereal, and some other things. This includes spinach, cabbage, onion, garlic, Lucerne leaves, crushed turnip, barley, gram, wheat flour and mustard oil preparation.

3. Un-optimized, Energy and Protein Percentage Level in Feed

Scientific feed formulation with optimum M.E. (k. cal) and C.P. % is very essential for this region. Because of cold stress and high altitude, the ready-made feed available in the market would not be appropriate for birds of this region. Ascites becomes a major problem in fast growing broilers if energy and protein level in feed is not optimized. Depending on season (winter/summer), quality and type of day old chicks (fast/slow growing), scale of feeding (ad-lib/ restricted) and incidence of ascites, feed formulation needs to be adjusted and energy and protein level along with feed additives has to be optimized.

Feeding of Poultry

Feeding of poultry constitute major managemental practice in poultry production, since feed constitutes 60-70% of total expenditure. In cold arid hypoxic environment of Himalayas, poultry birds attain lesser body weight gain and consume relatively more feed to meet energy needs for endogenous heat production. On the other hand excess feeding leads to ascites and poor FCR. Therefore, feeding scale and nutrient density of the feed for broiler and layer should be selected appropriately. The major nutrients required for poultry are: (a) water, (b) protein, (c) carbohydrate, (d) fat, (e) mineral, (f) vitamin and (g) feed additives.

Carbohydrate and fat are principle source of energy. Fat provide 2.5 times more energy than carbohydrate and essential fatty acids like linoleic, linolenic and arachidonic acids. But the digestibility of fat and percentage inclusion in poultry ration is very low (4-8%). Protein is essentially required for body growth and for supply of essential amino acids which are not synthesized in required amount in the body. Poultry requires arginine, glycine, histidine, leucine, lysine, isoleucine, methionine, phenyl alanine, threonine, tryptophan and valine as essential amino acid and of these arginine, lysine, methionine, cystine and tryptophan are critical amino acids, since they are usually deficient in poultry diet. Lysine and methionine are considerd as limiting amino acids and must be included in diet. Cereal grains are usually low in critical and limiting amino acid. Thus, it is must to include a good proportion of animal proteins to ensure the inclusion of all critical amino acids in diet. Minerals and vitamins do not supply energy but regulate several essential metabolic processes in the body and in their deficiency several disease conditions arise in poultry. Essential minerals and vitamins required for poultry are calcium, phosphorus, sodium, copper, iodine, iron, manganese, zinc, selenium, vitamin A, vitamin D_3, vitamin E, vitamin C, pyridoxine, riboflavin, panthothenic acid, niacin, folic acid, vitamin B_{12} and choline. Balanced poultry ration should include all the above nutrients in right proportion for maintenance and productive function for different age group of birds. Commonly used poultry feed ingredients are given in (Table 2).

In lean period of winter months when availability of conventional feed becomes minimal, the farmers use locally available vegetables, pulses and human ration (wheat flour, rice etc.) for sustaining the feed requirements of birds. These are given in (Table 3).

All these less commonly used ingredients are high in fibre content. So care should be taken that the total percentage of fibre in diet should not exceed 6-8% of ration.

Table 2: Commonly used poultry feed ingredients

Feed	CP%	ME (k cal)	Maximum level of % inclusion chicks (broiler)		Maximum level of % inclusion growers/layers	
			0-4 wk	418 wk	4-20 wk	>20 wk
A. Energy source						
Maize	8.5	3,350	60	70	70	60
Wheat	11.5	3,120	20	25	20	20
Barley	11	2,640	10	15	20	15
Rice (Rough)	8.7	2,990	10	20	25	20
Pearl millet	12.7	2642	50*	-	50*	-
Oats	11.4	2,550	10	25	25	20
Finger millet			10*	-	20*	-
Sorghum	8.8-11	3,288	25*	-	40*	-
Rice polish	12.2	3,090	15	20	20	15
Rice polish (Deoiled)	14.1	2,235	20*	-	20*	-
Rice bran	12.9	2,980	15	20	20	15
Wheat bran	14.7	1,300	8	20	15	10
Bakery byproduct	10.5	3,862	10	15	15	10
Molasses	2.8	2,400	1	5	5	5
B. Vegetable Protein Source						
Ground nut cake	40.9	2,596	40*	-	40*	-
Ground nut cake (deoiled)	48.6	2,328	20*	-	20*	-
Soya bean meal	41.7*	2,230	25	25	30	20
Full fat Soyabean	44	2,230	15	20	20	20
Sesame til cake	43.8	2,210	8	10	15	10
Safflower cake	43	1,912	8	10	15	10
Linseed meal	29.6	--	20*	-	20*	-
Sun flower meal	32	1,543	8	10	15	15
Mustard cake	35.1	2,373	10*	-	10*	
Dehydrated alfalfa (lucerne)	17.5	1,200	5	10	5	5
Maize gluten feed	21	1,750	15	20	20	20
Maize gluten meal	62	3,720	15	20	25	20
C. Animal Protein Source						
Fish meal	43.1	1,834	5	5	8	5
Meat meal	2195	54.4	7	7	10	7
Blood meal	81.1	2,830	2	2	2	2
Meat cum bone meal	50.4	2,150	6*	-	6*	-
Feather meal	81	2,360	1	1	2	1
Vegetable oil	-	8,000	5	5	5	8

Source: Steven and Summers (1991); * Panda and Mohapatra (1998)

Table 3: Unconventional Ingredients for Himalayan region

Ingredients	CP %	Remarks
Berseem leaf meal	29.0	High in fiber, high in NFE and
Early stage, dehydrated	15.3	High in calcium
Cabbage leaves	29.6	High fiber content, high glucosinolates, cyanogens
Carrot leaves	14.9	
Cauliflower leaf -meal (dry)	26.9	
Bengal gram	14.0, 2496 k cal	Good source of energy and protein
Potato leaf meal	17.5	
Radish leaves (dry)	25.9	
Tomato waste	24.4	

Table 4: Requirement for chicken feeds. ISI, 1974-77

	Broiler starter (0-6 wk)	Broiler finisher (6-9 wk)	Chick (0-8 wk)	Grower (8-20 wk)	Layer (20-64 wk)
Metabolizable energy (k cal /kg)**	2,900	3,000	2,700	2,700	2,700
Crude protein** (%)	22	19	22	16	18
Crude fibre* (%)	6	6	7	8	8
Acid insoluble (ash %)*	3	3	4	4	4
Total sulphur amino-acid (%)*	0.75	0.75	0.5	0.5	0.5
Lysine (%)*	0.9	0.9	1.0	0.7	0.5
Methionine (%)**	0.35	0.35	0.35	0.25	0.25
Vitamine A (IU/kg)	6,000	6,000	4,000	4,000	8,000
Vitamin D_3 (ICU/kg)	600	600	600	600	1,200
Thiamine (mg/kg)	2	2	6	6	6
Riboflavin (mg/kg)	5	5	5	5	5
Pantothenic acid (mg/kg)	12	12	10	10	15
Nicotinic acid (mg/kg)	40	40	30	20	20
Biotin (mg/kg)	0.1	0.1	0.1	0.1	0.15
Vitamin B_{12} (mg/kg)	8	8	15	15	15
Alpha Tocopherol (mg/kg)	20	20	10	10	10
Choline chloride (mg/kg)	1,400	1,400	1,300	—	—
Linoleic acid (%)**	1	1	1	1	1
Salt (%)*	0.6	0.6	0.6	0.6	0.6
Calcium (%)*	1	1	1	1	2.75
Avail. Phosphorus (%)*	0.5	0.5	0.5	0.5	0.5
Manganese (mg/kg)	60	60	55	55	55
Iodine (mg/kg)	1	1	1	1	1
Iron (mg/kg)	40	40	20	20	20
Copper (mg/kg)	4	4	2	2	2
Zinc (mg/kg)	50	50	-	-	-
Moisture (%)*	10	10	10	10	10

*Maximum **Minimum

Table 5: Requirements of chick feed (90% dry matter); NRC, 1994

Nutrient	Broiler 0-3 wk	3-6 wk	6-8 wk	Layer (white-egg-Laying Strain) 0-6 wk	6-12 wk	12-18 wk	>18 wk
Metabolisable Energy (k cal ME_nkg)[1]	3,200	3,200	3,200	2,850	2,850	2,900	2,900
Crude Protein (%)	23	20	18	18	16	15	17
Aminoacids							
Arginine (%)	1.25	1.10	1.00	1.00	0.83	0.67	0.75
Glycine+ Serine (%)	1.25	1.14	0.97	0.70	0.58	0.47	0.53
Histidine (%)	0..35	0.32	0.27	0.26	0.22	0.17	0.20
Isoleucine (%)	0.80	0.73	0.62	0.60	0.50	0.40	0.45
Leucine (%)	1.20	1.09	0.93	1.10	0.85	0.70	0.80
Lysine (%)	1.10	1.00	0.85	0.85	0.60	0.45	0.52
Methionine (%)	0.50	0.38	0.32	0.30	0.25	0.20	0.22
Methionine+ Cystine (%)	0.90	0.72	0.60	0.62	0.52	0.42	0.47
Phenylalanine	0.72	0.65	0.56	0.54	0.45	0.36	0.40
Phenylalanine+ Tyrosine (%)	1.34	1.22	1.04	1.00	0.83	0.67	0.75
Proline	0.60	0.55	0.46				
Threonine (%)	0.80	0.74	0.68	0.68	0.57	0.37	0.47
Tryptophan (%)	0.20	0.18	0.16	0.17	0.14	0.11	0.12
Valine (%)	0.90	0.82	0.70	0.62	0.52	0.41	0.46
Fat							
Linoliec acid (%)	1.00	1.00	1.00	1.00	1.00	1.00	1.00
Macrominerals							
Calcium (%)	1.00	0.90	0.80	0.90	0.80	0.80	2.00
Non-phytate poshorus (%)	0.20	0.15	0.12	0.40	0.35	0.30	0.32
Potassium (%)	600	600	600	0.25	0.25	0.25	0.25
Sodium (%)	0.45	0.35	0.30	0.15	0.15	0.15	0.15
Chlorine (%)	0.30	0.30	0.30	0.15	0.12	0.12	0.15
Magnesium (mg)	0.20	0.15	0.12	600	500	400	400
Trace minerals							
Manganese (mg)	8	8	8	60	30	30	30
Zinc (mg)	0.35	0.35	0.35	40	35	35	35
Iron (mg)	80	80	80	80	60	60	60
Copper (mg)	60	60	60	5	4	4	4
Iodine (mg)	0.15	0.15	0.15	0.35	0.35	0.35	0.35
Selenium (mg)	40	40	40	0.15	0.10	0.10	0.10
Fat soluble vitamins							
A (IU)	1,500	1,500	1,500	1,500	1,500	1,500	1,500
D_3 (ICU)	200	200	200	200	200	200	300
E (IU)	10	10	10	10	5	5	5
K (mg)	0.50	0.50	0.50	0.5	0.5	0.5	0.5
Water soluble vitamins							
Riboflavine (mg)	0.01	0.01	0.007	3.6	1.8	1.8	2.2

Contd...

Pantothenic acid (mg)	0.15	0.15	0.12	10	10	10	10
Niacin (mg)	1,300	1,000	750	27.0	11	11	11
B_{12} (mg)	0.55	0.55	0.50	0.009	0.003	0.003	0.004
Choline (mg)	35	30	25	1300	900	500	500
Biotine (mg)	10	10	10	0.15	0.10	0.10	0.10
Folic acid (mg)	3.5	3.5	3.0	0.55	0.25	0.25	0.25
Thiamin (mg)	3.6	3.6	3.0	1	1.0	0.8	0.8
Pyridoxine (mg)	1.80	1.80	1.80	3	3.0	3.0	3.0

Table 6: Inter-relationships between nutrients in poultry ration; ISI, 1974-77

	Broilers		Growers			Layers
	0-6 wk	6-9 wk	0-6 wk	6-14 wk	14-20 wk	> 20 wk
Calorie:Protein (C/P)	139:1	160:1	145:1	180:1	241:1	190:1
Calcium:Phosphorus	1.4:1	1.4:1	1.4:1	2:1	2:1	4.5:1
Vitamin D_3	200	200	200	200	200	200
Arginine:Lysine	1.12:1	1.09:1	1.09:1	1.05:1	1.03:1	1.6:1
Methionine:Cystine	1.15:1	1.14:1	1.14:1	1.14:1	1.14:1	1.12:1
Phenylalanine:Tyrosine	1.14:1	1.16:1	1.16:1	1.1:1	1.16:1	-
Tryptophan(%)	0.2	0.2	0.2	0.16	0.12	0.11
Niacin (mg/kg)	27	27	27	11	10	10

Table 7: Feeding scale for broiler

Age (weeks)	Feeding scale (Minimum) g/bird /day	Feeding scale (Maximum) g/bird /day	Feeding scale (Optimum) g/bird /day
1	7	14	21
2	14	28	42
3	18	36	54
4	30	60	90
5	40	80	120
6	48	96	144
7	62	124	186
8	78	156	234

Source: FRL research data

Table 8: Feeding scale for growing and laying birds

Age (weeks)	Feeding scale g/bird/day	Type of ration
1	8	Layer starter
2	10	Layer starter
3	15	Layer starter
4	20	Layer starter
5	30	Layer starter
6	40	Layer starter
7	50	Layer grower
8	55	Layer grower
9-10	60	Layer grower
11-12	65	Layer grower
13-16	70	Layer grower
17-21	75	Layer grower
22-24	85	Layer
25-28	100	Layer
>28	120	Layer

Source: FRL research data

5. Feeding Management

In cold and high altitude areas it not advisable to feed complete ad-libitum to broilers. Broilers can survive in hypoxic and cold stressed condition only by reducing the rate of body growth. Complete ad-libitum feeding leads to ascites and other metabolic diseases. It also predisposes the birds, susceptible to other disease. Feed restriction is the only measure to curtail the mortality in broilers. In broilers feed restriction can be done by:

(1) Feed available for 8 h/d for 6 weeks.

(2) Feed available for 8 h/d during the first 3 weeks, then full feed for the remaining 3 weeks.

(3) Full feed for the first week, followed by 3 weeks of 8 h of feed availability, and then 2 weeks of full feed.

(4) Feeding calculated quantity of balanced ration per week as given in Feeding scale (Table 7-8).

All feed restriction treatments significantly reduces ascites incidence, when compared with the fully fed controls. Broilers subjected to restricted feeding, shows lower live weight, live weight gain and feed consumption at 28 days. However, no differences are found at 42 days, as a consequence of the compensatory gain during the refeeding period, if fed ad-lib at later stages. With scale feeding (minimum, optimum and maximum) broilers attain one kg body in 08 wk, 06 wk and 04 wk, respectively at the altitude of 1,000 feet from mean sea level at Ladakh. This restricted feeding greatly reduces the mortality in commercial broilers due to ascites.

In growing pullets (8-20 wk) feed restriction is most commonly followed to delay sexual maturity. This increases the egg production, extends the productive life and checks excessive weight gain in broiler pullets. It also reduces the feed cost. Recommended methods of feed restriction for growing pullets are (i) limited feeding time (ii) quantitative restriction (iii) qualitative restriction (iv) nutrient dilution (v) skip-a-day feeding.

Mash feeding, dry or wet is the most common method of feeding poultry. Pellet feeding in broiler is not suitable for this region, because of high nutrient density, it leads to ascites.

Broiler Production

Broiler farming is an intensive operation which starts from procurement of quality chicks and ends with marketing of uniform sized flock. 'All in' and 'All out' system is most commonly practiced in which chicks of same age, weight and of same hatchery are housed together and are reared till the age of marketing and all birds are sold at the same time. Some basic points in broiler production are as follows:

(a) *Procurement of day-old-chick*: All procured chicks should be disease-free or be vaccinated. At least parent stock should be free from mycoplasma. Only quality chicks of uniform weight, health and mobility should be purchased. The success of broiler production depends on quality and disease-free status of DOC. FRL advocates use of slow growing breed as it is beneficial in this area especially in winters, as has been established by research findings. All procured chicks should be of same hatchery or shed.

(b) *Housing*: At a time, chicks of same age group should be kept on farm or chicks of same age group/same breed should be housed in a shed. If birds of two different age groups/breeds are to be housed, there should be sufficient distance between two sheds.

Houses should face South-north to catch sun rays especially in winter. They should be at leeward side of mountains and sheltered from direct wind. The type of house and housing materials required for maintaining optimum temperature and ventilation depends on the farmer's affordability. The room should be properly ventilated in day time, especially in winter, to get rid of ammonia. The poultry house should not have large rooms and ceilings should be low. This facilitates easy warming and ventilation. Cleaning and washing of all cemented parts and windows should be done with any disinfectant (phenyl, attack, aldipol). After drying the house should be fumigated with formaldehyde and potassium permanganate in 2:1 ratio (40 ml: 20 g/100 cubic feet) 24-48 hrs before arrival of chicks and all windows, openings should be sealed. All windows should be open 3-4 hrs before placing chicks to remove all the toxic fumes and gases. Water and feeder should be cleaned in advance and should be moisture free. The litter should be fresh and well placed inside brooder, mixed with quick lime. Newspaper, waterer, feeder and bukhari should be placed and arranged properly at least 3-4 hrs before arrival of chicks to bring the room temperature at optimum.

(c) Brooding of chicks: Brooding is provided for supplementary heat to day-old-chicks until they no longer require exogenous heat (0-6 weeks). There are two general systems of brooding: (1) natural method, and (2) artificial method. Natural brooding is done by deshi hens/mother of chicks. Before arrival of chicks in brooder house, the house should be prepared in advance. The broody hen will provide all the warmth needed by the chicks. In a 2 feet square and 2 feet high packing box or wooden basket a deshi hen can brood 15-20 chicken. Small amount of chick feed mixed with water is kept under the box. Artificial brooding is handling of chicks without the aid of hen.

Brooding period for broilers is 2-3 weeks and for layers, it is 6 to 8 weeks. Brooding period for broilers may go up to 6 weeks depending on weather condition or at least in night. This period is very critical in life of birds because maximum mortality occurs in this period. Ventilation of brooder house should be judiciously done. There should be a balance between required room temperature and adequate ventilation. At least some part of window should be kept open always. There must not be too much moisture and dusty inside the house.

The litter can be of saw dust/paddy husk/coarse wheat bhusa etc. Its thickness may be 5-10 cm depending on whether condition. Litter should be stirred and cleaned frequently to prevent caking. Wet litter if any should be replaced immediately with dry litter. When Moisture is low, dust becomes a problem, and when high the litter cack, gives ammoniacal odour and is potent source of Coccidiosis. A good litter adheres slightly and breaks up when dropped from hand.

Brooding Temperature

Chicks are very susceptible to low temperature (cold stress) and fluctuation of temperature in early stage of their life. Too high or too low temperature slows down growth and causes mortality. During first week the temperature should be 95°F (35°C) which may be reduced by 5°F per week during each successive week till the temperature reaches 70°F. The temperature inside room should be 20°F below brooder temp crature. A maximum and minimum thermometer should be placed in brooder house to monitor the room temperature. Behavior of chicks like huddling in corners or coming very close to heat source tells about the temperature inside room/brooder. If chicks are evenly spread under hover/room, the temperature shall be considered optimum. When temperature is less than required, the chicks try to go closer to the source of heat and when temperature is too high, they scatter away from heat. If any draft air is coming then also the birds stray away heat and may be found sitting in corner or any one side of house. To provide artificial heat hover type electric brooder are most commonly used. But if there is problem of electricity the gas brooder or kerosene based bukharies can be used. In this region both electricity and availability of gas is a problem, so the most common source of heat is bukhari. Regular check up, cleaning and refueling of bukhari is must to avoid accidents due to fire and leakage of kerosene oil.

The brooder space per chick is 7 sq inch and if brooding is being done in whole house then the floor space required is 0.1 sq m till the age of brooding. If hover type brooder is used then the chick guard should be placed 60-70 cm from the edge of hover. In cold arid region, major mortality occurs in first 1-2 weeks of age because of very low temperature and less humidity. To balance the required temperature and optimum ventilation with bukhari becomes difficult. So judicious opening of windows at frequent interval or some part of window to keep open, even in night, is required. Poultry birds require the same standard of clean water as human beings. The waterer must be cleaned with potassium permanganate at least once a day. All left over water of previous day should be discarded and only fresh water be given each time. Luke warm water may be given in winter. Height of water pots should be maintained at 2.5 cm above the back height of the chicks to reduce spoilage. For the first 2 weeks, 50 cm of linear water space for 100 chicks shall be provided. The requirement of feeder space can vary from 2.5 to 3 cm per bird from 0 to 8 weeks of age. Chicks should be fed only after 3 to 4 hours after arrival in the brooder. To prevent wastage of feed, the feeder should not be filled more than one -third at any time. Like watere the feeder equipments should be cleaned daily. The watere and feeder should be arranged in a circle around the hover or bukhari and must be placed at equal distance, so that all chicks get equal opportunity for warmth, feed and water. As a thumb rule, the number of feeder and waterer should be such that all chicks must find space to drinks the water and feed at a time. With advancement of chicks age number of watere and feeder must be increased.

On day one, just before arrival of chicks some preparations are needed like:

1. All windows must be open 6 hr before for complete removal of gases/fumes from the house.
2. Well laid fresh litter should be covered with newspaper.
3. All feeder and watere should be cleaned, dried and arranged properly inside the house.
4. Hover or bukhari must be kept on 3 hr before to maintain the required room temperature.
5. Foot bath/lime must be placed at the entrance.
6. Fresh, clean luke warm water should be kept ready.

On Arrival of Chicks

1. All chick boxes be open immediately and each chick should be hand fed medicinal water.
2. For the initial 3 hr only water should be made available to chicks. This will reduce the stress and dehydration.
3. Crushed maize only or small amount of starter feed can be thrown/spread over the paper for first 5-6 hrs.

Following medicines/Anti-stress may be added in water for first 3 days

1. Zeetress powder
2. Ele ctrolyte
3. Glucose/crushed sugar
4. Anti-oxidant like E-CARE-Se
5. Liver tonic
6. B-complex liquid
7. Pa raffin liquid to reduce blocking of vent due to eating of particles of litter by the chicks
8. Any antibiotic

(d) *Lighting*: Artificial lighting is very essential for broilers throughout the night for first 2 days to allow them to locate feeder, water pots and get organized. Uniform distribution of light in shed makes the chicks to get scattered uniformly and little warmth they receive from light source. Electric infrared bulbs or lantern may be used for lighting. Electric bulbs in large no, if hanged one foot above the chick's height gives warmth and chicks prefer to huddle under the light source.

After 2 weeks of age regular monitoring of broiler's health is required. The condition of litter and amount of ventilation becomes very important. At this age relaxation in temperature control may be permitted because of developed thermoregulatory mechanism of birds. But birds require more ventilation and better hygiene condition. The diseases like air sacculitis, Chronic Respiratory Diseases (CRD), mixed infections of mycoplasma and *E. coli* start manifesting their symptoms at this period. Also, a chance of coccidiosis is very likely at this period. So the upper layer of litter should be removed/cleaned twice a week. Birds drooping in litter should be checked for coccidiosis. After 3rd week, birds start developing ascites like syndrome. Increased ventilation may cure this problem. To prevent infection, only one person should be allowed to enter inside the house and with changed slipper/shoes. One shoes/pair of slipper for that shed should be kept separately. In this region, no vaccination is required for broilers. Only the hatchery, born disease is the source of infection. If severe case of out, break is noticed then only vaccine is required.

Grower Management

Most of the management practices, *viz.*, housing, feeding, watering, sanitation and disease control are same as in broilers. The difference lies in intensity of care and supervision. As the birds attain 8 weeks of age, there is less need to maintain the room temperature. Floor, feeder and water space should be increased. Floor, water and feeder space recommended for a grower are 800-1,000 sq cm, 1.5-2.5 cm and 6.5-10.0 linear centimeters, respectively. When water consumption increases and the litter becomes wet. Care should be taken for cleaning of litter and not allowing accumulation of ammonia gas inside house. Duration of lighting should be reduced in decreasing order. Eight hrs of artificial light is needed for growers. Increased light intensity leads to cannibalism and vent picking. Vaccination may be required for chicks which are meant to be kept as grower and layer for specific diseases prevalent in that locality.

Table 9: Floor space, feeding space, watering space, feed and water requirement of different age group of birds on Litter system

Age (wk)	Category	Floor space/bird (sq cm)	Water space/ bird (linear cm)	Feeder space/bird (linear cm)	Required brooder/ room temperature (F)	Litter thickness (cm)	Amount of water/100 bird (l)	Amount of feed/bird (g)
1	Broiler	100-200	0.5	2.5	95/85	12.5	2-3	11
2	Broiler	250-300	0.5	2.5	90/80	12.5	2-3	21
3	Broiler	250-300	1.0	4.0	85/75	12.5	3-4	27
4	Broiler	250-300	1.0	4.0	80/70	12.5	3-4	45
5	Broiler	700-800	2.0	4.0	75/65	10.0	12-14	60
6	Broiler	700-800	2.0	4.0	70/60	10.0	12-14	72
7	Broiler	700-800	2.5	7.5	70/60	10.0	12-14	93
8	Broiler	700-800	2.5	7.5	70/60	10.0	12-14	117
9	Grower	900-1000	2.0	7.5	70/60	10.0	20-25	60
12	Grower	900-1000	2.0	7.5	70/60	10.0	20-25	65
13	Grower	1900-2000	2.0	10	70/60	10.0	35-40	70
20	Layer	1900-2000	2.5	14	60/50	10.0	35-40	85
21	Layer	2300 2800	2.5	14	60/50	10.0	40 45	110
64	Layer	2300-2800	2.5	14	60/50	10.0	40-45	110

Layer Management

The flock should be transferred to layer house at 18-20 weeks of age. In cold climate, it is preferred to keep the flock on deep litter. One laying nest for every 4 pullets is required. The floor space recommended for layer is 1,800-2,000 sq cm, feeder space of 14 cm and water space of 2.5 cm. Lighting influences egg-production. Therefore, lighting of layer house is very essential for optimal production. From 21 weeks onward the lighting should be increased gradually till it reaches 16-17 hrs per day and should be maintained at that level thereafter. Irregular light results in drop in egg production. One 40 watt bulb with a reflector hanging 2.1 m above the floor for each 9.29 m square of floor space would provide the recommended intensity of light. Light bulbs should be cleaned regularly. Laid eggs should be collected immediately from the nest in the morning to prevent the layer from eating its egg. Male breeder birds can be kept together or separately. They should be given extra calcium, manganese and vitamin E to ensure proper fertility.

Ascites Syndrome: A Major Problem of High-Altitude Broilers

Ascites syndrome refers to the accumulation of fluid in the body cavity and occurs when broilers fail to consume sufficient amount of oxygen to supplement their metabolic demands. The lack of oxygen stimulates the red blood cell proliferation to the vascular system causing an increase in hematocrit. An increase in hematocrit results in higher blood viscosity and leads to pulmonary hypertension, right ventricular hypertrophy, edema and ascites syndrome (Julian, 1993; Wideman, 2001). The propensity for broilers to develop ascites syndrome increases as ambient temperature decreases, because the broilers need more oxygen in order to maintain their body temperature (Wideman and Tackett, 2000).

Ascites is more prevalent in rapidly growing broilers and a decrease in growth rate is sufficient to make birds less susceptible to it (Jones, 1994). The incidence of ascites is inversely related the level of available oxygen for respiration (Odom *et al.*, 2004). A relationship exists between the partial pressure of environmental oxygen, as influenced by altitude, cold weather, as it increases oxygen consumption and ascites in broiler chickens (Cueva *et al.*, 1974). Thus, it is clear that conditions contributing to development of ascites syndrome are high altitude, rapid growth, cool temperature and poor ventilation. The break point for occurrence of ascites related changes in the absence of other stress factors (cold stress) is 19.6% oxygen (Beker *et al.*, 2003).

A right ventricle to total ventricle ratio (RV/TV) > 0.27 is considered as accurate measure of onset of ascites (Huchzermeyer and De Ruyck, 1986). Apart from environmental and managemental aspect, ascites has got genetic aspect too. Broiler strains can be highly susceptible to less susceptible to ascites. Through selection experiments, few researchers have developed the ascites resistant broiler lines in simulated condition of altitude and cold stress (Balog *et al.*, 2003). A gene has been identified for making the birds susceptible or resistant to ascitas/hypertension syndrome. This Bone Morphogenetic Protein Receptor type II (BMPR 2) gene maps to chromosome 7 of chicken (Cisar *et al.*, 2003).

Probable Measures for Controlling Ascites in Cold Stressed (5-15°C) Birds at an Altitude of >1000 Feet are as follows:

a) Lesser amount (0.15 -0.3 %) of common salt in ration. Common salt potentiates ascites.

b) Ration should be alkalinized by adding $NaHCO_3$.

c) Addition of selenium in diet helps in prevention of ascites.

d) Drinking water (spring) in this region is rich in salts. So it should be kept stored overnight and next day, upper scum layer should be removed before watering to birds.

e) Birds on high density nutrients/high energy, high C.P. percentage are more prone to ascites.

f) Feed a low protein/low energy diet during the first 14 days (Currie, 1999).

g) Nitric oxide helps in prevention of ascites.

h) L-arginine supplementation reduces the incidence of ascites by helping in synthesis of endogenous nitric oxide which causes vasodilation (Odom *et al.*, 2004).

i) Coenzyme Q_{10} has a beneficial effect in reducing ascites mortality @ 40 mg/kg of feed.

J) Frusemide (diuretic) at 0.001%, 0.005% and 0.01% significantly reduces ascites without lowering the final body weight (Wideman *et al.*, 1994).

k) Increasing dietary bicarbonate and reducing dietary chloride has potential as a low cost and effective method to reduce ascites in broiler chickens.

l) Naked-neck birds are more competent homeotherms and present superior levels of Hsp70 in the organs that are first affected by So Naked-neck broilers would be more tolerant to ascites syndrome.
m) A combination ascorbic acid and chromium supplementation can offer a potential protective management practice in preventing the detrimental effects of cold stress in laying hens.
n) A combination of vit C, vit E and zinc can provide protection to the endothelial lining of lung and heart and in turn some protection from onset of ascites.
o) Restricted feeding prevents Ascites.
p) In day time all windows should be kept open and at least two windows in night, depending on local weather condition to provide maximum ventilation. Better ventilation prevents Ascites.
q) All broilers attaining one kg weight should be soid immediately.

Table 10: Relationship between altitude, oxygen content of air and barometric Pressure

Altitude above sea level (ft)	(m)	Barometric pressure in. Hg	Reduction in air oxygen (%)
4000	1217	25.84	11.2
6000	1829	23.98	16.4
8000	2438	22.22	21.4
10000	3048	20.58	26.2
12000	3658	19.3	30.7

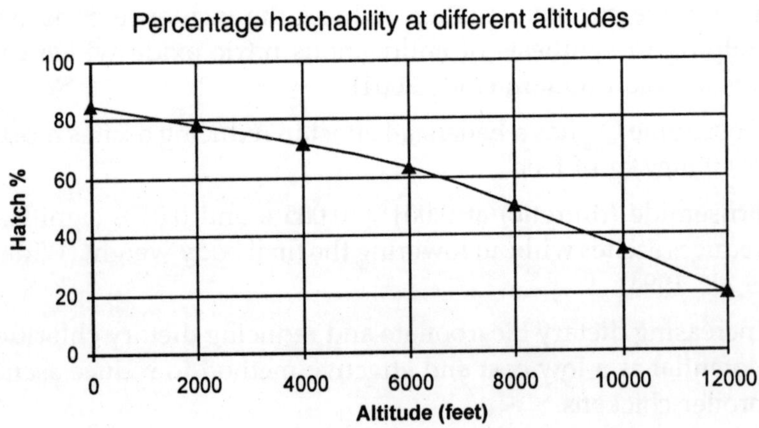

Fertility and Hatchability at High Altitude

High altitude is associated with reduced barometric pressure and reduced partial pressure of oxygen. External temperature and absolute humidity are lower. Thus fresh ventilating air tends to be colder and drier than at sea level. When air expands due to heating at high altitude, a cubic volume of air contains less oxygen.

Hatchability per se is adversely affected by hypoxia of high altitude (Lundy, 1969). Research has shown that hatchability of chicken eggs is reduced as the altitude at which they are incubated is increased. The cause of higher embryonic mortality has been reported to be lower hemoglobin level. Chicken requires adaptation period for normal fertility and hatchability at high altitude. If chicks are grown from day-old to onset of lay at high altitude, then the survivors will show better hatchability compared to adult hens brought from sea level (Smith *et al.*, 1955). Factors affecting hatchability are as follows:

- By increasing the air pressure inside incubator to sea level, it is possible to produce satisfactory results. Some hatcheries do provide pressurized atmosphere in the incubator room (Pressurerized Incubation).
- By injecting oxygen directly into the cabinet in which the eggs are to be incubated (Oxygen supplementation). Concentration of oxygen should be 23-23.5%. Oxygen is supplied into both setter and hatchery compartments by a tube from oxygen cylinders that have a pressure regulator valve and flowmeter. A gas analysis apparatus is used to determine the percentage of oxygen in the mixed air within the incubator cabinets.
- Hatchability fell by about 5% for each 1% (7.6 mm Hg) reduction in oxygen conc. below 21%. Hatchability falls down to 50% at 15% oxygen (114 mm Hg).
- Hatchability falls by 0.5 % for each 1% change in relative humidity from optimum.
- Temperature reduction of 1°F reduces the Relative humidity of air by 2.5%.
- Supplementation of oxygen into incubator increases hatchability of the order of 10 to 14% (Szopa and Mazanowski, 1963).
- Ells and Morris (1947), at 2195 m above sea level ($pO_2 \simeq$ 120 mm Hg), gave supplementary oxygen to eggs in forced draught incubators and reported that increasing pO_2 to 146 mm Hg

(19% O_2) throughout incubation increased hatchability by approx. 13%. Increased hatchability at altitude associated with oxygen supplementation was also obtained by Meshew (1949), Stephenson (1950) and Wilgus and Sadler (1954). Oxygen supplementation increased the haemoglobin concentration in embryos (Meshew, 1949), provides the evidence of physiological adaptation by embryos to low oxygen partial pressure.

- Higher incubation temperature in combination with increased oxygen tension decreased embryonic mortality during the 3rd and 4th weeks of incubation and resulted in higher hatchability. Christensen and Bagley (1988) suggested that hatchery managers at high altitudes should supplement oxygen and incubate turkey eggs at higher temperatures than at lower altitudes.
- At the altitude of 5,200 ft above MSL in Dirang (Arunachal Pradesh), Buragohain *et al.* (2006) achieved 17.86 to 76.04% (average 56.73%) hachability in Vanraja eggs brought from PDP Hyderabad. The cause of lower hatchability was reported to be transportation losses.
- Under hypoxic conditions, strain differences have been noted for hatchability (Stephenson, 1950).
- Resistance of chick embryos to hypoxia can be increased by selection (Smith and Abott, 1961).

Some facts about practical aspects of fertility and hatchability at high altitude are given below:

1. Incubation time is increased as the elevation increases, due to a decrease in the carbon dioxide concentration of the air rather than the change in the oxygen content.
2. Eggs produced by breeder hens kept at high altitudes seem to produce normal hatches if they are incubated at low altitudes.
3. Altitude has no effect on the fertility of eggs produced by the breeder flock.
4. Eggs from different strains have different hatchability.
5. Some of the hatchability may be restored over a period of time if genetic selection is practiced in the breeding programme, but it cannot be restored to normal.
6. At high altitude the hatchability and mortality will always be lower and higher from normal condition, respectively.

Table 11: Temperature, Relative humidity, O_2 and CO_2 concentration for optimum hatchability

	Temperature		Relative Humidity	O_2	CO_2
	°F	°C	%	%	%
0-19 Days (Setter)	99.53	7.5	50-60	21	0.3
20-21 Days (Hatcher)	98-99	36.7-37.2	65	–	–

References

Balog , J.M. *et al.,* (2003). Poultry Sci. 82: 1383-1387.

Beker, *et al.,* (2003). Poultry Sci. 82: 1550-1553.

Bhagat, U.C. (2003), Proceedings of workshops on "Devising strategies to make Animal- Husbandry & Poultry farming sustainable and economically viable in Ladakh". FRL, Leh, 27 sept 2003.

Buragohain, R. *et al.,* (2006). Indian Vet. J. 2006 ; 83 : 762-73.

Christensen, V.L. and Bagley, L.G. (1988). Poultry Sci., 67: 6, 956-960.

Cisar, C.R. *et al.,* (2003). Animal Genetics. 34: 475-476.

Cueva, *et al.,* (1974). Research in Veterinary Sci., 16: 370-374.

Currie, J.W.R. (1999). Avian Pathology 28: 313-326.

Ells, J.B. and Morris, L. (1947). Poultry Sci., 26: 635-638.

Gupta, A.L. (2003). Proceedings of workshops on "Devising strategies to make Animal- Husbandry & Poultry farming sustainable and economically viable in Ladakh". FRL, Leh, 27 sept 2003.

Huchzermeyer , F.W. and De Ruyck, A.M.C. (1986). Veterinary Record, 119: 94.

Jones, G.P.D. (1994). British Poultry Sci., 35: 97-105.

Jones, G.P.D. (1995). British Poultry Sci., 36: 123-133.

Julian, R.J. (1993). Avian Pathology. 22: 419-454.

Lundy, H. (1969). In: The Fertility and Hatchability of Hen's Egg , 143-176. Carter, T.C. and Freeman, B.M. Edinburgh (eds.) Oliver and Boyd.

Meshew, M.H. (1949). Poultry Sci., 28: 87-97.

North, M.O. and Bell, D.D. In: Commercial chicken production manual, 4 edition, Van Nostrand Reinhold Pub., New-York.

Odom, T.W. *et al.,* (2004). Poultry Sci. 82:835-841.

Panda, B. and Mohapatra, S.C. (1998). In: Poultry Production, ICAR.

Phuntsog, T. (2003). Proceedings of workshops on "Devising strategies to make Animal- Husbandry & Poultry farming sustainable and economically viable in Ladakh". FRL, Leh, 27 sept 2003.

Smith, A.H. and Abbott, U.K. (1961). Poultry Sci., 40: 1459.

Smith, *et al.*, (1955). Poultry Sci., 40: 1459.

Stephenson, A.B. (1950). Fm Home Sci., II: 65, 95-96.

Steven, L. and Summers, J.D. (1991). In: Commercial poultry nutrition, University Books.

Szopa, L. and Mazanowski, A. (1963). Zez. Nauk. Wyzsz. Szk. Roln., Wrocl. 52: 121-128.

Wideman, R.F. (2001). World's Poultry Sci. J. 57: 289-307.

Wideman, R.F. and Tackett, C.D. (2000). Poultry Sci. 79: 257-264.

Wideman, R.F. *et al.*, (1994). British Poultry Sci. 35: 633-667.

Wilgus, H.S. and Sadler, W.W. (1954). Poultry Sci. 33: 460-471.

□□□

15

Pashmina Producing Goat: A Unique Resource

F.D. SHEIKH, T.A.S. GANAI AND T. TSEWANG
Regional Agriculture Research Station
Sher-e Kashmir University of Agricultural Sciences and
Technology of Kashmir, Leh, Ladakh-194101 (J & K), India

One of the important livestock of our country is goat and it ranks first in the world, with its population estimated to be 120 million. Goats are known for excellent fibre, meat, milk, skin quality as well as for high prolificacy, therefore they contribute significantly towards the productivity, stability and sustainability of many farming systems of our country. The Pashmina, internationally known as "Cashmere", a fine luxury fibre, is the prince of speciality fibres obtained from domestic goats known as "*Capra ibex*". The word "pashmina" is a persian script. Literally "pashm" means wool and "pashmina" means woolen in English language. The cashmere breeds of goat are distributed throughout the world. The important breeds are, Tan goats of China, Markhore and Raini of Iran, Kurdi of Iraq, Vatani of Afghanistan, Don, Orenberg and Altai mountain of Russia, Feral goats of Australia, Chegu and Changthangi goats of India. The three major producers of cashmere wool are China, Mongolia and Iran in this order. India has one of the best animal genetic resources in the form of Changthangi (Changra) goats of Ladakh region and Chegu goats of Himachal Pardesh which are considered among the finest pashmina wool of the world. In our country, the Changthang region alone produces around 30,000 kg of raw pashmina fibre every year from about 1.51 lakh of Changthangi goats, which form 80% of the total pashmina production of the country. The Changthangi goats are reared by the Changpa nomads in Changthang region of Greater Ladakh. Ladakhi pashmina fibre is relatively longer and finer, which aids the Kashmiri weavers to spin easily for preparing various winter garments. This fibre has three times insulating values of the finest wool on a comparable weight basis and also posses 90% of the strength of merino wool and 60% strength of mohair. The importance of this valuable fibre was discovered by the people of Europe during the 17th century. It is believed that Emperor Napoleon presented a pashmina shawl to Empress Josephine. There is a rumour that pashmina garments are as soft as baby skin and provide enough warm to hatch an egg. The craftsmen in Kashmir developed their own technique to process the raw material and use it for the manufacture of shawls. Thousands of artisans are involved there for preparing shawls and the pashmina business has sustained the economy of thousands of artisan families, as the possibilities for agriculture in Changthang is quite low. People of Changthang rear huge herds of goats and remain in migration in specific routes in search of better pastures for their livestock. The available pasture was sufficient for the number of livestock reared in Changthang by the Changpas till 1959, but during the Chinese invasion of 1962 a number of western Tibet farmers migrated to Changthang along with their livestock. They settled in Changthang and ultimately this extra livestock feeding on limited pastures caused damage to the pasture land due to overgrazing. However, Ladakh is a world leader in the production of the finest pashmina and is a repository of germplasm with potential to exploit the superior traits for improvement of a whole range of pashmina producing goats in the world. Pashmina goats are of great importance for vitalizing the economy of poverty stricken region of Changthang and Leh districts of Ladakh.

Home Tract

Leh and Kargil districts of Ladakh are the home tracts of Changthangi goat. The Nyoma block in Leh district of Changthang subdivision is the main pashmina producing zone of Ladakh. It has 1.18 lakh of Changthangi goats distributed in 22 villages/hamlets while Durbook block comprising 12 villages, specifically Chushul, Kargium and Phobrang, has a population of 40,000 Changthangi goats (All India Livestock Census 2004). Besides, the Khaltsi block particularly Lingshet, Fotoksar and Tia villages, Nobra block, in the Diggar, Tangyar, Khardong and Panamic villages of Leh district are also supplementing the pashmina production of Ladakh. The total area of Changthang is about 20,000 sq. km, which constitutes about one third of the total area of Jammu and Kashmir State. This area is abundant in pasture lands but vegetative cover is very thin. Its distinct flora and fauna make it one of the most unique places in the world. The area has an excessive surface water system but due to rugged terrain, a very insignificant fraction of water is utilized for irrigation. Thus, a poor agriculture has been possible along the water channels which have perennial supply of melted glacier water. The whole Ladakh region is interspersed with small village hamlets which look like oasis in the desert. The Changthangi goat of Ladakh is highly adapted to the environmental conditions of the area. The animals are very stout and can walk very long distances every day in search of pasture grasses and herbs. It is believed that this goat evolved as a result of unplanned crossing between wild caprines and their crosses over a period of time. Some people also believe that ibex goat has contributed some genetic material information of the breed. This doubt has been cleared by the fact that the horn pattern and body colour of Changthangi goat looks like that of ibex goat. The Chegu breed of pashmina goat are mainly found in Lahaul, Spiti, and Chini valley of Himachal Pradesh and Pithoragarh district of Uttarakhand.

Breed Description

No information is available about the descendance of Chegu breed but on the basis of horn pattern and physical characteristics Chegu breed is considered to be descendent of *Capra falconeri* (Markhore) and *C. ibex* (Himalayan ibex) existing in the higher ranges of Himalayas in wild state. The animals are sturdy and have small to medium size body. Males are bigger and heavier than females. In Chegu breeds the ears are mostly long, leafy and pendulous while in Changra breed it is mostly medium size and sometimes stumpy. The horn characteristics can be classified into two main groups, resembling those of markhors in majority of the animals and the ibex in few animals. Horns are simply long, cork shaped, directed backward and upward. In both the breeds, head is slightly elongated, triangular and held high above the back line with a strong and well-set neck. In both the breeds of goat variable colour pattern is found. The most prevailing colour is white but black, grey, brown, fawn and mixture of these colours is also found. Generally, both the breeds of goat are intelligent, alert, graceful, swift-footed and have a majestic gait and quick movement. Females are docile compared to males and both sexes are beared.

Management

The Changra goats are mostly being reared by a special group of people called "Changpas". The Changthangi animals are hardy, agile and well adapted to the migratory life needed under difficult conditions. The Changpas always lead a migratory life with thousands of different livestock like yak, sheep and goat and they live in their traditional "Rebo", a circular tent with a heating arrangement in the centre. The flocks are taken out everyday for grazing, made to stay there through out the day and returned to the villages or encampment by night. No shelter is provided during the night encampment and the whole flock is supervised and guarded by "changkhi", a typical shephered dog breed of this region. The flocks remain migratory through out the year and the grazing areas for summer and winter have been properly earmarked. The important pasture species found in the area are grasses such as *Poa pratensis, Festuca rubra, Bromus macrostachy, Elymus sibiricus, Agropyron lange-aristatum* etc; legumes such as *Astragalus tribulifolius, Oxytropis microphylla, Cicer microphyllum* etc; bushes such as *Caragana vesicolor, Myricaria germonica, M. prostrata* etc. It is believed that the harsh conditions of nature induce the animals to produce an undercoat to insulate themselves against chilly weather. The animals are well adapted to survive on poor and sparse diet while preferring scrub to richer grasses. Normally goats molt in the beginning of summer season and then combing is done by the end of winter with the help of a combing device to get maximum yield of pashmina. It is an interesting fact that sexual cycle is dependent on photoperiodism where the cycle is triggered with the decreasing day length and animals become sexually active from June onwards. Thus, breeding normally takes place during the month of November and December and kidding takes place during April-May. During this time 98% of kiddings is completed and the kid mortality has been found to be very low, i.e. 3%. For mating, one buck is sufficient to breed 40-50 does. Weaning of the young animals is normally practised at the age of four months. The average daily milk production was observed to be nearly 700 ml/animal with a lactation length of 5 months. In State Govt. farm the animals are normally kept in open air with wire fencing. Normally high land grazing is followed during the month of July to September while pasture grazing at low land area is done during May to June and October to December. During the remaining four months, goats are specifically confined to stall feeding due to harsh winter weather and snow fall. On stall feeding amount of concentrate given to individual animals daily was 500 g for bucks, 400 g for does and 300 g for young animals while daily fodder given to each individual was 1.5 kg to bucks and 1 kg each to does and young ones.

In farms, 88.12% of the goats were of white colour and 11.88% had light brown colour. The colour of the head of the animals varied from white to black while some animals had grey head. This is due to the continous selection for white coat colour, practised over a period of 20 years. At the time of birth, the length of the animal was 26.01 ± 0.18 cm while the height at wither and girth at chest was estimated as 27.71 ± 1.99 cm and 29.92 ± 1.60 cm, respectively. The whole body was covered with a fine undercoat of pashmina and long hair. The face and muzzle were devoid of hair. The ears was small, erected and stumpy while the horns possessed a typical character of being curved first upward then backward, downward and onward. The body was straight and heavy.

Production Parameter

The birth weight of male kid was found to be ranging from 1.40 to 2.50 kg while its female counterpart was in the range of 1.30 to 2.40 kg. Above all, the average birth weight of male and female kids were estimated as 2.11 ± 0.26 g and 2.06 ± 0.19 kg, respectively (Table 1). The average weaning weight of male and female goats were estimated as 13.80 ± 1.22 and 13.06 ± 1.08 kg, respectively. The average body weight at 300 days was found to be 20.00 ± 2.09 and 18.68 ± 1.87 kg in male and female animals, respectively. The pashmina yields of bucks, does, male hogget and female hoggets were estimated as 401.79 ± 18.81 g, 248.33 ± 14.32 g, 254.87 ± 12.39 and 279.91 ± 16.02 g, respectively. Performance of different clipping has been presented. While considering year-wise yield, the pashmina production during third year gave the highest value. The length of pashmina fibre was found to be 4.25 ± 1.20 cm in male and 4.02 ± 1.50 cm in female. However, overall average staple length of fibre was estimated as 4.135 ± 1.29 cm. The fibre diameter in male and female goat was calculated as 12.86 ± 2.59 μ and 12.97 ± 3.02 μ, respectively. The overall fibre diameter, irrespective of sex, was estimated as 12.92 μ. The number of primary and secondary follicles per sq. mm area were found to be

4.00, 5.67 and 29.67, 38.56 in male and female Changthangi goats, respectively. The estimated percentage of down and scouring yield in male and female goats are 52.80 ± 18.65, 48.63 ± 18.60 and 85.08 ± 8.75, 84.32 ± 8.73, respectively.

The reproductive traits of animals are very important to maintain the farm systematically and scientifically and make the farming enterprise a profitable business. The age at first tupping, age at first kidding, gestation period, litter size and kidding interval of Changthangi goat were 667.73 ± 18.43 days, 829.90 ± 12.69 days, 151.71 ± 0.11 days, 1.03 ± 0.01 days 396.66 ± 6.96 days, respectively. The kidding percentage estimated in this goat breed was 73.3% which varied from year to year. The percentage of female kids born was found to be 49.65. The twining rate was very low in this breed, i.e., 0.3%. However, the majority of the kidding was completed between March and April, when normally natural service is preferred by the farmers where one buck is sufficient to breed 40 does.

Pashmina Fibre Production

Most of the research work on Pashmina goats has been concentrated on biometry and yield of Pashmina fibre as it is main produce of these goats. Average age at first combing is 11.69 months and is affected by season and year of birth. Male has lower age at first combing than female under semi-intensive management. Pashmina is being collected once in a year. Different reports have indicated average pashmina yield, length and diameter of the fibre ranging from 75.3 to 227.6 g per combing, 2.5 to 2.6 cm and 10.7 to 12.7 μm, respectively. Females produce more pashmina fibre per unit body surface, however, total yield is more in male because of larger body surface. Pashmina yield depends on number of hair follicles per unit body surface. Thigh region contains highest number of follicles, followed by flank, back, shoulder, neck and midbelly. Number of secondary follicles is more than that of primary follicles and the ratio is affected by sex and body area of the animal. Total pashmina yield is significantly affected by year of collection, age of animal and physiological status of the animal. Dry female and castrated male have higher yield than female and non-castrated male, respectively. Quality (length and diameter) of the pashmina fibre depends on age and sex of the animal and region of the body from which fibre has been collected. Younger animals and females produce finer fibre. Longest fibre is at neck, shortest at the back and finest fibre is found at back. Pashmina does not grow on head, lower part of leg and tail of the animal.

Female Reproduction

Reproductive age of pashmina nannies is upto 12 years. Average age of goat for first kidding in Cheghu breeds is 737 days and this is lower than that of Changthangi goats (988 days) in their natural habit. Body weight of doe at kidding ranged from 22 to 24 kg. Length of the cycle and duration of oestrus is 20.39 ± 0.39 days and 24.83 ± 1.67 hours, respectively. Weak heat symptoms are evident in females with scanty oestral discharge from genetalia and no firm pattern under microscope. If bucks are allowed in the flock, pregnancy rate increases. Though oestrus cycle occurs throughout the year, most of females in the flock exhibit heat during autumn and summer.

Kidding interval, gestation period and service period ranges from 272 to 378, 152 and 197.7 to 218.4 days, respectively. Pasturition process is studied in detail which takes 177.7 ± 37.8 minutes in primipara and 136.4 ± 14.3 minutes in multipara nannies. Twining is not very common, maximum occurrence being 3.25%. Mortality of twins is very high (69.82%) and birth weight of twin kids is smaller. Efforts to multiply superior germplasm rapidly, supervoulation, oestrous synchronization and embryo transfer are being carried out at Indian Veterinary Research Institute, Mukteswar. Superovulatory response was higher with FSH-P than PMSG and HCG. Response was higher in older animals (4-6 years) as compared to younger. Prostaglandin analogue (Carboprost Tromethamine) was found to be useful for synchronizing oestrous in low doses (18, 19). Overall conception rate under Embryo Transfer Technology (ETT) is 35.09%.

Male Reproduction

Testicular and sperm biometry and semen quality of pashmina bucks is studied. Scrotal circumference is 23.6 ± 0.44 cm and weight of both test is 133 g. Total length of spermatozoa is 59.8 ± 0.2 μ; dimension of head being 8.68 μ x 4.34 μ. Volume, pH, concentration of sperm, initial motility, percent of live and abnormal spermatozoa ranged 0.62 to 0.84 ml, 6.4 to 6.8, 2.94 to 3.52 million per sq mm, 60 to 61%, 71.7 to 80.6% and 33.2 to 33.4%, respectively.

Birth Weight

Average birth weight is 2 kg. Years and period of birth and sex of the kid significantly affect birth weight. However, two reports indicated non-significant effect of sex of kid.

Carcass

Pashmina goats are also reared for meat purpose. High carcass yield over local goat and tasty meat of pashmina goat at Mukteswar is being reported. However, reports on pashmina goats at Ladakh indicated low carcass yield and poor grade meat. Mean dressing percentage on live and empty weight basis ranged from 39.0 to 43.7 and 47.4 to 48.3%, respectively. Meat: Fat: Bone ratio is 69:4:27. Age of the animal has significant effect on quality and quantity of carcass yield.

Milk

Milk yield of pashmina goat ranged 157 to 700 ml per day and lactation length is 187±15 days (24, 27). Specific gravity, total splid, total ash, fat and lactose content of milk is 1.032 ± 0.001, 13.7 ± 0.3%, 0.80 ± 0.05%, 3.3 ± 0.2% and 3.85 ± 0.03%, respectively.

Blood Protein Polymorphism

Slow and fast type of alkaline phosphatase, Tf^A, Tf^B and Tf^C, type of transferring are found in pashmina goats. Unlike other goat breeds, Tf^A has been more in pashmina goats. No polymorphism was reported in albumin, haemoglobin, amylase and cerulloplasmin. Na^+ and K^+ concentration was found to be similar to other Indian goat breeds. Indian pashmina goats have a restricted distribution in higher altitude of Himalayas and represent unique genetic material. Little scientific information is available on these goats in their home tract. Acclimatization at lower altitude at Mukteswar indicated the possibility of propagating this germplasm at medium altitude. Local hill goats do not produce fine fibre. Cheghu is heavier than local goats. However, breeding and scientific management practices along with environmental physiology and nutritional aspects of Pashmina goat rearing need to be studied in detail before introducing these goats at lower altitude. Cheghu is an important regional pashmina goat breed in India; conservation and multiplication of this fine germplasm is an utmost important task. Selective breeding and application of modern techniques of biotechnology will help in preserving and multiplying these animals at faster rates.

Economics of Pashmina Production

As the profession of people in Changthang is pastoral farming, a major share of pashmina goats is reared in Changthang region of

Ladakh. Almost 90% of the population in Changthang area survives on the income from goat and sheep, which is nearly $8.4 million, annually. Ladakh is producing more than 80% of the total pashmina yield of the country. The pashmina prices dwindle and swing between extremes. The price of raw pashmina was $63/kg during 1993-94, $25/kg during 1996-97 and $20/kg in the year 2009. This trend of price is mainly due to serious involvement of middle men. Unfortunately, local industry does not consume more than 5% of the whole production and the remaining pashmina is being sold outside Ladakh through middle-men. Thus, the real benefit is going to the middle-men while the end users and poor growers are being exploited. The highest yield of pashmina per animal was found to be from breeding bucks and does while the returns, *i.e.*, benefit : cost ratio was highest in castrated bucks, *i.e.*, 1.49 (Rs. 765.02) followed by does, *i.e.*, 1.37 (Rs. 653.99) and breeding bucks, i.e., 1.14 (Rs. 254.87). More than 3,000 tons of cashmere is produced annually worldwide, the majority comes from Mongolia with smaller amounts from Iran, Afghanistan, Australia, New Zealand from which Scottish cashmere processors traditionally import. There are currently around 50 producers in the UK with a herd of around 2,500 goats. The UK would need a breeding herd of around 2 million breeding females to be self sufficient in cashmere. Feral goats produce small quantities (50 g) of high quality cashmere (less than 16 microns). Through a specialised breeding programme cashmere stock now produce more than to 300 g. One of the important criteria for price of cashmere in the world market is fibre diameter along with other characteristics. The finer the diameter, the higher the value. A price of between £80-100 can be expected for dehaired hosiery fibre (less than 16.5 microns) and 60-70% of this price for fibre of 16.5-18.5 microns. Scotland processes over 1,000 ton of imported cashmere (60% of world production) per annum. A premium is generally paid for fibre of less than 16.5 microns of good length and for white cashmere over coloured. Demand for cashmere garments exists world-wide but is particularly strong in West Europe, the USA and Japan (although these markets are affected by many economic down-turns). They are less affected by fashion trends and consequently prices are more stable than for mohair. Scottish Cashmere Producers Association operates a fibre pool through which producers can market their fibre. A meat co-operative has also been formed to market both culled and prime stock.

Disease Prevalence

Occurrence of diseases was found to be quite low in Changthangi goats. Most of the problems which arose were related to growth where stunted and retarded animals were common. Sometimes, animals were found to suffer from diarrheoea which might be due to parasites or poisonous grasses. In kids, tapeworm, roundworm and coccidia infestation was also observed. The occurrence of pox, CCPP and FMD was not common. In big farms, vaccination was followed for FMD, pox and clostridial infection. Gid was a common incidence in Changthangi goats. Tick infestation was sometimes found. Deworming with anthelmintics and dipping with some antiparasitic drugs two times a year was being followed in some of the organized farms. However, these medical activities were not so common among the poor small village farmers, which has resulted in comparatively higher mortality in the animals. Sometimes, genetic defects like cryptorchidism, stumpy ear, short legs, prognathism and mixture of body colour was found in some animals.

Future Strategy

- Health coverage of goats
- Assured value of pashmina
- Selection and breeding for improving both quality and quantity of fibre
- Selection and breeding for other important economic traits like meat, milk, skin and adaptability
- Improvement of local goats by grading up
- Pasture development and fodder production
- Establishment of pashmina fibre processing units
- Establishment of quality and control unit
- Control of middle-man by establishment of co-operative society
- Ban on individual selling of pashmina fibre
- Study at molecular level, like development of genetic marker for economic traits to improve selection and acquire high accuracy
- Encourage the selection of superior breeding stock
- Provide advice and assistance to new cashmere goat farmers
- Provide a central marketing operation for the collection, grading, processing and marketing of cashmere fibre and its products

Other Important Strategies to be Adopted in the Field of Research

- A research project which aims to determine the inheritance of production characteristics of Changthangi and Chegu goats should be carried out. The strong phenotypic and genetic correlations between down weight and length of down indicate that the latter may be used as an indirect indicator of down weight for preliminary selection.
- The rate of improvement in fleece weight will be reduced, if fiber diameter is kept to an acceptable level and fiber length is maintained. This is undesirable response for cashmere producers due to their return based on cashmere fleece weight. Optimal economic selection index needs to be developed for cashmere producers of India.
- The value of cashmere is determined by a whole range of characteristics which in the past were evaluated by an expert with many years of experience in the trade. The task was, and still is, to assess the amount of usable cashmere available in the undercoat, to determine the quality and the potential end use of cashmere and to estimate the cost of processing it into the finished product. Characteristics of cashmere which are assessed include:
    - Diameter
    - Length
    - Colour
    - Lustre and crimp (style and handle)
    - Strength and
    - Yield (amount of cashmere after dehairing and scouring the raw fibre).

The variability in each of these components is also important as is the character of the guard hair which should be long and coarse for easy and effective removal and to protect the undercoat while *in situ*.

- Objective measurements for many of these traits have been established and developed into IWTO (International Wool Testing Organisation) specifications (the main exceptions being colour and lustre) which buyers rely upon increasingly to purchase fibre. Objective measurements also provide the means by which the requirements of the trade can be translated into goals for the producer and breeder.

- The differential in the price paid for cashmere is variable but is always in favour of fine white cashmere with high crimp and low lustre and the production of such high quality fibre must be the ultimate goal of our breeders. However, the breeder has the additional task of increasing annual production per head since financial returns are determined not only by price/kg, but also by weight produced. Fibre traits such as fibre diameter and annual production, in cashmere goats are very strongly inherited but there is, as in other fibre producing species, a positive relationship between fibre diameter and fibre weight which prevents concurrent improvement in annual production and fibre quality. Devising a strategy to overcome this problem is the current challenge for researchers and breeders alike, but most breeders or even breeding groups in Europe currently operate on a small scale limiting the amount of information available and progress which can be made. One important aim of this Thematic Network, which is to be commended, is to establish common protocols for fibre analysis so that performance records from herds across Europe can be merged into a common database for the eventual benefit of all participants.
- In any breeding programme, the rate of genetic progress is faster with selection for fewer traits in larger numbers of animals. In the past, the hand separation of patch samples to determine cashmere yield and estimate annual production, and the projection microscope technique for the measurement of fibre diameter, was expensive in terms of time and money. It is limited not only the number of traits which could be measured but also the number of animals which could be tested. With the use of OFDA, we can not only measure fibre diameter more accurately, but we can also measure yield in a fraction of time, sparing resources to measure other traits such as length, colour and crimp, which can be included in the cashmere database.
- With the best Changra cashmere goats, currently producing 500 g of hosiery or 600 g of weaving quality cashmere, and the average goat producing closer to 260 g, selection for increased annual production of cashmere under 15 mm, using measurements of yield and diameter, is likely to be a priority for the foreseeable future. However, information in the database

will enable us to develop efficient selection strategies for other traits, if these become important. The first step is to decide what we want to measure and to establish a common protocol and standard of measurement. The data is only as good as we make it.

- Goat is always mistakenly regarded as misfortune for ecosystem because of some prejudices to it for it's biological characteristics. If goat production systems were not managed and assessed in terms of ecology, economics, animal science and grassland; livestock production efficiency, and herdsmen's income and regional economic development were not promoted, ecosystem construction and maintenance would not last long and the development of goat industry would be seriously restricted.
- In the Nucleus Herd of Caidamu Goats, cashmere production of 150 goats could make an annual income of 20,000 – 35,000 Yuan for the herdsman. In contrast, the Tibetan herdsman living in Yushu prefecture, with same number of goats, could only make 500 Yuan a year. Haixi herdsmen became rich on cashmere goats because of applying scientific technology and improving the quality of goats.
- Kids are kept in even worse hygienic condition. They are put in the shabby and tight pens where they can not even move. Hence mortality is highest during the first six months of life.
- In Changthang, there are some problems in harvesting the down. Firstly, the down is not harvested in time and it is always delayed. The herdsmen generally think that goat is less cold resistant, so they harvest the down from May to August. Secondly, the method of harvesting is not scientific. In Gonghe Basin and Huangnan Prefecture, when the down off the skin to middle and upper part of the hairs, herdsmen cut both hair and cashmere together, so called "Strip Down", which affect yield and quality of cashmere. While in Yushu, herdsmen harvest the down by combing when the down fiber forms balls at the end of the hair. It is called "Rusty Ball Down", which results in poorest quality and lowest production. Also, there are many problems with disease control, supplementary feeding, rearing and shift of breeding goats, etc.

- In conclusion, raising cashmere goat is a systematic program. Obviously, improvement of cashmere goat is an important way to improve economic efficiency of cashmere goat production. However, if feeding and management is not good, the benefit of the breed improvement will not show up.

- Since in the present situation there is a great scarcity and denudation of pasture due to over grazing in Changthang region goat industry should be adjusted for conservation of ecological environment and efficiency improvement of animal husbandry in pastoral areas. Only by making best and proper uses of grass resources can goat industry continue to contribute to local economic development and increase in farmers' income.

- Changra goat and non-traditional goats surrounding Leh district should be further selected and improved. Though selection is more difficult than crossbreeding, fineness and length of cashmere fiber should be major objectives of selection while increasing cashmere yield. The herdsmen should be provided with more training and guidance. Herd size should be determined by grass supply. Quality and efficiency should be emphasized. Number of goats should be adjusted according to the changes of grassland.

- Production strategies must therefore rely on dual-purpose meat and wool/fibre breeds, as well as retaining separate meat and fibre breeds within flocks, in order to reduce risks and maximise incomes. To a large extent, Central Asian producers are aware of and already practice these strategies, but policy and research action is required to increase their income from wool and fibres.

- Pashmina, as a valuable fibre being produced by Changthangi goat, should be marketed properly to help the farmers in terms of economic benefit. Fodder banks may be established to supply fodder to the farmers during lean period. A proper and well planned breeding programme should be adopted to enhance the performance of Changthangi goat. Field progeny testing should be taken up to produce good proven bucks which can be used to improve the local goat population.

Table 1: Performance of Changthangi goats

Characters	Mean ± S.E.
Birth weight of male kid (kg)	2.11 ± 0.26
Birth weight of female kid (kg)	2.06 ± 0.19
Weight at weaning of male kid (kg)	13.80 ± 1.22
Weight at weaning of female kid (kg)	13.06 ± 1.08
Mature body weight of male at 300 days of age (kg)	20.00 ± 2.09
Mature body weight of female at 300 days of age (kg)	18.68 ± 1.87
Pashmina yield by buck (g)	401.79 ± 18.81
Pashmina yield by doe (g)	248.33 ± 14.32
Pashmina yield by male hogget (g)	254.87 ± 12.39
Pashmina yield by female hogget (g)	279.91 ± 16.02
Staple length of pashmina in male (cm)	4.25 ± 1.20
Staple length of pashmina in female (cm)	4.02 ± 1.5
Diameter of pashmina fibre in male (μ)	12.86 ± 2.59
Diameter of pashmina fibre in female (μ)	12.97 ± 3.02
Twining rate (%)	0.30

References

Acharya, R. M. and Sharma, V. D. (1980). Note on pashmina production and its quality from Changthangi pashmina goats. Indian J. Animal Sci. 50: 586-587.

Anonymous (2000). Introduction of pashmina. Proceedings of National Level Seminar on "Development of Pashmina Wool in Ladakh", Leh, Ladakh, 19-21 June, 2000, pp 9-14.

Bhattacharya, T.K. (1999). Genetic characterisation of Indigenous breeds of cattle by PCR-based methods. Ph.D. Thesis, Indian Veterinary Research Institute, Izatnagar, Bareilly, U.P., India.

Darokhan, M.D. and Tomar, N.S. (1983a). Studies on body weight and measurements at birth in Changthang pashmina goats of Ladakh. Indian Veterinary J. 60: 564-569.

Darokhan, M.D. and Tomar, N.S. (1983b). Studies on pashmina yield of Changthang goat. Indian Veterinary J. 60: 650-653.

Koul, G.L., Biswas, J.C. and Somavanshi, R. (1987). Follicle and fibre characteristics of Indian pashmina goats. Research in Veterinary Sci. 43: 398-400.

Kumar, R. (2000). Present status of pashmina production in Ladakh. Proceedings of National Level Seminar on "Development of Pashmina Wool in Ladakh" held at Leh, Ladakh, 19-21 June, 2000, pp 57-63.

Mir, A. (2000). Pasture development for sustainable pashmina development in Ladakh. Proceedings of National Level Seminar on "Development of Pashmina Wool in Ladakh", Leh, Ladakh , 19-21 June, 2000, pp 89-94.

Misra R.K., Singh, B. and Jain, V.K. (1998). Breed characteristics of Changthangi pashmina goat. Small Ruminant Research 27: 97-102.

Samphel, T. (2000). Future of pashmina in Ladakh. Proceedings of National Level Seminar on "Development of Pashmina Wool in Ladakh", Leh, Ladakh, 19-21 June, 2000, pp 49-52.

Shareif, M. (2000). Factors affecting pashmina production and growth. Proceedings of National Level Seminar on "Development of Pashmina Wool in Ladakh", Leh, Ladakh, 19-21 June, 2000, pp 82-88.

Wani, M.H., Wani, S.A., Wani, G.M., Gall, C.F. and Gelderman, H. (1999). Economic viability of pashmina goat production in Ladakh-the cold arid zone of J&K. Appl. Biol. Res. 1: 129-132.

□□□

16

Changthangi Sheep: A Valuable Genetic Resource

F.D. SHEIKH, T.A.S. GANAI, T. TSEWANG,
S.S. MISRA AND S. AZMAT
Regional Agriculture Research Station
Sher-e Kashmir University of Agricultural Sciences and Technology of Kashmir
Leh, Ladakh-194101 (J&K), India

In the forbidden land of "*Changpas*" besides the *Changthangi* pashmina goat, there exists a potential dual purpose breed of *Changthangi* sheep also known as "*Changluk*". This breed has population strength of about 74,180 heads and is famous for its wool and mutton production. This breed of sheep sustains the economy of *Changpas'* (*Changthangi* people) to a greater extent but next only to pashmina goats. Until now no specific and scientific breeding strategy has been adopted for upgradation of this sheep either by the farmers or the State Government. This is evident from the fact that no breeding and research farm for improvement of this sheep exists in the whole of Ladakh region. The economy of many farmers in the breeding tracts of this breed is largely dependent on animal rearing in general and sheep rearing in particular. Massive crossing of the local sheep, followed as the state Sheep Breeding Policy to give boost to fine wool production, has reduced their number drastically. The number presently is very low and restricted mainly to Changthang area where crossbreeding proved a failure. Therefore, immediate need of initiating improvement programs for better utilization and conservation of this breed of sheep is essential. The base line data gathered on demographical, geographical distribution, production and reproduction performances of this breeds and identification of the shortcomings in the rearing of these animals are discussed in this chapter. The information generated will be immensely helpful in devising improvement plans for this breed. The generated information can be utilized to educate farmers regarding better feeding, management, health care practices, and proper breeding techniques to be followed which will ultimately lead to the development of a sustainable and profitable rearing system for this breed of sheep. The economic conditions of the farmers rearing the sheep shall be improved at the end and these gene pools shall get conserved for use in posterity too.

1. About the Breed

In modern literature the breed is generally described as *Changthangi* sheep as per the name of its breeding tract. The other name which is used by the local farmers of the area is Changluk meaning sheep of northern plateau (Chang = Northern, Thang = Plateau / Plain, Luk = Sheep). According to Moorcroft "Sheep of Ladakh are of two distant kinds. The tall black-faced "*Huniya* used chiefly of carrying burdens and pretty diminutive sheep of *Purik,* which is used only for food. The predominate (common) sheep is *Huniya* which, with the exception of *Purik* breed is almost only kind of sheep to be found throughout the Tibet and this breed is much larger than any of the Indian breeds, and the height averaging from 27 to 30 inches. Moorcroft was so impressed with this breed that he recommended crossing of this breed with common small sheep of hill provinces of England." In earlier times, nearly the whole of the traffic of Ladakh was dependant on these sheep. They were food, clothing and carriage for the people of Ladakh. The *Huniya* was therefore, much prized, and men wealth was generally estimated by the number of his sheep. The *Huniya* breed of earlier times is *Changthangi* breed of present times. The breed is reared for quadruple purpose: mutton, wool, manure and dung energy. The *Changpa* use the dried dung of sheep along with dried dung of other animals they rear as fuel for their specifically made iron stoves (*Thab*).

Theses stoves keep them warm as well as they cook their food on them, this is the only source of fuel energy in the area. The wool is used to make rugs and cloth. Use of these pelts as flooring material keeps the *Changpas* tent houses warm in the freezing temperature. The pelts of the animal are also used for clothing material. *Changpas* consume the mutton of these sheep and surplus animals are sold to mutton dealer of Leh market.

2. Breeding Tract

Distribution of native tract in terms of longitude and latitude extends between 32°-3 to 34°-0 N and 77°-5 to 79°-10 E. The geographical area of the breeding tract is about 24,000 squares km. The breeding tract of the Changthangi sheep comprises of various villages located in main and collateral valleys of Changthang area. It includes valleys of Rong, Nyoma, Kuyul (all these three valleys are basin of the river Sindh), Anley with some collateral valleys, Puga and Rupsu, Kharnak and Tso Kar of Nyoma block of Changthang. The villages of Cheshule, Pangong and Krygam valleys are the remaining portion of the breeding tract and these areas falls in Durbuk block of Changthang. These valleys are mostly of about 30 kms or so in length and of varying breath. The breeding tract lies in Leh district. This area falls in cold arid region of western subcontinent between the Korakaram and the Greater Himalayas interwoven with barren and rugged mountains forming some big and small side valleys. On the North-East side of the breeding tract is the Tibet and China and its southern side is bounded by Chamba and Lahul-Spiti Districts of Himachal Pradesh. The terrain is mostly mountainous and denuded. The Changthang area is divided into two rural developmental blocks- Durbuk and Nyoma. The Changthang is derived from a Tibetan word meaning northern plains (Chang = northern, thang = plains). Whole of the breeding tract was divided into 12 strata for the survey and documentation and the details of the villages included in each stratum is presented in Table 1. The elevation of the breeding tract varies from 3,340 to 4,560 m above MSL and these elevations correspond to Liksy and Korzook villages, respectively. The physiography of the area is mountainous with long valleys between them. The animals are reared extensively in the pastures up to 5,200 m above MSL. The detailed altitudes of the breeding tract are presented in the (Table 1).

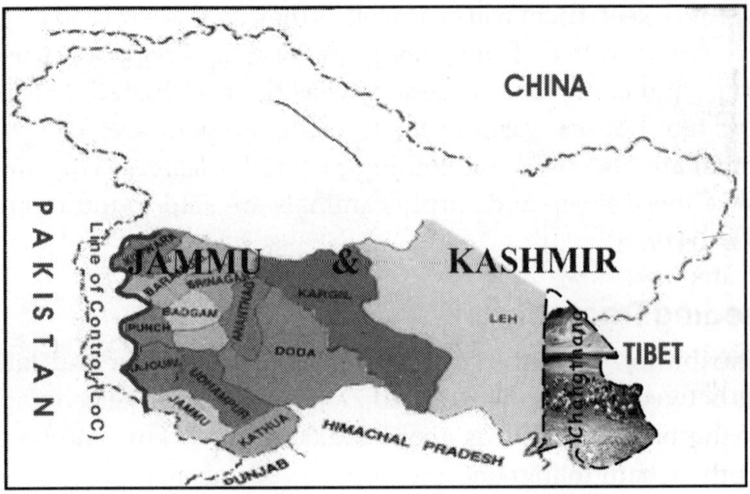

Fig. 1: Map of Jammu and Kashmir indicating breeding tract of *Changthangi* sheep

Table 1: Name of stratum identified in the breeding tract to characterize the *Changthangi* Sheep

S.No.	Name of stratum	Elevation (m above MSL)	Name of villages
1.	Liksey	3,340-3,510	Shara, Liksey, Tulka, Tulka Pho, Trachit
2.	Chumathang	4,000-4,050	Chmathang, Keria, Ney, Skithmong and Kesar
3.	Chushul	4,310	Chushul valley
4.	Pan gong Tso	4,300-4,420	Man-Merik, Spangmik, Kakstl and Phobrang
5.	Kargyam	4,320-4,500	Satuu, Lungburma, Kharbulu, Chibru, Tibetian settlement
6.	Durbuk	3,890-4,500	Tangste, Durduk, Shyok, Shashakul, Tharuk and Chelam
7.	Samad	4,560	Samad val ley, Rahchun and adjoining areas
8.	Khamak	4,560	Khamak valley and adjoining areas
9.	Kurzook	4,420-4,490	Kurzook, Samdho, Ankum and Monastery
10.	Nyoma	4,170	Nidder, Noyma, Mud, Tibetan settlement
11.	Anley	4,200	Rango, Anley, Khaldo, Pongog
12.	Koyal	4,200	Koyal and Demjok and adjoining area

3. Native Environment

The breeding tract of *Changthangi* sheep falls under cold arid climatic zone. In this zone, there is harsh climatic dryness (with very low rainfall) and temperature varies from hot and extreme cold. Because of these climatic conditions the area is know as cold Himalayan desert and it has very low thermal and hydric indexes. The temperature in the area shows wider seasonal and diurnal fluctuations. July and August are the hottest and the January is the coldest. The minimum and maximum temperature fluctuates between - 40° C to 40° C in the Changthang area. The rainfall in the area is very meager. The average relative humidity varies between 13 and 90% with the highest humidity corresponding to the lowest temperature. The snowfall in the area is very scanty and when it falls it is dry and powdery. Wind velocity in the area is very high and the fallen snow gets accumulated in the gorges and this remains there up to summer. In summer, melting of this snow causes regeneration of various grasses in the area where the water of these snowmelts flow through or creeps in. Most part of the breeding tract remains close form November to April due to closure of passes connecting the area with rest of the state.

4. Population Dynamics of Changthangi Sheep

The *Changpas* have mixed sheep and goat flocks. The population status of the sheep and goats in the breeding tract of *Changthangi* sheep, as per the survey conducted during the project, are presented in (Table 2 and 3). Total population of the *Changthangi* sheep is 74,180 heads and sheep to goat ratio in the flock is almost 1:2. The population of the *Changthangi* sheep is lower than the *Changthangi* goat or Changra goat. The first report on the population of *Changthangi* sheep (*Huniya*) was published by Cunningham (1857). According to him total population of this breed (*Huniya*/*Changluk*) was 2,25,000 in 1847 and number had drastically reduced to 73,875 (Census, 1971), 66,822 (Sheep Husbandry Department, J&K, 2000-2001) and slight increase in 2003 AD with population of 98,968 heads (the increase is due to accounting of animals reared by the Tibetan refugees which was not accounted for earlier). The total estimated sheep population under the project in 2003 was 74,180 heads (Table 3).

Table 2: Population status of *Changthangi* sheep in different years

S.No.	Year	Population	Reference
1.	1847	2,25,000	Cunningham (1857AD)
2.	1971	73,875	Livestock Census 1971
3.	2000-2001	66,822	Sheep Husbandry Deptt
4.	2002-03	98,968	Sheep Husbandry Deptt
5.	2003	74,180	Estimated in the project in 2003

Table 3: Break up of estimated population status of sheep and Goat of *Changthang* the year 2003

Species	Name of the Breed	Name of the Block Nyoma		Durbuk		Total
		Male	Female	Male	Female	
Sheep	*Changthangi/*	25,294	32,743	6,729	8,972	74,180
Goat	Changra4	6,379	75,082	14,139	22,891	1,58,491
Overall		51,673	1,07,825	20,868	31,863	2,32,671

5. Housing

Sheep during night area are kept in paddocks. The paddocks have walls of stones piled in an orderly fashion in such a way that air may pass through it. The animal house has no roof. The animals are let out for grazing early in the morning and comeback to paddocks in the evening. There are no *kaccha* or *pacca* animal houses anywhere in the breeding tract. The animals are kept in all the seasons under the sky. The floor of paddocks has 3-4 inch thick sandy soil. There is no system of drainage in the paddock and there is not even need for it as the environment is arid and therefore there is no chance of damping of floors if the paddocks is there. Housing space of these animals varies during summer and winter. A space area of 9 x 9 m is sufficient to accommodate 300 animals during winter and only 200 during summer. Animals of all ages are housed in same paddock and the sheep are housed with the goats. *Rongpas* (stationary *Changpas*) some times provide a *kaccha* type of house in extreme winters.

6. Feeding Practice

The breed is exclusively maintained on natural grazing during summer. On melting of the snow, nutritious clover crops up in the pastures but in winter the dried surface grasses and some under sand herbs and shrubs, after removing the top soil by hooves, are eaten by the sheep. Only during severe winter, if there is snowfall and the grasses are covered by the frost; barely, pelted concentrate feed and some dried grasses are fed to the animals otherwise the flock remains on surface grazing only. The concentrate feed is provided by Husbandry Department on subsidy rates. Fodder and feed resources are short, in general, in the area and improvement of pastures is essential for sustaining the breed and other animals in the area.

7. Grazing

The animals remain in the pasture throughout the day from morning till evening. The grasses growing naturally in the area are extensively grazed by the animals. No cultivated grass is fed at the lower elevations. The straw of only grown crop barley is given to the animals during winter especially in January–February when the pastures get covered with the frost. During autumn and winter as mentioned earlier animals dig out the roots of the pasture grasses and shrubs by their hooves and eat them.

8. Type of Pasture

The *Changthang* area is one of the highest inhabited areas in the world. It falls in the Greater Himalayas and the pastures are obviously mountainous desert type. The elevation of the pasture is up to 5,000 m above MSL. The various types of vegetation cover of the pastures are as under:

- Grasses : The important species of pasture grasses found in the area *Poa pratensis, Festuca rubra, Bromus macrostachy, Elymus sibiricus, Agropyron lange-aristatum,* wild clover, *Loma, Pangsa, Nergyal (Astragalus* sp.), *Kibu* etc.
- Legumes : *Astragalus tribulifolius, Oxytropis microphylla, Ciecr microphylum*
- Bushes and trees : *Caragana vesicolor* (*Toma*-a leguminous bush about 1-2 feet high). *Myricaria germonica, M. prostrate* (a small tree of 2-6 feet high)

- Major fodder shrubs (the local names are given): *Sehyoah, Troia, Aabry, Somoede, Burwak, Yakzes, Ambang, Kyandon, Taksha, Shyyoung, Snalo, Tseery, Sarolo, Talo, Chyut, Umbo, Sirma, Shaya, Penmaa, Sonam shepa, Loma*
- Source of dry fodder : Wild pasture grasses (names are given above) and straw of barley
- Seed and grain feed : Barely (only during winter crisis)

9. Livestock Migration

As already said in whole of the Changthang area livestocks (sheep, goats and yak) are reared under migratory system of rearing. The livestock remains in migration through out the year with defined routes to tread and camping spots to stay. The duration of stay at a given spot depends upon the availability of grazing on the land surface and also the migration pressure of other flocks. In most of the Changthang, the mixed flocks of sheep and goat are reared by the shepherds or persons deputed by the villagers turn by turn in the vicinity of the permanent villages or pasture near by in radius of 10-15 km of villages except in revenue village Phobrang, where livestock migrate to farther distances. Three villages of Changthang are almost totally migratory and almost all the families of these villages move with their livestock. These villages are Kharnak, Samad-Rukchan and Korzok.

The migration pattern is (are) mutually agreed by the animal rears (villagers). In each migration cycle the leader flock(s) is flocks of monastery followed by the flock of "*Amchee*" (traditional village physician). Flocks of the other villagers follow these flocks in the migration route. The stay and movement in the routes are mutually decided by the villagers and the areas to be grazed are also decided by different flocks in and around the migratory route(s).

10. Physical Characters

Coat colour in *Changthangi* sheep is of complex nature, and may be controlled by multiple alleles. The colour ranges from complete black to white and brown. Most of the animals has a mixture of these three colours however the white is predominant.

The head is convex in majority of the cases (80%). However, 13% of the animals were having straight head and only 7% of the animals were with concave type of head. In general, the *Changthangi* sheep has got a typical convex head with a long tapering face. Both the ears are pendulous. Both horned and polled animals are there in *Changthangi* breed. The horns, when present, are two in the number. Wattles were observed in 2% of these cases and majority of the animals did not have any wattle. The tail in *Changthangi* sheep was from medium or short. In 82% of the cases the tail was very short and only in 8% of the cases tail is short. The tail is thin and straight in all the cases.

11. Body Weights and Body Measurements at Different Ages

Body weights and body measurements of *Changthangi* sheep at different ages are presented in the Table 4-9.

At birth: Overall average birth weight is 2.50 ± 0.65 kg. The male lambs are heavier than female lambs. The sex-wise averages of chest girth, height at wither, body length are presented in (Table 4).

Table 4: Birth weights (kg) and measurements (cm) in *Changthangi* sheep

Trait	Sex	N	Mean±SE	Range
Birth weight (kg)	Male	333	2.54 ± 0.67	1.10-3.90
	Female	296	2.464 ± 0.63	1.15-3.90
	Overall	**629**	**2.50 ± 0.65**	**1.10-3.90**
Height at wither (cm)	Male	333	32.99 ± 0.29	24.00-46.00
	Female	296	32.83 ± 0.31	24.5-46.50
	Overall	**629**	**32.91 ± 0.21**	**24.0-46.00**
Chest girth (cm)	Male	333	27.3158 ± 0.48	24.00-49.00
	Female	296	26.6500 ± 0.42	24.00-47.50
	Overall	**629**	**26.9744 ± 0.32**	**24.00-49.00**
Body length (cm)	Male	333	31.16 ± 0.27	25.00-45.00
	Female	296	30.94 ± 0.28	25.00-44.00
	Overall	**629**	**31.06 ± 0.19**	**24.00-49.00**

At three months age: Average overall body weight at 3 months of age in *Changthangi* sheep is 10.51±0.04 kg and it varies between 6.00 to 12.80 kg. Males are heavier than females. The sex-wise averages of chest girth, height at wither, body length are presented in (Table 5).

Table 5: Three months weights (kg) & measurements (cm) in *Changthangi* sheep

Trait	Sex	N	Mean±SE	Range
Body wt. (kg)	Male	327	10.59±0.06	6.00-12.15
	Female	294	10.44±0.06	6.00-12.80
	Overall	**621**	**10.51±0.04**	**6.00-12.80**
Body length (cm)	Male	327	45.02±0.21	40.00-55.00
	Female	294	44.64±0.20	40.00-54.00
	Overall	**621**	**44.83±0.14**	**40.00-55.00**
Chest girth (cm)	Male	327	46.88±0.36	39.00-59.00
	Female	294	48.28±0.37	39.00-59.00
	Overall	**621**	**47.59±0.26**	**39.00-59.00**
Height at wither (cm)	Male	327	43.92±0.25	38.00-55.00
	Female	294	44.40±0.26	38.00-55.00
	Overall	**621**	**44.1664**	**38.00-55.00**

At six months age: The overall average body weight at 6 months of age is 15.68±0.04 with no significant variability between the two sexes; however body measurements were higher in males than the females and it ranges between 12 and 18 kg. The sex-wise averages of chest girth, height at wither, body length are presented in (Table 6).

Table 6: Six months weights (kg) and measurements (cm) in *Changthangi* sheep

Trait	Sex	N	Mean with error	Range
Body weight (kg)	Male	328	15.69±0.06	12.00-18.00
	Female	352	15.67±0.06	12.00-18.00
	Overall	**680**	**15.68±0.04**	**12.00-18.00**
Body length (cm)	Male	328	51.25±0.40	40.00-70.00
	Female	352	50.23±0.39	40.00-69.00
	Overall	**680**	**50.72±0.28**	**40.00-70.00**
Chest girth (cm)	Male	328	59.58±0.55	42.00-80.00
	Female	352	58.94±0.52	42.00-80.00
	Overall	680	59.25±0.38	42.00-80.00
Height at wither (cm)	Male	328	51.45±0.39	39.00-67.00
	Female	352	51.34±0.37	39.00-67.00
	Overall	**680**	**51.39±0.27**	**39.00-67.00**

At nine and twelve months age: The overall average body weight at 9 and 12 months of age are 23.86 ± 0.25 and 24.52 ± 0.20 kg, respectively. Animals do not grow too much between 9 and 12 months of age in spite of the fact that it is the growing phase of the animals. This is due to the extreme winter condition which corresponds with this phase of the age. Males are either equal or better than the females in weight as well as body measurement traits. The sex-wise averages of chest girth, height at wither, body length of 9 and 12 months of age are presented in (Table 7 and 8), respectively.

Table 7: Nine months body weights (kg) and measurement (cm)

Trait	Sex	N	Mean with error	Range
Body weight (kg)	Male	261	23.86±0.36	15.00-29.00
	Female	316	23.86±0.36	15.00-28.00
	Overall	**577**	**23.86±0.25**	**15.00-29.00**
Body length (cm)	Male	261	60.93±0.48	46.00-80.00
	Female	316	60.41±0.59	44.00-80.00
	Overall	**577**	**60.06±0.49**	**44.00-80.00**
Chest girth (cm)	Male	261	64.85±0.50	46.00-81.00
	Female	316	63.71±0.51	48.00-80.00
	Overall	**577**	**64.30±0.36**	**46.00-81.00**
Height at wither (cm)	Male	261	52.71±0.43	43.00-70.00
	Female	316	51.92±0.47	41.00-68.00
	Overall	**577**	**52.33±0.32**	**41.00-70.00**

Table 8: One year above body weights (kg) and measurement (cm)

Trait	Sex	N	Mean±SE	Range
Body weight (kg)	Male	280	24.52±0.20	17.00-30.00
	Female	365	24.08±0.24	17.00-30.00
	Overall	**645**	**24.31±0.15**	**17.00-30.00**
Body length (cm)	Male	280	60.93±0.48	51.00-84.40
	Female	365	60.06±0.49	51.00-81.70
	Overall	**645**	**60.51±0.34**	**51.00-84.40**
Chest girth (cm)	Male	280	64.85±0.50	51.00-87.50
	Female	365	63.71±0.51	53.00-88.00
	Overall	**645**	**64.30±0.36**	**51.00-88.00**
Height at wither (cm)	Male	280	52.71±0.43	44.00-77.70
	Female	365	51.92±0.47	41.00-77.80
	Overall	**645**	**52.33±0.32**	**41.00-77.80**

Adult age: There exists a great variability in the body weight of adult animals and it varied between 25 to 69 kg. The animals are of better body weights in the flocks of higher altitudes. The average overall body weight of the adults is 40.96 ± 0.46 kg with males significantly heavier than females. The superiority of males over females does exist in body measurements too. The sex-wise averages of chest girth, height at wither, body length of adult animals are presented in (Table 9).

Table 9: Adult weights (kg) and measurements (cm) in Changthangi sheep

Trait	Sex	N	Mean ±SE	Range
	Male	341	44.38±0.56	25.00-69.00
Body weight (kg)	Female	318	35.60±0.62	20.00-54.00
	Overall	**659**	**40.96±0.46**	**20.00-69.00**
	Male	341	78.18±0.45	46.00-80.00
Body length (cm)	Female	318	71.91±0.52	60.00-89.00
	Overall	**659**	**75.73±0.36**	**60.00-99.00**
	Male	341	90.49±0.64	52.00-115.00
Chest girth (cm)	Female	318	81.89±0.78	55.00-104.00
	Overall	**659**	**87.14±0.52**	**52.00-115.00**
	Male	341	72.58±0.42	50.00-96.00
Height at wither (cm)	Female	318	66.26±0.45	50.00-86.00
	Overall	**659**	**70.12±0.33**	**50.00-96.00**

12. Peproductive Traits

12.1 Reproductive Performances

The reproductive performances of the males and females are given in (Table 10).

Table 10: Reproductive performances of the males and females of Changthangi sheep

1. Males	
a. Age at first mating	2.00-2.50 years
2. *Females*	
a. Age at first oestrus	15-17 months
b. Oestrous cycle duration (days)	21-22 days
c. Estrous duration (hours)	36-48 hours
d. Seasonality	July to December
e. Age at first lambing	28 months
f. Lambing interval	One year
g. Litter size	One (twinning less than 1%)
h. Lifetime number of matings	6-8

12.2 Reproductive Disorders

The frequencies of abortion and still birth are very rare (<1%). Infectious metritis and pyometra is rarely reported. However, reproductive disorders due to nutritional deficiencies occur when the pastures are not good in the area.

12.3 Reproductive Management

a. *Mating method* : The ewes are mated by the natural service and the ratio of male to female in breeding is kept 1:25-40. No artificial service is practiced for breeding of these sheep.

b. *Breeding season* : In the *Changthangi* sheep animals are seasonal breeders and breeding season is primarily from July-November the farmer of high altitude, bred their sheep in July and August but farmers living comparatively at the lower altitudes bred their animals in September to November. The flock breeding is normally practised and the rams 3-4 per 100 breedable ewes are utilized. No breeding record is maintained by the farmers.

c. *Lambing season* : The lambing starts in the month of December at higher altitudes and goes upto March. However, main lambing season at high altitude is from December to February though a few lambs come thereafter too. The farmers prefer the extreme winter condition for lambing as they have low mortality of the new born lambs during this period. This low mortality during this season is attributed to better body condition of ewes and the minimum fluctuation of temperature during this part of the year. The lambs, in particular, are highly sensitive to fluctuation of temperature because of lowly develop thermoregulatory system. The lambing percentage is between 60-70% in the flocks of *Changthangi* sheep.

d. *Mothering ability* : No milk is taken from the sheep however the sheep have sufficient milk for their lambs.

13. Wool Production

a. *Utility of wool* : The wool is used to make rugs and cloth of special type. They also make quilts and mattresses of special type by this wool and are locally called as *Rstkotul* and *Rstdhan*, respectively. These items are prepared and processed generally by the women. These items get good price in the local market. Changpas consume the mutton of their dead sheep and surplus animals are sold to mutton dealer of Leh market. The pelts of the animal are used as flooring material by the Changpas to keep their tents (*Rebo*) warm.

b. *Wool cover* : *Changthangi* sheep has wool on neck, body and on hind quarter. Faces and legs have no wool growth. In some of the animals wool do not grow even on neck.

c. *Shearing season* : *Changthangi* sheep are usually shorn once in a year the shearing season is in May–June. At low altitude animals are shorn twice a year in May–June and September-October.

d. *Greasy wool production* : There is higher wool production from males than females and with increasing age production increases up to 2.5 years and there after is static or a bit low. Wool production potential of animal varies from 0.50 kg to 2.65 kg. Age-wise and sex-wise wool production by the *Changthangi* sheep is presented in the (Table 11). The male produced significantly higher fleece than female coevals. The canary coloration has not been observed in the fleece of the *Changthang* sheep.

Table 11: Sex-wise and age-wise greasy wool production in *Changthangi* sheep

	N	Mean± SE (kg)	Range
Sex-wise			
Male	164	1.53±0.05	0.50-2.65
Female	145	1.31±0.04	0.50-2.50
Total	**309**	**1.42±0.03**	**0.50-2.65**
Age (years)			
1.5	75	0.67±0.02	0.50-1.60
2.5	53	1.69±0.59	0.50-2.65
3.5 and above	181	1.66±0.03	0.90-2.60
Total	**309**	**1.42±0.03**	**0.50-2.65**

e. *Fleece quality* : The wool samples were collected at the time of shearing from the mid side region, these samples were properly labeled and put in the polythene bag and carried to Fleece Testing Laboratory of the University, for ascertaining the quality of the wool produced. The mean length of the staple and the average number of crimps per cm of wool, fiber diameter and medullation percentage were recorded. Keeping in view the staple length this breed has the highest one among the Indian breeds of sheep. Based on the quality the wool, it is good for use in making rugs & local woolen products. The various wool quality attributes studied in the present study area are presented in the (Table 12 to 14).

Table 12: Wool quality traits of *Changthangi* sheep (sex-wise)

Traits	Sex	No. of Animals	Mean + SE	Range
Staple length (cm)	Male	201	13.79±0.44	6.80-20.10
	Female	190	11.25±0.15	5.50-19.30
	Total	**391**	**12.86±0.26**	**5.50-20.10**
No. of crimps	Male	201	6.88±0.23	6.48-7.50
	Female	190	6.48±0.26	5.96-7.00
	Total	**391**	**6.69±0.17**	**5.96-7.50**
Average crimp/cm	Male	201	0.53±0.01	0.49-0.57
	Female	190	0.55±0.01	0.51-0.59
	Total	**391**	**0.54±0.01**	**0.49-0.59**
Fibre length (cm)	Male	201	20.90±0.42	7.35-35.40
	Female	190	17.92±0.40	6.20-31.40
	Total	**391**	**19.45±0.30**	**6.20-35.40**
Medullation %	Male	201	5.41±1.04	1.08-12.50
	Female	190	4.17±0.50	0.80-11.80
	Total	**391**	**4.60±0.05**	**0.80-12.50**

Table 13: Age-wise wool quality of *Changthangi* sheep

Traits	Age(in years)	No. of animals	Mean	Range
Staple length (cm)	1	114	10.69±0.34	5.50-20.10
	2-3	164	13.73±0.52	6.80-19.30
	Adults	113	12.37±0.31	6.42-19.20
	Total	**391**	**12.26±0.39**	**5.50-20.10**
No. of crimps	1	114	6.48±0.23	2.40-23.60
	2-3	164	6.66±0.30	1.90-27.20
	Adults	113	6.94±0.34	1.90-23.60
	Total	**391**	**6.64±0.34**	**1.90-27.20**
Crimp average	1	114	0.60±0.01	0.26-1.87
	2-3	164	0.50±0.02	0.11-2.10
	Adults	113	0.54±0.02	0.20-2.09
	Total	**391**	**0.54±0.01**	**0.11-2.10**
Fibre length	1	114	16.45±0.62	6.20-31.60
	2-3	164	21.05±0.43	7.35-35.40
	Adults	113	20.17±0.43	7.50-31.40
	Total	**391**	**19.45**±0.30	**2.35-35.40**
Medullation %	1	15	**6.05**±1.01	**1.10-12.50**
	2-3	13	3.60±0.52	**1.10-8.30**
	Adults	12	3.88±0.83	**0.80-11.80**
	Total	**40**	**4.60**±0.50	**0.80-12.50**

Table 14: Sex-wise and age-wise average fiber diameter of *Changthangi* sheep

Age group	Sex	N	Average diameter (μ) with mean	Range (μ)
1 Year	Male	66	33.95 ± 1.46	28.11-51.48
	Female	62	34.78 ± 1.03	27.64-49.44
	Overall	**128**	**34.46 ± 0.84**	**27.64-51.48**
2-3 Years	Male	50	34.08 ± 1.81	27.67-59.40
	Female	62	36.67 ± 1.94	28.31-63.36
	Overall	**112**	**35.68 ± 1.38**	**27.67-63.36**
4 Years and above	Male	54	31.21 ± 0.22	29.22-32.86
	Female	56	30.99 ± 0.36	29.10-33.32
	Overall	**110**	**31.03 ± 0.19**	**29.10-32.86**

f. *Staple length* : Male had more staple length than their female counterpart. Short staple in females than male is probably due to double production in females (lamb+wool) than males (only wool) wherein ewes have to divide the body resources between lamb and wool.

g. *Number of crimps* : The crispiness in *Changthangi* is very less and it is depicted by long staple length too, as there exits negatively correlation between these two traits.

h. *Medullation percentage* : It is less in this breed of sheep therefore the wool is not well suited for quality carpets. However, the rugs and other locally made articles of this wool are in good demand in the local market.

i. *Fibre length* : Fibre length is higher in males than females. Lesser fiber length in females is also evident from lower staple length in females than males. Based on fiber length, the *Changthangi* sheep produce wool with longest fibre among the Indian sheep breeds.

14. Disease and Mortality

14.1 Diseases and Parasites

The occurrences of the diseases in this breed of sheep in its native tract are very less. Common bacterial diseases affecting the *Changthangi* sheep are Diarrhea and Clostridial diseases and occasional incidence of Brucellosis are also reported. The commonly reported viral diseases are Foot and Mouth disease, Ecthyma, Sheep pox. The sheep are sometimes vaccinated against Foot and Mouth disease, Pox and Clostridial disease (Lamb dysentery, Struck, Malignant oedema, Pulpy Kidney, Black's disease). For protection against the Clostridia disease multi-component Clostridial vaccine is administered.

14.2 Parasitic Status

Gastrointestinal Protozoa : The only species of protozoa recorded was *Eimeria* (Coccidia).

Cestodes : The only cestode eggs observed in the fecal samples of *Changthangi* sheep were of *Moniezia* species and the positive cases were 11.48% with low (+) and moderate (++) infection cases of 9.55% and 1.93% cases of mild infection in case of *Nematodirus* species.

14.3 Measures Against Diseases

Vaccination : Sheep farmers vaccinate against FMD, CCPP and Clostridial diseases.

Dosing : Dosing with anthelmentics in these animals is done from time to time in accordance with season and parasitic loads in the flocks.

Dipping : Dipping on large scale against mange and ectoparasites is done periodically and Ectomin is the dipping material of choice.

14.4 Mortality in Changthangi Sheep

The mortality for 0-12 months decreases with increasing altitude from 15.00 to 24.83 per cent. The chief cause of mortality is respiratory diseases followed by nutritional reasons and attack by wild animals. Average mortality is 3.04-12.62% in the animals aging more than one year of age. The mortality rate increases with increase in the elevations.

15. Carcass Characteristics

Age at Slaughter

The average age at slaughter is between 20-60 months. This age is 20-32 months for male and 32-60 months for females.

Carcass Weight (kg)

Average carcass weight is 16.78 kg for males and 15.86 kg for female (Table 15).

Dressing Percentage

The over all dressing percentage in *Changthangi* sheep is 45.41 ± 0.25% with values varying from 40.00 - 51.85% (Table 15).

Pelt Characteristics

The over all skin weight (wet) in *Changthangi* sheep is 2.99 ± 0.05 kg with values varying from 1.00-5.00 kg (Table 15). The over all skin weight (dry) in *Changthangi* sheep is 1.81 ± 0.02 Kg with values varying from 0.50-3.50 kg.

Table 15: Carcass traits of *Changthangi* sheep

Trait		Total no. of animals	Mean ± SE	Range
Carcass weight (kg)	Male	62	16.78±0.54	7.00-27.00
	Female	95	15.86±0.33	6.00-25.00
	Total	**157**	**16.23±0.29**	**6.00-27.00**
Dressing %	Male	62	45.47±0.40	40.00-51.85
	Female	95	45.38±0.34	40.00-51.72
	Total	**157**	**45.41±0.25**	**40.00-51.85**
Skin wt. (wet)(kg)	Male	62	3.25±0.08	2.00-5.00
	Female	95	2.82±0.7	1.00-4.00
	Total	**157**	**2.99±0.05**	**1.00-5.00**
Skin wt. (dry)(kg)	Male	62	1.92±0.07	1.00-3.50
	Female	95	1.74±0.05	0.50-2.70
	Total	**157**	**1.81±0.02**	**0.50-3.50**
Skin length (cm)	Male	62	101.30±2.43	60.00-150.00
	Female	95	95.34±1.64	60.00-130.00
	Total	**157**	**97.69±1.40**	**60.00-150.00**
Skin width (cm)	Male	62	58.06±1.61	30.60-99.00
	Female	95	56.73±0.91	32.00-82.00
	Total	**157**	**57.25±0.84**	**30.60-99.00**

The overall skin length in *Changthangi* sheep is 97.69 ± 1.40 cm with values varying from 60.00-150.00 cm. (Table 15). The over all skin width in *Changthangi* sheep is 57.25 ± 0.84 cm with values varying from 30.60 to 99.00 cm.

16. Elite Animals

16.1 Criterion for Elite Animals

- Animal should be regular breeder
- Body weight should be more than 50 kg in males and more than 40 kg in females

- Greasy wool production by the animals should be more than 2.00 kg

16.2 Elite Flocks

All the sheep breeders of high elevation have some animals of elite type in their flocks. So, instead of giving name of the breeders, name of the villages having such animals is presented below:

- Korzok
- Samad
- Anely
- Koyul
- Nyoma

17. Important Features of Breed

1. It is the only breed of sheep adapted to cold arid condition low hydric and hypoxic conditions with good returns to farmers
2. It has the longest staple wool fiber among the Indian sheep breed
3. The breed is the tallest of Indian breed
4. The breed produces the strongest wool among the Indian breeds
5. Great variability is there in the body weights and thereby heralding great chances of improvement in meat production
6. Variability exists in wool production and wool quality, therefore, if required, these traits can also be set at desirable level (but within physiological limits of the breed)
7. Breed produces good quality mutton greatly relished by the local people of the area.

Recommendations

- Ram exchange programme among the farmers needs due consideration to avoid in-breeding and to have enhancement in returns from this sheep
- Improvement programmes increasing bodyweights/carcass weights should be initiated
- Stress on wool production needs to be avoided. (The above mentioned two factors shall increase the economic returns to the farmers and enhance the interest of the farmers in rearing this sheep. In return, the breed shall get conserved automatically).

- Advantages of crossbreeding of this sheep with hill sheep must be explored for increasing body weights of the small hill sheep
- Breed Society or Breed Cooperative must be established to conserve and improve the breed
- Pasture improvement programme in potential areas of Changthang must be taken in hand and the existing program must be encouraged.

References

Ganai, T.A.S. (2003). *Changthangi* Sheep Unit of ICAR- NBAGR Network Project

Razdan, H.K. (1995). Chemical pools of m/n cation in cold Arid zone of Ladakh as influenced by soil proportions M.Sc. Agri. Thesis SKUAST.

□□□

17

Yak Husbandry : An Overview

GAUTAM MONDAL
Regional Agricultural Research Sub-Station, SK University of Agriculture Science & Technology, Kargil - 194103 (J&K)

Yaks are large ruminants and one of the highest dwelling animals in the world. High hills of the Himalayan range at high altitude of 3500 m or more above the mean sea level where the coldest and extreme weather prevails are the natural habitat of the yak. The common saying that 'what moves Tibet and Mongolia does so on yaks' highlights the importance of yak in high altitudes. Wild yak is normal inhabitant of North Tibetan desert which lacks of trees and bushes at a minimum elevation of 5.0 km. They are habitant of coldest, wildest and most desolate mountains of both arctic and desert condition. Colour of wild yak is a uniform blackish brown with a little white on muzzle. This had probably been domesticated thousand years B.C. by the Tibetans and much smaller than its wild ancestors. The colour of domestic yak is brown, reddish grey and coat resembles the wild ones, although in most cases, they have longer stomach hairs. The horns are weaker and even hornless animals may be observed at the ratio of 1:100. The utility of yak in high altitude for the people living there is invaluable, i.e., it is used for carrying loads which can not be taken by other animals, agricultural use; its milk which has a high nutritive value due to its high fat content than cow, and it is also used for meat purpose during extreme winter when no vegetable is grown in the region and road is closed for more than six months from adjoining areas like Kashmir valley or Himachal Pradesh.

Yak rearing is closely associated with the social and cultural life of the people in the Tibetan plateau and other parts around Himalayan range. The yak, however, a component of religious practices and manifestations of Tibetan Buddhism, though there is no large farm maintaining the yak rather, small no. of yaks are reared by individual family. In Bodkharboo, a Govt. maintained yak farm is there but no. of yak is very less. In some areas yak population is decreasing due to mechanization in agriculture and transportation and poaching etc. but its use in the everyday life of an average farmer is unlimited.

Under Indian situation, Arunachal Pradesh has highest no. of yak population and Indian Council of Agricultural Research (ICAR) has set up National Research Centre on Yak. Breeding, feeding and other management practices are the main research criterion for the scientists working there but under Ladakh condition, no research work has been conducted so far. It is an important animal in the region as far as milk, meat and drought is concerned and research should be conducted on cross breeding with cattle, its efficiency to produce milk and meat under confinement as well as in rangeland, nutrient requirement for different physiological stages and crosses. Its conservation is also necessary as some places are prone to poaching and people replace it with other ruminants like cattle.

Zoological Classification

Phylum	Chordata
Class	Mammalia
Order	Artiodactyla
Family	Bovidae
Genus	Bos
Species	*Bos grunniens* (domesticated)
	Poephagus grunniens (wild)

Local name of wild yak is *Tibetan dong* or *brong dong* and domestic yak is *pegu* while in Hindi wild and domesticated yak are called as *chour gau* and *ban chour*, respectively. Like other domestic ruminants, i.e., cow, goat the yak has same chromosome no. (2n = 60).

Origin, Domestication and Distribution

It is a native of Tibet and surrounding areas in Central Asia. It is distributed along with Tibet, China, India, Pakistan, northern Afghanistan, Bhutan, Mongolia and in some Russian states. In India, Ladakh, Spiti valley, some North eastern states like Arunachal Pradesh, Sikkim and Nagaland have maximum no. of yak. Unequivocal evidence to link the modern yak to its earliest ancestors is not available. Fossil evidences suggest that yak were extensively distributed in North-eastern Eurasia in the late Tertiary period (2.5 million years ago) and that these are the ancestors of wild yak found as Pleistocene fossils in northern China, Inner Mongolia (China), eastern Siberia and northern mid-Asia and on a line roughly connecting these locations (Olsen, 1991). The present domestic yak is descended from wild yak, which may have been domesticated by Tibetans residing in a dry and cold area that covers more than half of Tibet.

The outstanding achievement of the Tibetan culture was taming of the wild beasts for domestic purposes. Sheep and goats had already been tamed successfully and this led to the taming of yak, horse and other herbivores and the development of a society based on animal husbandry. Domestication of yak led to progress in terms of economic development of the people because of the value of the yak and its products of milk, hair, hides and meat, dung as a fuel and fertilizer. Yak expanded outward to East to West, and from North to South for multipurpose use in day to day life of hilly people.

Today, yaks are found extensively in alpine and sub-alpine regions of western China at the height of 2-5 km with a cold and semi humid climate. In Mongolia, majority of yak population are found in Hangay and Hovsgol mountains and in the high altitudes, on the western and northern side of the country. Yaks are distributed in Commonwealth of Independent States in the narrow mountain area on the borders with China and Mongolia and these are introduced in high alpine areas for drought as well as meat purposes. Yak of Nepal and Bhutan are on the southern slopes of the Himalayas while those of India are distributed in the high altitude northern provinces (Ladakh, Spiti and Garhwal) and North Eastern States like Arunachal Pradesh, Nagaland and Sikkim. Other pockets of yak populations are in alpine areas of Afghanistan and Pakistan, adjacent to the Qinghai-Tibetan Plateau. In modern times, for research purpose and to acclimatize to other environment, some yak has been exported to different places but population is negligible.

Population

Yak distributed mainly in the coldest higher mountains of Asia, and Commonwealth of Independent States (former USSR). In the Asian countries, over 90% of yak are distributed in China and Tibet (Xu Shangzhon, 1998), whereas rest are found in Mongolia, India, Pakistan, Nepal, Bhutan and Afghanistan. Total population of wild yak is 15,000 but declining rapidly due to hunting by people, sharing of habitat and disease transmission by domestic yak (Namgyal, 2006). Wild and domestic yak in the world is around 13.75 million. Though there is no census on yak population in India but its crosses with cattle is counted. Among the yak rearing states, Arunachal Pradesh is in leading position by constituting over 70 percent of population. Ladakh province of Jammu & Kashmir which is known for its coldest and most dry climate in India have the animals as follows (FRL-9):

S. No.	Livestock	Population
1.	Cattle (native and crossbreds)	84, 362
2.	Goat	4, 63, 661
3	Sheep	2, 62, 473
4	Equines	2, 055
5	Yak and its crosses	36, 238
6	Camel (double humped)	200

Distinctive Characters

a. Characteristics

In the Tibetan language, yak is pronounced as 'yag', mean yak bull and *dri* as the female yak. Other languages follow this name closely. The ancient Chinese people called the animal *Ya Niu* and is written in a bigger picture- denoting its large body, outstretched horns, long hair and big tail. In time, the name was reduced to a word pronounced as 'ya' and thereafter name and pronunciation kept on changing.

Presently, *Li Niu* was said to live in the mountains and denoted the wild yak, while *Mao Niu* was used to denote the domestic yak (Li Ruimin, 1986). Now a days, different countries or even different locations in the same country uses different names including both male and female. Appearance, hair colour and tail of wild yak do not resemble to those of the domestic yak; wild yak has the larger body size as compared to domesticated one. But in course of time, due to mating of wild and domesticated yak, hair colour has become same except grey white hair which is found in domestic yak around the mouth. Miller and Steane (1996) stated that millions of wild yak in the Tibetan plateau are also mingling and mating with domestic one and resemblance in hair colour suggests about the ancestral appearance.

The wild yak is massive and of strong built with thick, long hair covering the whole body, drooping head, high humped shoulders, a straight back and short limbs. The colour of the hair is jet brown or jet black and coarse hair hang from flanks. A golden brown colour may also be seen among wild yak in the Aru Basin of North-western Tibet (Schaller, 1998). A colour line down the back of the body and behind the withers is silver grey and there are grey-white hair around the muzzle. Dewlap is absent and the back has 14 pairs of ribs instead of 13 in all other oxen. The skull is bulky, broad and well proportioned to the length of frontal and nasal bone (Banerjee, 1998). The horns are round and very thick, arch is open and the head shape has a fierce appearance. The wild yak has a very acute sense of smell, is highly alert and timid; it tries to escape immediately on sensing or seeing people or other animals. Wild yak dislikes heat but are highly tolerant of cold and starvation. Wild yak bulls often wander off individually during the non-breeding season in hilly areas away from the high mountains. Such males are known to attack people on remote roads. Wild female yak does not normally cry, it will let out squeaks and cries during oestrus and in the breeding season and if it meets other wild beasts, just as does the domestic yak. Domesticated yak is much shorter than its wild ancestors. These are brown, reddish grey in colour in addition to black and black/ brown ancestors. These also make frequent grunting sounds and withstand much lower cold and starvation than wild one. Under this condition, it has been also observed that the yaks are grazing on fields which are covered with thick snow during peak winter.

State Animal Husbandry Dept. (Jammu and Kashmir) maintains a yak farm at Bodhkharboo which has small no. of animals and physical measurement of yak bull and cow are as follow (Table 1):

Table 1 : Physical measurement of bull and cow

Parameters	Bulls (cm)	Cow (cm)
Height of wither	111-121	100-116
Chest depth	56-72	52-70
Chest width	30-42	25-41
Oblique length	120-149	110-133
Heart girth	170-198	150-171
Cannon circumference	15-21	14-20
Hip width	30-43	29-37

b. Adaptation to Environment

Yak has many characteristics for adaptation to environment like extreme cold, high altitude with low oxygen tension in the air and high solar radiation, steep mountain with short grazing period, low availability of fodder/feed and difficulty to move from one place to another by the people. Yak may perform at an ambient temperature of 5-15°C but can perform at -55°C (Drass block which is 2nd coldest inhabited place in Kargil district of Ladakh). These animals have some special physiological characteristics by which they can withstand the climatic harshness. They withstand cold by conserving heat and require less food. Heat loss is decreased by thick layer of hair. Again, they accumulate a subcutaneous fat which helps to conserve the body heat. Yak can adapt in a area where atmospheric oxygen is low as experienced in Ladakh, due to large chest (14-15 pairs of thoracic ribs), large lungs and a large heart relative to their body size. Solar radiation is resisted by highly pigmented skin and black hair over their coat. It is also reported that yak has large rumen relative to their body size and can graze very short herbage like sheep which help them adapt the condition like Ladakh.

Utility

Yak is domesticated long ago by the Tibetans and now these animals are used for different purposes like milk and meat. Initially, these animals were used for carrying of loads from one place to another which were not easily carried by the people. But gradually people understood the utility in various field like milk and meat particularly in places like Ladakh, which are detached from the rest of the country for more than 6 months due to thick snow. The main utility of the animals are as follows :

a. A single adult animals can carry up to 150 kg of load in steep mountain route which otherwise is not transportable.

b. Initially the animals were used for drought purpose but gradually people understood the other production potential of the animal. Milk production is lower than cows/ lactation (around 500 kg) but it is of high nutritive value (fat content 7-11%) (Banerjee, 1998) and comparable to buffaloes. Though milk quality and quantity is highly dependent on availability of feeds, altitude and management practice of the farmers. It is well known to the farmer that crosses of cattle and yak produces more milk than both its parents and for the purpose no. of yak is diminishing in the area. Crossbred (F_1) males are always sterile and called as 'tzo' in Ladakh.

c. Under conditions like Ladakh, yak is slaughtered during winter when road is closed for more than 6 months and availability of vegetable is rare to the common people in the region. Market survey shows that everyday during summer months in Kargil city itself 8.0 q of sheep, goat and poultry is being slaughtered. It is reported that the dressing percentage of yak is around 50-55% and shoulder meat is found to contain water 68%, protein 20%, fat 7% and ash 5.1%, Ca 1.283%, phosphorus 1.08% (Banerjee, 1998).

d. Fine hair and coarse hair from a yak may be around 300-600 g and 1-2 kg, respectively. Fine hair is used for making of cloths whereas later is used for carpet making, bag or rope preparation. Even today, skin of the animal is used as such by aged people as winter cloth after drying thoroughly. Fibre production is dependent on different factors like age and size of the animal, sex and with the method of harvesting. Proportion of down fibre is high in calves and declines as the animal gets older. Down fibre grows as additional protection for the yak over winter and has to be harvested prior to being shed in the early summer. The down fibre is much valued for textiles. Hair of the animals are also used for making waterproof *Rebo* tents (Namgyal, 2006).

e. Manure available from yak is used by the farmers in Ladakh, as soil of the area lacks organic fertilizers. But the nutrients present in dung are to be explored for the exact calculation of nutrient supply to the soil. Different apple/apricot growing places have the age old practice to supply dung as nutrient and mulch to the plants.

f. Now a days, farmers are using tractor or other machines for the agricultural purpose, bullocks can be used for ploughing of field or carrying of loads from one place to another.

Reproduction

Female: Yak is seasonal breeder and the time oestrous is affected by climatic factors, grass availability, both altitude and latitude and nutritional status of the animal. With the higher altitudes, breeding season is delayed which allows calves to be born in somewhat warmer weather or closer to the onset of warm weather to allow the dam and offspring to get fresh herbage. From that point of view, a delay in the ability to breed must be regarded as a sensible adaptive response. The yak that mated late in the season have less chance to be remated that year and conception may be failed than those who mated in early season and chances of conception is more. Also, calves that are born late in the year have insufficient time to get into good body condition to improve their chances of surviving the extreme cold of first winter.

First oestrous may be observed in 2-3 years but depends on body development and nutritional status of the animal. Though this also varies with the altitude, season of birth as well as availability of fodder. Sign of oestrous is like that of cow and can be well detected by the farmers. Mounting on to other animal is less pronounced than cow but swelling of vulva or mucous discharge from the vagina may be seen. In general, female shows qestrous in the morning or late in the afternoon. Length of qestrous and duration of qestrous varies from 18-22 days and 1-2 days, respectively. Research showed that qestrous can be induced which can improve the reproductive efficiency of the yak.

Gestation period of yak is around 258 days (range 253-278 days) and this is helpful in less stressed condition of yak and that gives birth of a smaller calf, and less O_2 in the environment helps to sustain both dam and calf. Conception rate varies with the management system, disease control, nutritional status and hormone profile of the animal. Post-partum anetrous period is around 125 days though it varies from 80-200 day under this condition.

Male: Anatomical and histological structures of the male genitalia are very similar to other bovines except small and hairy scrotum in yak due to adaptation to the environment. It has been seen that the age of puberty is 3 years and ejaculate volume, concentration and motility of sperm in semen rose steadily from age 3 to 9 and then declines (Zhang Yun, 1994). Yak bulls fight each other for possession and strongest bull has the maximum mates.

Productive Characters

Growth: Rapid growth is required for survival of the animal in first winter. The offspring born in early warm season have more chances of survival than those born in comparatively colder months. After getting birth of the offspring these are to be cared as in case of calf and fed colostrums initially and then reared on milk. Calves are initially not allowed to graze with the dam but after certain period of time (weaning) is allowed even at night. Growth pattern of male and female calves is different like other mammals and dependent on season of birth, milk availability and management practices of the farmer. Local farmers said that during winter months loss of body weight occurs due to unavailability of quality feed which may go up to 20-25%. Growth of yak takes place up to 6-8 yrs of age and after that no growth takes place till death (20-25 yrs). Meat production from yak is dependent on age of the animal though young one is preferred by the people. Again, age of the animal has an effect on carcass quality and yield, though pure and crosses of yak have the difference on yield and quality of meat. Supplementary feeding with maize, urea and bone meal along with grazing have an additional effect on growth and hence on meat production of the animal. Supplementary feeding with concentrate during winter has beneficial effect on meat production but farmers claims that it is not economically feasible under this condition.

Milk production: Milk yield as well as lactation period is low as compared to cattle. Hybrids, i.e., cross of yak and cattle are more productive than yak though the hybrid male offsprings are sterile. Milk production in a single lactation is low as compared to cow though the milk from yak is of high quality with i.e. moré fat (Table 2). Milk yield varies depending on various factors, i.e., stages of lactation, nutrient availability during maximum growth of the foetus or advanced stage of pregnancy, early lactation and management practice of the farmer. Though there is normal diurnal variation in yield as well as composition like other animals. Lactation period varies with the management practices of the farmer and it does not readily let down milk due to strong sphincter muscles in small teats as compared to cow.

Table 2: Comparison of milk composition of different ruminants

	Water	Total solid	Fat	Lactose	Protein	Ash
Yak	81.85	19.15	10.90	–	4.06	0.98
Sahiwal	86.70	13.30	3.60	5.00	3.30	0.66
Buffalo	83.20	16.80	7.40	4.90	3.80	0.78
Sheep	83.70	16.30	5.30	4.60	5.50	6.90
Goat	88.00	12.00	3.50	4.60	3.10	0.79

Others: In addition to milk and meat, it produces so many valuable items like hair and horn etc. which have been used by the people of the region. Productivity of the materials have been discussed earlier.

Management of Yak

a. *Farm:* Management is a term that includes art as well as science of combining ideas, available resources and labour to produce and market a worthwhile product successfully and profitably. There are different functions of management which are dependent on a particular situation but following five are major the functions : planning, organizing, co-ordination, direct and control. Under Ladakh condition, there is no large scale farm maintained and 0-5 yak or cow crossbred along with few crossbred cow is reared by the farmers. Though there is sheep and goat also maintained in every family for meat, fibre as well as milk. Animals are maintained primarily for domestic use of milk, meat and fibre and extra is marketed at lower rate. Proper management practices are not followed by the farmers and the animals are kept only for domestic use. No farm records, no budgeting (comparative, annual and partial) is maintained by the farmers. Two types of management is practiced in Ladakh, i.e., summer or hill grazing and winter stall feeding. During stall feeding period mainly alfalfa hay, straws from oats, barley and wheat is given to the animals. If the animals are in productive level are given grains or some time compound cattle feed from the market. During hill grazing period the animals are allowed to graze on hill tops or a specific area allotted for the each village. This time animals feed on wild grasses, shrubs and other wild plants. No extra feed is given to the animals in this period.

Care of cow and calf: Domestic yak or its crosses should be taken care of during parturition. Close observation should start 10 day before the expected date. Sign of parturition is like that of other bovines, i.e., swelling of udder, swelling of vulva and drooping away ligament around the tail head. Generally, it does not require any assistance unless the calf is reasonably healthy and in vigorous state. The calving pen should be well ventilated, clean and disinfected properly. Normal parturition takes place within 2 hr and then exterior of genitalia should be washed with clear water with few crystals of $KMnO_4$. Just after birth all the mucus is to be removed from the nose and mouth of the calf and naval should be detached 2 inch away from body, allowed to drain and some antiseptics to be applied. Colostrums is then fed to the calf which allows the calf to absorb immunoglobulin present. Weaning is not commonly practiced in Ladakh and calf is allowed to graze along with dam. Farmers rear very small no. of large ruminants and the calf born in late season slaughtered if optimum growth does not takes place.

Housing: Housing or shelter should give comfort to animals by protecting it from cold and hot weather and natural enemies. It should be planned and sufficient floor space should be there for the animals. In Ladakh, as weather is extreme (-30 to +37°C) housing should be good enough to protect the animals. Farmers who rear small no. of animals, have no separate shed for the animals. On the ground floor animals are kept along with poultry and on the upstairs people used to stay. This is due to the concept that the heat generated by the animals will warm the building. During summer all the animals are allowed for hill grazing and no shed is required. In majority of places, small ruminants are allowed for hill grazing so for yak and other large ruminants housing is necessary. To warm the room, some people developed 'green house' in winter months. It is the extension of concrete animal shed (south facing) which is covered with UV film as a result during day time inside of the room becomes warm and comfortable. But with the onset of summer, this UV film is replaced with straws or other grasses to protect from intense solar radiation.

b. *Feeding and nutrition:* yaks are ruminant and survive only on grass like other ruminants. Fermentation takes place in rumen and produces acetate, propionate and butyrate. Among which propionate is the glycogenic and others are ketogenic volatile fatty acid. Microbial population in the rumen is responsible for fermentation and not much research has been conducted in this regard. Like other ruminants bacteria, protozoa and fungi are there but their classification or metabolic pathways during fermentation process is not fully understood. Microbes degrade protein into small peptide, amino acid and ammonia which are used by themselves and forms microbial protein. This protein when comes to the intestine is digested by host enzyme and becomes available to the animal. Ammonia concentration in the rumen may vary according to the protein quality and quantity fed and when the concentration goes above a limit, produces toxicity. Carbohydrate in the feed for ruminants is in the form of cellulose, hemicellulose and pectin etc. from plant origin and fermented in the rumen to simpler sugar then into volatile fatty acids. Fat from the feed on fermentation hydrolysed and biohydrogenated to form saturated fat or long chain fatty acids which may be used by the microbes or goes to the intestine and digested by host enzyme. Under this condition, as discussed earlier, animals are fed mainly alfalfa hay along with some straws but there are a no. of plant sp. eches to be used by the farmers. Sharma *et al.* (2000) reported that the region has a no. of forage plants and tree/shrubs which are used for animal feeding (Table 3) but few are very important.

Table 3: Common forage plant, tree/ shrubs found in Ladakh

Name of tree/ shurbs	Name of forages
Populus balsamifera	*Aconogonum tortnosum*
P. euphratica	*Agropyron repens*
P. ciliata	*Artemesia sieversiana*
P. nigra	*Artiplex crassifolia*
P. sylvestris	*Astragalus frigidus*
P. alba	*Astragalus zanskarensis*
Hippophae rhamnoides	*Avena fatua*
Eleaganus angustifolia	*Avena sativa*
Salix alba	*Cicer microphyllum*
S. angustifolia	*Dactylaus glomerata*
S. babylonica	*Echinochloa colonum*
S. excelsa	*Fertuca pretese*
S. sclerophylla	*Hordeum vulgare*
S. tetrasperma	*Lepidium latifolium*
S. denticulate	*Medicago sativa*
Rosa webbiana	*Melilotus officinalis*
Myricaria germanica	*Pisum arvensis*
Tamarix gallica	*Poa pratensis*
Ulmus wallichiana	*Trifolium repens*
Morus alba	*Triticum austivum*
Betula utilis	*Zea mays*
Caragana pygmea	
Pinus wallichiana	
Juniperus macropoda	
Robinia pseudoaccacia	

Table 4: Common fodders and their nutritive values.

Ingredients	DM (%)	CP	NDF	ADF	HC	C
Aconogonum tortuosum	88.87	8.63	51.04	38.22	12.82	24.59
Avena fatua	90.09	6.79	38.35	28.96	9.39	24.34
Cicer microphyllum	91.40	15.30	36.38	32.34	4.04	26.44
Ephedra gerardiana	91.81	9.27	45.23	42.48	2.75	21.20
Hippophae rhamnoides	91.09	23.93	22.72	20.21	2.51	11.74
Iris lactea	90.37	10.62	47.28	42.35	--	--
Lepidium lactifolium	91.74	24.62	11.20	--	--	--
Medicago falcate	73.6	15.85	20.86	--	--	25.8
M. sativa	74.6	18.22	20.65	--	--	25.9
Melilotus officinalis	89.95	2.01	20.94	18.53	2.41	13.83
Prangos pabularia	89.43	12.66	35.78	32.69	3.09	27.57
Salix babylonica	91.03	18.12	36.35	24.68	11.67	15.40
Trifolium pretense	89.68	17.56	39.14	32.56	6.58	26.35

DM- dry matter, CP- crude protein, NDF- neutral detergent fibre, ADF- acid detergent fibre, HC- hemicellulose, C- cellulose. Values are is on DM basis.

Table 5: Major diseases of yak

Disease	Causative organism	Common symptom
Anthrax	*Bacillus anthracis*	High rise of temperature, oozing of blood from natural orifices and death
Brucellosis	*Brucella abortus* *B. melitensis*	Abortion in last trimester of pregnancy and bulls loose sexual desire, becomes infertile
Haemorragic septisemia	*Pasteurella multocida*	High rise of temperature, profuse salivation, conjunctivitis, diarrhoea or dysentery, edema in neck, dewlap or brisket region and in acute case death within 24 hr
Botulism	*Clostridium botulinum*	Partial or complete paralysis, profuse salivation, tongue protrude from mouth and recovery seldom occurs
Calf scour	*Eschericia coli*	Depression, inappetance, stiffness, rough hair coat, profuse loose feces and yellowish brown to grey white feces
Mastitis	*Staphylococcus* sp. *Streptococcus* sp. *Mycobacterium bovis* *Mycoplasma bovis*	Swelling of udder, pain, drop in milk production, thin milk and flakes in milk, fever and chronic case fibrosed udder with low milk production
Leptospirosis	*Leptospra* sp.	Fever, anaemia, haemoglobiuria, icterus abortion and reproductive disorder
Tetanus	*Clostridium tetani*	Restriction in movement, muscular stiffness, difficulty in walk, hypersensitivity to any external stimuli and lock jaw. Lateral recumbence and bloat
Salmonellosis	*Salmonella* sp.	Enteric form characterized by anorexia, depression, high rise of temperature and profuse watery feces. In the initial stage, semisolid feces followed by diarrhoea with blood. Septicaemic form generally found in young animals and symptoms like high rise of temperature, dullness, ataxia and recumbence. Feces are putty coloured and may contain blood streaks. Signs of pneumonia may also be developed
Viral diseases FMD	Picorna virus- OAC, Asia I, SAT I, SAT II and DAT III strain (RNA virus)	Mortality less but morbidity is 100%. High rise of temperature, generalized malaise, depression, dejection, and apathy to food and water. Reduction in milk production. Vesicles appear in oral mucosa, interdigital space and udder and profuse salivation. Post FMD sequel is panting, anaemia, overgrowth of hairs and sometime diabetes mellitus
Rinderpest	Paramixo virus (RNA virus)	The virus has got tremendous affinity to lymphoid tissue and epithelial cells of gastro intestinal and respiratory tract. High rise of temperature, dull, anorectic. Necrosis of mucous membrane in oral cavity. Watery dark feces with blood, temperature subsides. Respiratory distress and nasal discharge may contain blood

Contd...

Parasistic diseases*		
Ticks, mites and lices		Ticks: Cutaneous injury which paves the way for infestation of fly and bacteria. It sucks blood and may cause extensive damage to the leather Mites: scale, alopecia, itching and hyperkeratosis. Scab or nodule formation may be seen Lice: itching, irritation, formation of erythamatus macules, dermatitic lesion and anaemia
Babesiosis	*Babesia* sp.	High rise of temperature, anorexia, depression, and weakness in the last stage-severe jaundice, haemoglobinurea, paralysis of different muscles
Coenurosis Liver fluke	*Coenurus cerebralis* *Fasciola* sp.	Constipation followed by diarrhoea, loss of appetite, pot belly, bottle jaw, emaciation and death
Nematodes	Round worm	Progressive loss of body wt, unthriftiness, diarrhoea, pot belly, anaemia
Miscellaneous		
Contagious bovine pleuro pneumonia	*Mycoplasa mycoides* var. *mycoides*	High rise of temp., general depression, panting, nasal discharge, edematous swelling on throat and dewlap
Vitamin and mineral deficiency	Dietary deficiency of different fat soluble vitamins and minerals as vegetable grown only in summer months.	Requirement of these nutrients are very but deficiency may cause interference in cellular activity, growth and metabolism which may lead to death of the animals
Milk fever	Depression of ionic calcium in the blood which is associated with dietary and hormonal factors	Hyperexcitement and slight elevationof body temperature and hind limbs becomes rigid initially but then muzzle becomes dry, lateral kink of neck or head, atony of rumen, subnormal body temperature, lower productivity and lateral placement of body and can not stand. Response to calcium therapy
Bloat	Excessive production of gas or physical obstruction of eructation of gas from the rumen either due to primary or secondary cause	Dull and depression, anorexia, distended rumen with fluid or gas, atony of rumen, nervous symptoms, gasping and inextreme cases death may occur

* The organisms reported in cattle and assumed that the same are responsible for yak.

Nutritive value of different feedstuffs is also estimated under this condition. No. of tree leaves, hay or grain has been tested for its chemical composition and fibre fractions FRL[9] as follows:

But the farmers are not well aware of these things and they feed their animal according to the availability of ingredients not to the requirement. There is also a practical problem of storing the hay or other feed ingredients in the house; common practice is that the hay is packed in jute bag and stored. Maize is not grown by the farmers and there is a immediate need to introduce new feed for the animals otherwise productivity will remain low. Some unconventional feeds may be used for the purpose and most important one is apricot oilcake. As apricot enjoys the main fruit crop in the region, cakes available from wild variety remain unused due to presence of cyanide. Now a days, some people detoxify it by heating the cake by 1.5 hr by adding 3 times water to it and that has no detrimental effect. It contains about 30% crude protein but quality is not known.

c. *Health and disease*: For optimum production in any farm, animals should be disease free and healthy. Shed and surroundings should be clean and disinfectants should be applied at a regular interval. Drinking water should be clean and fresh so that no contaminants can come in contact with the animals. There are so many diseases found in the yak but few are important as listed below with its causative organism and common symptoms:

Herd should be properly vaccinated against the diseases found in the area and ailing cases should be treated in nearest veterinary hospitals. Sometimes does with anthelmentics should be done to get a healthy herd.

Conclusion and Future Research

Yak or its crosses are one of the most valuable animals in Ladakh and play an important role in transportation of goods in steep mountain, milk, meat production. Fibre and skin are also used for clothing and carpet making. Moreover, yaks play an important role in social and cultural life of Buddhists in the region. Its crosses are more productive and people are interested in rearing of the crosses of yak and cattle. Due to mechanization in agriculture and transport and even poaching of the animals, number is declining rapidly. Research in different fields like physiology, nutrition, management, disease prevention and control and processing of meat and milk is not done under this condition. Intensive research in the specified field as well as conservation of the animal is essential for the region to improve the economy of the farmers.

References

Banerjee (1998). A Text book of Animal Husbandry (8th Edition). Oxford and IBH Publishing house Co. Pvt. Ltd. Calcutta. pp. 1060-1063.

FRL-9. Field Research Laboratory (DRDO), Leh. Chaurasia, O.P., Ballabh, B.,Verma, A., Ahmed, Z. and Raut, A. Potential fodder plants of Ladakh. Extension folder no. 9.

Li Ruimin (1986). Records of bovine species in ancient Chinese history. In: Zhang Zhongge and Zhu Xianhuang (ed), *Proceedings of papers about history of animal husbandry in China*. Beijing, China Science Press. pp. 146-149.

Miller and Steane (1996). Conclusions. In: Miller D.G., Craig S.R. and Rana G.M. (ed), Proceedings of a workshop on conservation and management of yak genetic diversity, *at ICIMOD, Kathmandu, 29-31 October 1996*. ICIMOD (International Centre for Integrated Mountain Development), Kathmandu. pp. 191-209.

Namgyal (2006). Wild yak: the bison of the east. The Magpie, 10, July. pp. 5.

Olsen (1991). Confused yak taxonomy and evidence of domestication. Illinois State Museum Scientific Papers. 23: 387-393.

Schaller (1998). Wild yak. In Wildlife of the Tibetan Steppe. University of Chicago Press. 125-142.

Sharma, J.P.; Mir, A.A.; Singh, B.; Chaurasia, O.P. and Dwivedi, S.K. (2000). Plant genetic resources of cold arid region, their conservation and utilization. In: Dynamics of cold arid agriculture. J.P.Sharma and A.A. Mir. ed. Kalyani Publishers. pp. 121-143.

Xu Shangzhon (1998). Development strategies for the beef industry in China in the twenty-first century. Forage and Livestock Supplement: 1-6.

Zhang Yun (1994). The relationship between season and age of stud yak bull in Damxung. Proceedings of the first international congress on yak. J. Gansu Agricultural University (Special issue June 1994). pp. 303-307.

❏❏❏

18

Double Humped Camel : The Endangered Animal

CHAMPAK BHAKAT AND K.M.L. PATHAK
National Research Centre on Camel, Jorbeer, P.B No. – 07, Bikaner Rajasthan 334001, India

Double-humped camel (*Camelus bactrianus*) is an important component of the cold desert ecosystem and to the people inhabiting in region of Nubra valley of Ladakh (Jammu & Kashmir) where the remoteness, inaccessibility and extreme climatic conditions of the area spell slow economic development. The camel is a source of superior quality of hair, meat and milk in addition to load carrying/ haulage utility for people of cold arid region. The breed available in India is *Ladakhi* which is originated from Ladakh. The *Ladakhi* camels are short in height, compact with well-built robust muscular body, long thick hair cover on body particularly at neck, shoulder, hump and mid-side region. The double-humped camel originated from Gobi desert of Mongalia. *Camelus bactrianus* comes under the Order Artiodactyla, Sub-order Tylopoda and Family Camelidae. The total World population is estimated at about 2.2 millions (FAO, 2005) most of which are confined in high land of central Asian countries like China, Mongolia, Turkmenistan, Kazakhstan, Uzbekistan and

Afghanistan. A small population of about 105 camels is confined to Nubra valley of Ladakh (Sahani, 2002). The birth weight of bactrian camel varies from 30 to 40 kg. The body weight in adult camels varies from 350 to 650 kg and males were found to be heavier than females. The body colour varies from brown to dark tan. The body length varies from 170 to 178 cm in male and from 152 to 156 cm, in female. The height at withers varies from 168 to 176 cm in male and from 152 to 156 cm in female. The heart girth ranges from 194 to 208 cm in male and from 192 to 199 cm in female (Sahani, 2002). In earlier times, double-humped camel has served as a vital means of transport on the 6,400 km historic silk route. It was used to carry tea, silk and porcelain of China. Business men from Central Asian countries also used the camels to carry their merchandise to Ladakh and adjoining desert regions via the Karakoram pass. Double-humped camels can easily thrive on natural resources which might otherwise be unusable.

Sustainable Livelihood through Double-Humped Camel

One important aspect which is common to many places where camel inhabits is the utilization of its hair. Although draught use is the primary utility of Indian double-humped camel but hair is also being utilized by local communities just as in case of dromedary camels. Hair is used in village cottage sector for preparation of common utility items, *viz.*, rope, blankets, floor rugs, bags and mattresses etc. in Rajasthan and Gujarat (Sahani and Khanna, 1993). The hair and its products of this species can be an important source of additional income for camel keepers. The handicraft articles made up of camel hair, provide work to rural women in the field of grading of hair, tops preparation, spinning of hair, weaving, embroidery with 100% speciality hair and blending with sheep wool, goat hair, cotton and other products. The hair of camel is durable, strong and has low conductivity (Mukasa, 1981). It has been estimated that a camel hair fabric of 620 gm weight will be as warm as a pure wool fabric of 900 gm weight (Khanna and Rai, 1991). The main body sites of hair coverage in bactrian camels are shoulders, mid portion of body, neck and hump regions.

This species gives return in terms of cash by draught, at the same time it gives sufficient milk to provide calories to the family of camel keepers and also protects them by providing all sorts of warm clothing and household items. The cost to the camel keepers for such items would otherwise be very high and out of reach. This requirement is being fulfilled by this animal in such a small form, which otherwise would go unnoticed. Small and marginal farmers in their own village and in their own house prepare blankets from the hair of adults and shawl from the hair of calves. Some of the traditional weavers charge Rs. 50 per kg for yarn making and Rs. 80 per kg for making blankets and shawl. These carpets and blankets are very cheap and durable. The life of a carpet is approximately 20 to 30 years and that of a blanket is 30 to 40 years. Camel hair is also used for making ropes, which are subsequently used for tying the animals and in making cots. Thus, this species beautifully adjusts in the economics of a camel keeper in cold desert region.

Distribution Status

The archeological survey indicated that the Persians and their northern and eastern neighbours were familiar with the double-humped camels, however, they were inclined to adopt the dromedary when they met it in their conquest of the West. As per previous record of double-humped camel have been found in northern Persia and Mongolia, Mesopotamia, on the Iranian plateau and in Turkistan. The double-humped camel had a wide distribution as a wild animal in Central and North western Asia in pre-historic times. Russian travellers have reported existence of double-humped camel as wild animal and recent observation of a Russian expedition has confirmed this view that wild double-humped camels are still in existence.

Rearing Systems

The double-humped camels are left open for grazing in forest of Nubra valley near the banks of river Shayok and Siachen round the year. The close inbreeding activities are going on since more than last few decades. Presently, these camels are reared under extensive system of management. They breed in natural condition with browsing and grazing in the rangeland all over the year. Nobody provides any supplementary feeding except for the few working camels and females during parturition for a short time period. Since salt availability is limited, so it should be provided for the well being of the species. As per the practice no herd men follow the camels in rangeland and they are supervised only according to their requirements. The lucerne/ alfalfa locally called "Ole" is the main and only cultivated legume fodder, which also grows wild. Even people harvest certain wild plants and dry them properly for winter use. Legumes and grasses dominate the fodder plants. In addition, the fodder, *viz.* oat (*Avena fatua*), Zamak (*Agropyron repens*) etc. are grown and generally utilized for feeding purpose.

Double-humped camels are seasonal breeders. Breeding occurs during winter month from December to March. The camel attains puberty at the age of 3 to 4 years in female and 5 to 6 years in male. The reproductive life usually continues up to 16 to 18 years of age. Generally, parturition occurs once in 1.5 to 2 years. During sexual arousal males and females seek the company of each other. Male smell and bite the vulva, rubs the female along the back with the under side on his neck or rubs his head under her abdomen and forces her to a crouched position. Once the female sits down, the male mounts her, takes a grip on the female, erects his penis and copulates. Coitus lasts 1 to 6 minutes. After copulation, the male will lift one of the fore legs over the female to the other side and knee on the ground with fore legs and then gets up. Conception rate in double-humped camel is around 60 to 70%. In normal practice hair clipping is done during onset of summer months. In case of wild animals the shedded hair is collected by the farmers from April to June. In draft camels wooden nose peg are inserted at around 3 to 4 years of age and then camels are trained for work. The working camels are used for transportation of fodder, fuel, wood, stones, manure and building materials *etc.* on hilly tract.

Production of Hair

The hair of double-humped camel and its products can be an important source of additional income for farmers inhabiting the cold arid region. The annual hair production of adult double-humped camels varies from 3 to 4 kg per Indian animal, whereas annual hair production is 6 to 8 kg from bactrian camels of Mongolia and Kazakhstan. In China, with a population of around 6.0 lakh double-humped camels, the total annual production is around 1,500 tonnes out of which about 600 to 800 tonnes is exported to other countries (Sahani, 2002). The longest, fine and densest fibers are from the shoulder region of the body of camel. The number of wool fibers per square cm of skin of a camel in good physical condition varies from 4,000 to 4,800, while the number in an animal in poor condition is only a half of this number. The total yield of the male is about 30% higher than that of female bactrian camel. Bhakat *et al.* (2002) found lesser production in dromedary camel. The quality of the wool of the young animal is finer as compared to adult camel.

Utilization of Hair

The hair is spun by camel keepers by using traditional simple techniques. The coarse variety is spun by hand dheryai. The other method of spinning fine quality camel hair is by hand charkhas. In the recent times machine spinning is becoming popular for carpets, durries and clothing. The preliminary results of camel hair blends with wool, silk waste and polyester have shown encouraging results (Gupta *et al.*, 1987, 1989). Blended products may also be prepared with sheep wool, goat hair and cotton. Patni and Dhillon (1988) reported that it is worthwhile to blend camel hair with polyester, wool or silk waste. Hair is one of the important production traits of bactrian camel, which is extensively used for manufacture of various items like shawls, coats, caps, sweaters and hand gloves etc. in village cottage sector of this region. Double humped camel's wool is soft, elastic and has good insulation properties too. The fineness of the wool compares to that of sheep but is less dense, with fewer scales, the crimps are irregular and the diameter of the individual fibre is uneven, so the spinning quality is inferior as compared to sheep wool.

Quality of Hair

The fibre diameter of pure hair among three sites of double-humped camel calves varied from 12 to 19 m, hetero fibres from 20 to 26 m and hairy fibre 30 to 55 m, while in adult double-humped camel pure hair diameter varies from 20 to 32 m, hetero 30 to 46 m and diameter of hairy fiber varies from 45 to 68 m. Bhakat *et al.* (2001) found comparatively more hair diameter. The staple length of bactrian camel is the highest (16.13 cm) in the hump region of calves which varies from 16 to 18 cm, while staple length of shoulder and mid side is 10 to 11 cm and 5 to 6 cm, respectively in adults. Yadav *et al.* (2000) reported lesser staple length. The mean fiber diameter in calves varies between 19 to 26 m while in adult from 25 to 43 m. The mid-side region fibre was found to be finer both in calves (17 m) and adults (25 m) as compared to hump region (26 m for calves and 43 m in adults). In case of adult bactrian camel the mean fibre diameter values show more variance in hump and shoulder regions than the mid-side region. However, the fibre fineness trend is same, *i.e.*, the mid-side region fibers have the highest fineness (30 m) in comparison to 45 and 46 m of shoulder and hump region, respectively.

As far as the elongation percent property is concerned, higher elongation per cent value has been observed in mid, shoulder and hump regions. It shows that in adult camel too the mid-side region fibres not only possess lowest mean fibre diameter, highest fibre strength but also comparable elongation percent with the fibers of other regions. The fibres from mid-side region have much high elongation percentage, *viz.*, 38% as compared to 16% of those from the hump region at the same gauge length, *i.e.*, 50 mm indicating that fibre from mid-side region are more suitable to be used for textile purpose, in blends with wool and other fibres.

Management of Milk Production

The milk of bactrian camel is rich in vitamin C. It is an important source of milk supply for people living in the Gobi desert and steppe area of Mongolia. Camel's milk consumption in these areas makes up about 35% of the total food ration. Milk production potential of Indian bactrian camel varies, generally it is considered to be low in milk production (1 to 1.5 l/day) as compared to Indian dromedaries whereas the data of milk production of bactrian camel of Kazakhstan is as high as 6 to 7 l/day. The lactation length varies from 13 to 19 months and lactation yield varies from 500 to 1254 l. The peak milk yield is obtained at around 3 to 4 months after calving.

Health Hazards

The calf mortality is the major health hazard in this area. The higher mortality is mostly due to accidental ground such as drowning etc. in nearby rivers of Nubra valley in the new born calves as compared to adults. The common diseases prevalent in double-humped camels are contagious cough, vaginal and prepucial myiasis, fungal disease caused by *Trichosporum capitatum, Herpes encephalitis,* diseases caused by *Thelazia leesi, Dictyocaulus cameli* and other common helminths, osteoporosis, senile plaques, influenza virus, multiple lipomas of liver and sub-clinical mastitis (Sahani, 2002). As compared to single-humped camel skin affections and ectoparasite are not common in bactrian camel.

Female Reproductive Features

Generally, gestation period of this species varies from 380 to 410 days. Double humped camels are seasonal breeders. The breeding season commences from mid December and extends up to March. In females puberty can occur before physical maturity, The female can be bred at three years of age and calving can occur at four years of age. The triggering factor for initiation of cycling during breeding season appears to be increasing daylight in this cold desert region. Like dromedary camel bactrian too is an induced ovulator. Once a normal follicle starts growing, it usually takes up to 19 to 24 days from initiation of development to onset of regression.

Parturition Behaviour

Parturient female shows signs of marked restlessness and tends to seek isolation from the main herd. The female gets agitated, stops feeding, lies down and gets up repeatedly, lifts the tail and walks around. The signs of approaching parturition also include swelling and oedema of vulva, softening of pelvic ligaments and udder enlargement. Swelling and oedema of vulva occurs 10 to 40 days before parturition. Swelling of udder begins 1 to 1.5 months before. The udder becomes enlarged 10 or more days before. It is filled with milk from 1 to 6 days before delivery. The teats become swollen up but colostrum does not ooze from the teat orifice till the delivery of foetus. Parturition process takes place in 3 stages. First stage is characterized by intermittent restlessness of the female. During this particular phase, parts of the amnion and the fore feet of the foetus are forced through the dilated cervix into the vagina. Second stage is rapid and comparable to that of mare. It is of 10 to 60 minutes duration. Third stage varies from 91 to188 minutes. In bactrian camel whole foetal membrane leaves the uterus without eversion. The placenta is diffuse and epitheliochorial type. Foetus is covered by a pale and transparent epidermal membrane being characteristic of camelids. Uterine involution is completed within 40 days.

Male Rutting Behaviour

The sexual desire in male is called rut. The rutting continues from December to April month. Individual breeding activity lasts for 50 to 100 days. There are major changes in behaviour as soon as breeding season commences. Sexually excited males develop fighting instinct, their lips are smeared with saliva, the soft palate is frequently exposed and loud noise is frequently uttered. They grind their teeth continuously, make some peculiar metallic sound from mouth, seek air, belch, draw the head back, rub the poll or lash the tail. They often adopt a squatting posture with the croup lowered, the hind legs placed widely apart, the tail beating up and down and frequent urine sprayed to form a typical urine mark on rear hump. The male in rut loses condition due to loss of appetite and increased physical activity. The abdomen is tucked up and the hump gradually decreases in size. The male becomes emaciated after rutting period. The poll gland secretes copious colourless and transparent secretion, which adheres on the hair, and becomes dark brown in colour. It has a typical smell also. The puberty and rut appears at 3 to 4 years of age especially in the male of good body condition. Sexual activity to a desire level is attained by 5 to 6 years of age. Physiological capacity may increase and remain at a more or less constant level until 15 to 20 years of age, then decreases gradually.

Opportunities and Limitations

The major opportunities of double-humped camels are utilization of this species as load carrying for short distance transportation (5 to 10 km) in hilly tract of cold desert area. This species is used for transportation of various items such as fodder, manure, sand, fuel wood and stone etc. They can carry an average load of 150 to 200 kg. Very few camel carts are introduced for short distance transportation under cold arid region. As per literature available on draught power of bactrian camel from China, it is indicated that it can carry load around 40% of the live weight, *i.e.*, 150-200 kg. with a maximum of 250 kg. Carrying this load, the camel can cover 35 to 40 km within 24 hours. As riding camel, it usually maintains a speed of 5 to 8 km per hour in summer and autumn season. The speed is of 7 to 10 km in winter and spring season. In long distance travel camels can cover 40 to 50 km per day for a month. Double-humped camel can plough 0.2 hectare of land in 8 hours.

The bactrian camels are short in height which may be considered as one of the limitations of the species. Presently rearing and maintenance of small population of double-humped camel has become unprofitable and the survival of this species in this region was endangered because of fast mechanization and expansion of roads. Some other limitations are scarcity of grazing land, non-availability of fodder and lack of bactrian camel product's market. But the wide prospects of double-humped camel can be towards five dimension (meat, skin, bone, dung, sport) which are as follows:

Utilization as Meat Animal

Bactrian camel meat may be another use of the species. It is not a common practice to eat camel meat in India, although it is used in substantial quantities in several other countries. Dressing percentage of camel carcass ranges from 40 to 60%. Camel meat contains about 22% protein and only 1% fat. It has been estimated that an average carcass weighing 210 kg would yield 10 kg fat, 160 kg meat and 40 kg bones and would thus yield 35 kg protein and 9973125 KJ energy which will be sufficient to meet the protein requirement of about 35 peoples (Wilson, 1984). The optimum age for camel slaughter is around 3 to 4 years. Liu Zongping (2003) reported that in China bactrian camel are providing meat, hair, wool, hides along with indispensable means of transport.

Utilization of Skin

The skin of bactrian camel may be used for making various type of leather goods and fancy items of public interest, *viz.*, shoes, sandals, purse, bag, stool cover, rope, jewelry box, toys and various decorative items etc. as utilized in case of dromedary (Bhakat and Sahani, 2004). In earlier era, camel's hides have utilized to make containers for storing and carrying water, oil and ghee etc. This hide is well suited for application of colour and in Rajasthan there is a tradition of making elaborately painted small container (*kuppi*) by using camel hide. It is having good export potential.

Utilization of Bone

Bactrian camel bones may be used as important raw material for the manufacture of jewelry in some parts of the world. In India, since ivory has been banned, it may be replaced by camel bone. Now a days ivory is replaced by dromedary camel bone and use for making toys, jewelry and camel long bone may be used for various decorative items. A few samples of various items are kept at camel museum, NRCC, Bikaner. It has potentiality as good as ivory but cost is less as compared to that of ivory. The bone of double-humped camel is also used as bone-meal and fertilizer at cold desert region.

Utilization of Dung

Dung ranks as an important by-product. Camel dung is good source of organic manure and dry dung is also used as fuel. For farmers inhabiting cold desert barren territories, it can be the most important source of fuel and fertilizer. This dung can be sold by camel herd owner at good price. Camel keepers used to compensate the land owners by cash or by providing camel dung for overnight stays on their fields. Brick factories are also using camel dung as fuel.

Utilization As Sport Animal

In recent times, certain amount of money is involved and the economic importance as sport animal is increased. The safari and riding on double-humped camel is very much popular among tourists. In India, camel dance is major attraction in this area. The tourist importance of double-humped camel is in increasing trend. The use of camel in marriage and other social programme have a great socio-cultural importance.

References

Bhakat, C., Yadav, B. and Sahani, M.S. (2001). Effect of certain factors on hair quality attributes in Indian camel managed in an organized farm. Indian J. Animal Sci. 71 (10): 992-994.

Bhakat, C., Mehta, S.C. and Sahani, M.S. (2002). Studies on hair production attribute in Indian camel managed in an organised farm. Indian J. Animal Sci. 72(3): 275-276.

Bhakat, C. and Sahani, M.S. (2004). Pattern of utilization and potentiality of camel skin. Indian Farming 54 (2): 3-5 & 23.

FAO (2005). Animal Production Yearbook. Food & Agricultural Organization, Rome, Italy.

Gupta, N.P., Pokharna, A.K.; Arora, R.K. and Sugumar, S. (1987). Mechanical processing of wool, speciality hair and their blends. Annual Report 1987, CSWRI, Avikanagar.

Gupta, N.P., Patni, P.C. and Sugumar, S. (1989). Properties and processing of camel hair in India. Indian Textile J. 99 (4): 180.

Khanna, N.D. and Rai, A.K. (1991). Camel rearing in the Indian arid zone. Annals of Arid Zone 30(1): 1-10.

Liu Zongping (2003). Studies on the haematology and trace element status of adult Bactrian camels (*Camelus bactrianus*) in Chaina. Veterinary Research Communications 27 (5): 397-405.

Mukasa, E. (1981). A Bibliographical Review. International Livestock Centre for Africa, P.O. box 5689, Addis Ababa, Ethiopia.

Patni, P.C. and Dhillon, R S. (1988). Areas and prospects of utilization of camel hair and hide. National Seminar on perceptions and potentials of camel research in Indian, 9-10 October 1988, NRCC, Bikaner.

Sahani, M.S. (2002). Indian double humped camel. A hand out, National Research Centre on Camel, P. no – 07, Bikaner, Rajasthan.

Sahani, M.S. and Khanna, N.D. (1993). The camel fibre and its prospective utility. National seminar on production and utilization of animal fibres, Bikaner, 93, December, 13-14.

Wilson, R.T. (1984). Studies on the livestock of Southern Darfur. V. Notes on camels of Sudan. Tropical Animal Health and Production 10: 19-25.

Yadav, B., Mishra, B.P., Bhakat, C. and Sahani, M.S. (2000). Hair quality attributes of *Camelus dromedarious*. Indian J. Animal Sci. 70(2): 211-212.

□□□

Sahani, M.S. (2002). India's double humped camel. A [illegible], National Research Centre on Camel. P. no 4-37, Bikaner, Rajasthan.

Sahani, M.S. and Khanna, N.D. (1998). The camel fibre and its prospective utility. National seminar on production and utilization of animal fibres, Bikaner, 03 December 1-14.

Wilson, R.T. (1984). Studies on the livestock of Southern Darfur V. Notes on camels of Sudan. Tropical Animal Health and Production 10:19-25.

Yadav, B., Mishra, B.P., Bhakat, C. and Sahani, M.S. (2000). Hair quality attributes of [illegible] dromedaries. Indian J. Animal Sci. 70(2): 211-212.

19

Potential of Plant Biotechnology

TSERING STOBDAN[1], SANJAI K DWIVEDI[2], ZAKWAN AHMED[3]
[1]Defence Institute of High Altitude Research, Leh, Ladakh
[2]CEPTAM (DRDO), Metcalfe House, Delhi - 110 054
[3]Defence Institute of Bioenergy Research, Haldwani, Uttarakhand

High Himalays represent a blend of temperate and arid climates. It is a unique region of the world where crop plants have to experience conspicuously different environment as compared to their traditional niche. The region is characterized by extreme temperature (-40°C in winter to +40°C in summer), low precipitation, thin atmosphere, high wind velocity and sparse plant density. Such circumstances demands adaptation of plants to meet the requirement of this agro-climate. Conventional breeding for crop improvement has its limitation because of the variability in agronomic traits at different period during the life cycle of a plant. Most of the breeding traits are of multigenic nature and therefore difficult to include in a single plant. Besides, there is limitation in exploring the desirable traits from other living organisms because of reproductive barrier. Recent advances in plant molecular biology and genetic engineering have allowed plant biotechnologists to aim at overcoming the difficulties faced by traditional plant breeders. Two main avenues to improve crop plants are: (i) through genetic modification, by introducing new genes into already existing elite genotype, and (ii) through the smart exploitation of genetic diversity using molecular markers in breeding program.

Abiotic Stress Tolerance

Abiotic stresses, especially drought and frost, are the primary cause of crop loss in cold arid region. The improvement of crop abiotic stress tolerance by classical breeding is fraught with difficulties because of the multigenic nature of these traits and the large variability in stress sensitivity at different periods during the life cycle of a given plant. Abiotic stress tolerance mechanisms are caused by cellular and molecular events that lend themselves to biotechnological manipulation. In contrast to plant resistance to biotic stresses, which is mostly dependent on monogenic trait, the genetically complex responses to abiotic stresses are multigenic, and thus more difficult to control and engineer. Plant engineering strategies for abiotic stress tolerance rely on the expression of genes that are involved in signaling and regulatory pathways or genes that encode proteins conferring stress tolerance or enzymes present in pathways leading to the synthesis of functional and structural metabolites. Current efforts to improve plant stress tolerance by genetic transformation have resulted in several important achievements (vanocur and Altman, 2005). Several successful approaches the achieving tolerance through the genetic engineering of specific genes have been used.

Cold tolerance transgenics : Temperature is a major environmental factor that changes from season to season and undergoes daily fluctuation and short, erratic lows and highs. In cold arid region, there is a drastic fluctuation in temperature within a short period due to which plants suffer frost injury. Also, frost is almost certain during second or third week of September when weather is calm and clear, i.e, cool, clear night and no wind movement (Sharma, 2000). During freezing of plant tissues, ice typically forms in the extracellular compartment reducing its water potential and leading to loss of water from the cell by osmosis. Thus, cellular dehydration is a major component of freezing stress. Freeze-induced dehydration have a number of effects that result in cellular damage, such as the denaturation of proteins, precipitation of various molecules and formation of multiple forms of membrane lesions (Thomashow, 1998).

In nature, some plants have increased freezing tolerance. Winter wheat has freezing tolerance of about -20°C and that of winter rye is about -30°C. The mechanisms responsible for freezing tolerance are not well established. Classical genetic studies have demonstrated that the ability of plant to tolerate frost is a quantitative trait involving the action of many genes with small additive effects. In recent years, three major approaches have been taken to determine the nature of genes with roles in freezing tolerance: the isolation and characterization of cold-induced genes; the isolation and characterization of mutants affected in freezing tolerance; and QTL mapping using molecular probes to identify freezing tolerance loci (Thomashow, 1999). A growing number of genes have been shown to be induced during cold acclimation. Some of these genes encode proteins with known activities that could potentially contribute to freezing tolerance. Most, however, encode newly discovered proteins with unknown function. Because, the cold-induced genes are easily identified by differential screening, they have been characterized by several groups and have been given different names such as *COR* for cold regulated, *LTI* for low-temperature-induced, *CAS* for cold-acclimation-specific and *RD* for responsive-to-dessication genes. Using transgenic approaches cold-induced genes appear not to confer useful cold tolerance when expressed individually. No measurable improvement in freezing tolerance has been observed in *COR15a* transgenic lines at the whole-plant level (Jaglo-Ottosen *et al.*, 1998). The *COR* genes have been used in investigating the signal pathways that are associated with cold acclimation using molecular-genetic approaches. Gene-fusion studies have uncovered *cis*-acting elements in the promoter region of *COR* genes. These elements are named either dehydration-responsive-elements (DREs) or low-temperature-responsive-elements (LTREs). The DRE/CRT elements have been used to isolate DRE/CRT-binding proteins. These proteins are called DREBs (DRE-binding proteins) or CBFs (CRT-binding factors) which function as transcriptional factor to activate the expression of *COR* genes. Transgenic approach involving over-expression of a single transcriptional factor has shown that it is possible to improve freezing tolerance to well beyond that which can be achieved naturally by the normal mechanisms of cold acclimation. Transgenic plants containing a cauliflower mosaic virus 35S promoter *(35S)::CBF3* construct displayed high levels of *CBF3* transcript and an induction of COR polypeptides to a level 3-5-fold higher than those found in fully acclimated wild type plants (Gilmour *et al.*, 2000). Similarly Liu *et al.* (1998) have independently demonstrated that induction of the CRT/DRE-regulon not only increases the freezing tolerance but also increases tolerance to drought. This finding provides strong support for the notion that a fundamental role of cold-inducible genes is to protect plant cells against cellular dehydration. Kaye *et al.* (1998) made spinach CAP85 and CAP160 (cold acclimation protein) proteins in plants. The results suggested that both proteins have a detectable effect on freezing tolerance. Expression of dehydrins (DHN) genes in transgenic *Arabidopsis* was shown to confer enhanced tolerance to freezing stress (Puhakainen *et al.*, 2004). In light of the success achieved in improving freezing tolerance through transgenic approach, it may be possible to improve important cereal crops of cold arid region against freezing injury. The possibility of using the *Arabidopsis CBF* genes (or orthologs from other plants) as "master switches" to manage activation of freezing tolerance regulon and thereby improve freezing tolerance in broad range of plants looks promising.

Drought tolerant transgenic : In cold arid zone, water deficit is attributable to high evapo-transpiration rate, low water holding capacity of soil and mearge rainfall. Development of drought tolerant transgenics of important crops of cold arid region will have far reaching consequences in higher survival of plants in their habitat and economic utility. The use of transgenics to provide enhanced drought tolerance is still experimental, but progress is being made. Research on dehydration tolerance has established the basis of at least four signal transduction pathways that function under conditions of abiotic stress and has identified binding elements that brings about stress-inducible gene expression. Wu *et al.* (2003) expressed several candidate genes for abiotic stress tolerance in rice that clearly increased the accumulation of biomass under water deficit. Currently many research groups worldwide are attempting to demonstrate the success of marker assisted selection (MAS) in breeding for drought tolerance in various cereal crops (Toenniessen *et al.*, 2003). A key challenge faced by these groups is determining the quantitative trait loci (QTLs) that enhances performance across varying combination of water-stress conditions, plant growth stages and environment. In pearl millet, improved drought tolerance has been achieved by integrating QTL molecular markers for drought tolerance in breeding programs. On QTL, which explained 23% of yield under water deficit has been achieved. In sorghum, QTLs for drought tolerance was found to be consistently associated with the 'stay green' traits (Kabede *et al.*, 2001).

Domestication of Forest Trees

Cold arid region witnesses presence of very few timber species. Willow and populus sp. are the main timber species and hence the choice of timber tree production is limited. There is a growing demand for wood for timber in the region which needs research on the domestication process of tree breeding of forest trees. Breeding of forest trees for quality and yield is a slow process mainly because of the long generation intervals. Biotechnology provides avenues to accelerate improvement of quality and yield of the forest trees either through genetic modification or through marker-aided selection. *Populus* has been adopted by the scientific community as the model plant of choice for forest tree genetic improvement. It offers numerous advantages, such as fast growth, facile vegetative propagation, interspecific hybridization, amenability to tissue culture and genetic transformation and a small genome size (~500 mb). The genome sequence of *Populus trichocarpa* was released and large expressed sequence tag (EST) collections have been assembled. The small genome size of popular and its high transformation efficiency have allowed the construction of gene trap, enhancer trap, and activation tagging libraries for gene discovery (Boerjan, 2005). Since *Populus* is a common forest tree in cold arid region, it offers advantages in improving the species and other forest trees of high altitude both for quality and yield.

Microarray and reverse genetics experiments have led to identification of genes that control traits relevant to tree domestication. Studies have provided sets of marker genes for different stages of xylem and phloem differentiation. Transcriptional factors for dormancy induction has also been identified which further accelerate the potential use of biotechnological tools. The largest efforts in genetic engineering has been devoted to modifying the amount and composition of lignin in trees to improve lignin extractability during pulping (Baucher *et al.*, 2001). Overexpression of ferulate-5-hydroxylase (*F5H*) and down-regulation of 4-coumarate CoA-ligase has led to reduction in lignin amount and increase in cellulose content in poplar. Besides, quality and yield of poplar has also been increased through genetic modification. Overexpression of a cytosolic pine glutamine synthase (GS), a key enzyme involved in nitrogen assimilation, increased height by 41% and stem diameter by 36% in three-year old field grown popular. Over expression of the *Arabidopsis* endoglucanase *CEL1* in popular increased height, leaf size, stem diameter, and cellulose/ hemicellulose content (Shani *et al.*, 2004). Similarly, constitutive overexpression of a *Aspergillus* xyloglucanase has led to enhanced growth and cellulose production and reduction in lignin. Success achieved in forest tree improvement through genetic engineering can be followed for improving the timber tree adapted to the cold arid region. Yield enhancement by introduction of transgenes will solve the large gap in demand and supply of timber in the region. Marker-aided selection has been used for efficient exploitation of genetic diversity in forest trees. Over the past few decades, genetic maps have been made for many tree species. QTLs have been mapped for a range of agronomically important traits with the aim of using genetic markers linked to QTLs to follow the trait in breeding programs (Brown *et al.*, 2003). However, the use of this strategy in forest tree breeding remains limited because of the decrease in linkage between the traits and the marker with each generation owing to genetic recombination.

Nutritional Enhancement of Cereals

Cereals are the most important staple crops and, thus, are an important source of micronutients for the people living in cold arid region. But, cereals in general contain low level of micronutients, most of which are lost during processing for food or feed. To solve the problem of micronutrients deficiency, biotechnological approach via genetic engineering offers significant potential to enhance the capacity of cereals to acquire micronutrients from the soil and to store them in seeds and improve grain nutrient quality. An example of this approach is the production of 'Golden Rice', which has enhanced levels of the carotenoids β-carotene in its grain (Ye *et al.*, 2000; *Bayer et al.*, 2000). Progress has been reported in enhancement of vitamin E. Over expression of the barley genes for homogentisics acid geranylgeranyl transferanse in maize resulted in a six fold accumulation of tocotrienol and tocopherol antioxidants, two members of the vitamin E-family (Cahoon *et al.*, 2003). Similar approach can be used for nutritional enhancement of cold arid staple crops like wheat and barley which are poor in vitamins, minerals and rare amino acids. Cereals are generally poor in sulfur containing amino acids. In a novel biotechnological approach, a codon-altered chimeric transfer RNA for lysine ($tRNA^{Lys}$) was expressed in rice, to enrich lysine in rice seed storage proteins without changing their quality or quantity (Wu *et al.*, 2003). The success achieved in rice offers avenues for enhancement of essential amino acids in other cereals as well. Storage proteins are important for baking quality in wheat. Genetic engineering of gliadins, storage proteins responsible for baking quality, in varieties with low baking quality but high in micronutrients content can make this genotype suitable for consumption. Screening of traditional germplasm for enhanced micronutrient content should be combined with genetic engineering strategies to improve nutritional quality in cereals. Some traditional rice varieties were found to have two-fold higher level of Fe and Zn than the modern rice breeding varieties. Similarly, large differences in protein content have been found in different rice varieties (Kennedy and Burlingame, 2003). Similar characteristics may also be found in traditional high altitude germplasm of cereals crops. It offers opportunities to select local varieties for genetic modification for other micronutrients with minimum transgenes. Nutritional enhancement of cereal via genetic engineering is still in infancy. Little is known about the physiological and biochemical mechanisms that control micronutrient accumulation in plants. The bioavailability of these nutrient is another area where there is a need for better information. The form in which nutrients are absorbed or how they may be metabolized within the body is often unclear. The use of genetic modification approach has met with considerable consumer resistance amid concern for its safety, stringent controls and the provision of more public information are necessary to assure and encourage public acceptance (Tucker, 2003).

References

Baucher, M.; Petit-Conil, M. and Boerjan, W. (2003). Lignin: genetic engineering and impact on pulping. Crit. Rev. Biochem. Mol. Biol., 38: 305 350.

Bayer, P.; Al-Babili, S.; Ye, X.D.; Lucca, P.; Schaub, P.; Welsch, R. and Potrykus, I. (2000). Golden rice: introducing the β-carotene biosynthesis pathway into rice endosperm by genetic engineering to defeat vitamin A deficiency. J Nutr. 132: 506S-510S.

Boerjan, W. (2005) Biotechnology and the domestication of forest trees. Curr. Opin. Biotechnol. 16: 159-166.

Brown, G.R.; Bassoni, D.L.; Gill, G.P.; Fontana, J.R.; Wheeler, N.C.; Megraw, R.A.; Davis, M.F.; Sewell, M.M.; Tuskan, G.A.; Neale, D.B. (2003). Identification of quantitative trait loci influencing wood property traits in loblolly pine (*Pinus taeda* L.). III. QTL verification and candidate gene mapping. Genetics 164: 1537-1546.

Cahoon, E.B.; Hall, S.E.; Ripp, K.G.; Ganzke, T.S.; Hitz, W.D. and Coughlan, S.J. (2003). Metabolic redesign of vitamin E biosynthesis in plants for tocotrienol production and increased antioxidant content. Nat. Biotechnol. 21: 1082-1087.

Jaglo-Ottosen, K.R.; Gilmour, S.J.; Zarka, D.G.; Schabenberger, O. and Thomashow, M.F. (1998). Arabidopsis CBF1 overexpression induces COR genes and enhances freezing tolerance. Science 280: 104-106.

Kabede, H.; Subudhi, P.K.; Rosenow, D.T. and Nguyen, H.T. (2001). Quantitative trait loci influencing drought tolerance in grain sorghum (*Sorghum bicolor* L. Moench). Plant Mol. Biol. 48: 713-726.

Kaye, C.; Neven, L.; Hofig, A.; Li, Q.B.; Haskell, D. and Guy, C. (1998). Characterization of a gene for spinach CAP160 and expression of two spinach cold-acclimation proteins in tobacco. Plant Physiol. 116: 1367-1377.

Kennedy, G. and Burlingame, B. (2003). Analysis of food composition data on rice from a plant genetic resources perspective. Food Chem. 80: 589-596.

Liu, Q.; Kasuga, M.; Sakuma, Y.; Abe, H. and Miura, S. (1998). Two transcriptional factors, DREB1 and DREB2, with an EREBP/AP2 DNA binding domain separate two cellular signal transduction pathways in drought- and low-temperature responsive gene expression, respectively, in *Arabidopsis*. Plant Cell. 10: 1391-1406.

Puhakainen, T.; Hess, M.W.; Makela, P.; Svensson, J.; Heino, P. and Palva, E.T. (2004). Overexpression of multiple dehydrin genes enhance tolerance to freezing stress in Arabidopsis. Plant Mol. Biol. 54: 743-753.

Shani, Z.; Dekel, M.; Tsabary, G.; Goren, R. and Shoseyov, O. (2004). Growth enhancement of transgenic poplar plants by overexpression of *Arabidopsis thaliana* endo-1,4-β-glucanase (*Cel*1). Mol. Breed. 14: 321-333.

Sharma, J.P. (2000). Breeding field crops for stress environments in cold arid region. In: Sharma JP and Mir A.A. (Eds), Dynamics of Cold Arid Agriculture. Kalyani, India, pp. 169-184.

Thomashow, M.F. (1998). Role of cold-responsive genes in plant freezing tolerance. Plant Physiol. 118: 1-7.

Thomashow, M.F. (1999). Plant cold acclimation: Frezing tolerance genes and regulatory mechanisms. Annu Rev. Plant Physiol. Plant Mol. Biol. 50: 571-599.

Toenniessen, G.H.; O'Toole, J.H. and DeVries, J. (2003). Advances in plant biotechnology and its adoption on developing countries. Curr. Opin. Plant Biotechnol. 6: 191-198.

Turker, G. (2003). Nutritional enhancement of plants. Curr. Opin. Biotechnol. 14: 221-225.

Vanocur, B. and Altman, A. (2005). Recent advances in engineering plant tolerance to abiotic stress: achievements and limitations. Curr. Opin. Biotechnol. 16: 123-132.

Wu, X.R.; Chen, Z.H. and Folk, W.R. (2003). Enrichment of cereal protein lysine content by altered $tDNA^{Lys}$ coding during protein synthesis. Plant Biotechnol. J. 1: 187-194.

Ye, X.D.; Al-Babili, S.; Kloti, A.; Zhang, J.; Lucca, P.; Beyer, P. and Potrykus, I. (2000). Engineering the provitimin A (β-carotene) biosynthetic pathway into (carotenoid-free) rice endosperm. Science 287: 303-305.

□□□

20

Extension in the Changing Agriculture Scenario

M.S. NAIN AND G.R. BHAGAT
Division of Agriculture Extension Education, Faculty of Agriculture
SKUAST, Chatha, Jammu (J&K)

Information is essential for people to survive. In order to be really instrumental, however, information must be transformed into knowledge. Individuals and communities are continuously transforming what they know into strategies and practices they can use for their own development. In this process of development tailored extension strategies can be instrumental by providing a flow of information and by bringing this information to the people concerned in such a way that they can choose to make use of it. The scope of analysis, of why innovations are adopted and how to ensure that future innovations will be incorporated into agricultural policy of the area, could offer interesting information that can be turned into useful knowledge which may prove to be most sustainable and 'the people centered development'. The people centered development is based on the conviction that people are capable of directing their own development process, consistent with their aspirations and stresses the importance of anchoring development programmes in local knowledge and skills. It implies that individuals must be empowered to participate in their own development process. Building on local knowledge and resources reduce the likelihood that the development intervention will de-skill people or increase their dependence on external experts (Korten, 1980), on the contrary it will empower people by increasing their self reliance (Van Vlendern, 1999).

On the other hand, sustainable agriculture embraces the management and conservation of resource base and the orientation of technological and institutional changes in such a manner as to ensure attainment and continued satisfaction of human needs for present and future generations.' Such sustainable development is environmentally non-degrading, technically appropriate, economically viable and socially acceptable. Thus, there is a wide length in sustainable development and sustainable agriculture which emphasise the modification of each and every aspect of agriculture production. In order to tackle the situation the agriculture extension system needs to be prepared and reoriented to assimilate the change. It is therefore imperative to understand that agricultural development process under cold arid zone can be enhanced through approaches which are system oriented, participatory, utilizing farmers' wisdom, emphasizing partnership between farming community and research/ extension professionals, farmers' problem identification as starting point and utilizing modern communication media. The foregoing discussion in the chapter explores issues like reorganization of institutional processes to make them flexible and responsive, to promote the farmer led extension, approaches and methods that encourages functional participation *vis-a-vis* empowerment of farmers' organisation especially the women farmers, use of new information and communication technology (ICTs) and the blending of farmers' knowledge into the extension system.

Farmer Led Extension

Low income household in rainfed farming areas rely heavily on agriculture and natural resources for their livelihood. New technologies and management practices are essential if the productivity of these is to be enhanced on sustainable basis. Yet Government research and extension services are at their weakest in these areas: extension message and demonstration of new package and practices are rarely based on the assessment of the farmers needs; necessary inputs are rarely available on time and affordable prices; there are major difficulties in the management of common pool resources such as forests or grazing land; Government field staff lack technical and logistic support and many of them regard remote rainfed areas as punishment posting resulting in frequent staff turnover and unfilled posts. It is now clear that the *green revolution* took place in a unique set of circumstances; dwarfing genes were introduced into wheat and rice at a time when fertilizer responsive varieties were being already cultivated under stable irrigated conditions where market and communication infrastructure were well developed. For the future technical change especially in highly variable conditions of rainfed areas will be location specific and incremental not revolutionary. A major effort will be required by research and extension staffs in order to find out farmers' requirement for technology change under these widely varying conditions and where possible to stimulate farmers demand on research and extension services. In addition many have argued that farmers themselves undertake experiment and spread the results of these among their neighbours so that a major opportunity for research and extension staff lies in supporting these local processes of experimentation and dissemination. It is likely that the future gains in agricultural productivity through technological innovations will have to be more incremental, location specific and geared towards specific farmers' constraints. This is particularly true for resource poor farmers operating in environment which can not be unified through irrigation and purchased inputs, which are remote from markets and political and urban centres and in which the natural resource base is fragile. The need for locally specific technological innovations means that if agricultural research and extension organisations are to be more demand driven then they were in the past. This calls for *farmer led extension* (FLE). The farmer led approach in agricultural extension can be defined as a process of extension where farmers are playing the major role in adopting the technology suiting to their own situation and ultimately the same technology is being adopted by a large group of farming community in the same locality (Young, 1995). Here role of extensionists, research institutions and commercial firms are negligible or very passive. In the process of farmer led extension the farmers opinion should not be simply replaced with expert knowledge but to combine farmers point of view and the professional knowledge together so as to improve farmer's ability to understand social development and to help farmers gradually to become the major decision maker and directly participating in the process of decision making for the village development. FLE may be broadly classified under three heads; (a) Farmer to farmer extension in which farmers are the extension agents and outsiders facilitate their work, (b) Farmers research activities supported by specialist researchers and professionals, (c) Collaboration with NGOs.

The people centered agricultural development works on the principles, (i) Motivate and teach farmers to experiment, (ii) Use rapid recognizable success in the experiments rather than artificial incentives or subsidies to motivate farmers to innovate, (iii) Use technologies that rely primarily on in-expensive locally available resources, (iv) Begin the process with very limited number of technologies so that the programme is focused and (v) Train village leaders as extensionists.

The farmer led extension primararily imply that the farmers do have the capacity to experiment. However, Government services often remain motivated by a desire to protect farmers from crops, varieties or cultivation practices deemed inappropriate to their circumstances and to promote particular crop or means of cultivation deemed to be in a national interest. Secondly, the farmer to farmer extension is feasible, low cost way of spreading the results of experiments and other technical information and the links between public extension workers and multiple sources of technical information contribute importantly to effectiveness along with that the strategies can be evolved for the sustainability of farmer to farmer approach. The Government in supporting farmer led extension needs necessary action such as :

- A reduced role of centrally determined plans and schemes in extension and increased freedom and skills to identify and respond to location specific requirements.
- A reduced role of centrally determined plans for research, and increased freedom and skills to develop problem focused and need based research agenda.
- The freedom and skills to create close two way issue focused working links between research and extension at local level.
- The ability to draw in and adapt technical information in support of public extension workers from multiple sources.
- Structural flexibility in extension services to allow public extension workers to link with staff at different levels.
- Decentralization of decision taking and control over resources to permit the necessary flexibility.
- Change in performance assessment criteria and methods.

New criteria for the lower level of extension services need to be designed around the question 'How adequately have officials identified and responded to farmers requirements and supported farmer-to-farmer extension?' Much of the assessment would be concerned with the uptake and impact of technical change but part of it would be concerned with the support they have given to the formation of farmers group and such farmer-to-farmer mechanism as cross visits. At the higher level of extension they would be based on the question 'How adequately have senior officials listened to and responded to requirements being articulated from the field?' Much of the same system of 'performance by results' would be applied to research institutes and their staff in relation to the needs of clients. Key question would centre on how adequately needs have been identified and responded to.

Issues and the Process of Integration of Indigenous/ Traditional Knowledge

Indigenous Knowledge is the object of increasing attention in the field of research and policy making. Indigenous Knowledge for say, includes far more than observations about the migrating pattern of wildlife; it also touches upon the spiritual dimensions of a person's relationship with the environment. Many people regard it as something intimate and personal which should not be discussed with outsiders or revealed in a written document to be circulated outside the community. Just as a beating heart cannot be replaced by a anatomy textbook there is a sense that a written word rarely capture the depth and power of that spiritual relationship and a central aspect of one's identity and existence is often reduced to a description of unrelated bits of knowledge. In addition to the sensitivity issue there is also a purely practical concern. Communities become suspicious when they hear about some companies (Pharmaceutical) and other using information from people to develop market and make profit without sharing those profit with the people who provided the information in the first place. The story of *Lehberry* is live example to discuss over here. Indigenous leaders are now calling for increased local control of research projects on indigenous knowledge and in some cases cost/ benefit analysis have led to the conclusion that the benefit of research are out weighed by the potential cost.

In recent years the number of projects devoted to the documentation and use of indigenous knowledge has increased greatly but at the same time concerns have been voiced about the appropriateness of such use and its limitations (Cruikshak, 1998). The objective of such projects is primarily to document indigenous knowledge, so that it can be disseminated more widely than is possible by traditional oral means. Such knowledge is used by people from outside the community and also serves as secondary means of passing on important information to future generations. The projects are also intended to expand the quality and quantity of information available for environmental research and management in order to ensure that management regulations more accurately reflect local perspectives on wildlife pattern and on the use of local resources. To some extent, the projects also sought to give indigenous people a greater voice in decision making and activities affecting their land and their resource.

The Process

With this background the concern is how can IK and scientific knowledge be integrated in the best interest of sustainable development. The process of integration of IK with scientific information: It has been argued that indigenous knowledge and scientific information should be combined in order to solve emerging problems of a practical nature (Salas, 1994; De Walt, 1994; De Kruij *et al.* 1995). However, little is known about the nature of the interaction between new and pre-existing knowledge among farmers. The typology of cognitive interaction between indigenous knowledge and scientific information and a model for information interpretation should be based on analysis of information dissemination among the farmers on the subject. The process involves the information flow which takes place when individuals who have certain knowledge and confronted with new information. Once the new information is interpreted and interaction takes place between the pre-existing knowledge and new information. This happens when farmers, extension workers and researcher exchange information during training, meetings, group discussions etc. This result in the formation of an agricultural information system (Rolling, 1990). Although cognitive processes are highly complex (Malins, 1994), four main types of interaction between farmer's knowledge and scientific information can be identified and are described (Ortiz, 1997) as follows:

Formative Interaction

Formative interaction occurs when new knowledge is formed which in some cases replace the previous belief by the individual. This happens when farmers has access to scientific information regarding a subject.

Modifying Interaction

Modifying interaction occurs when farmers' knowledge is slightly adjusted by scientific information, so that farmers are better able to understand the scientific and biological principles behind the things they observe.

Reinforcing interaction

Reinforcing interaction occurs when scientific information confirms the farmers own knowledge. Reinforced knowledge allows them to feel confident about their own observations and practices. For example, certain practices already known to farmers are also included in IPM such as crop rotation, use of repellent plants and the use of chickens to get rid of insect larvae. The practices are better implemented when farmers understand insect reproduction and need to interrupt life cycle.

Confusing Interaction

Confusing interaction occurs when there is a conflict between farmers' knowledge and scientific knowledge. This is often seen when scientific information is presented in an appropriate manner or when farmers lack knowledge of biological principles related to the concept.

The Information Interpretation Model

In order to simplify the complex cognitive processes by which the scientific information are interpreted by the farmers' and the factors that influence the interpretation may be presented in a modular form. Such a model may help to understand the principles which underlay the process and enable extension workers to design suitable approaches to farmers. There are three groups of factors which influence the interpretation of information. The first group includes those factors which are related to the quantity, comprehensiveness and sequence of information and the way in which it is presented to farmers. For example, we found that those activities that involved the direct participation of the farmers and the observation of phenomena in natural setting are usually a better way of acquiring suitable information. They facilitate informative, modifying and reinforcing interactions while conventional training/extension activities tend to produce confusing interactions.

The second group of factors influencing information interpretation is related to the presence of key knowledge elements on the farmers' cognitive map (*i.e.*, everything that is known by a person and retained in his memory). The first possible conclusion from the model is that where there is no appropriate pre-existing knowledge, new information either can not be interpreted or is likely to be misinterpreted, resulting in a confusing interaction and the formation of inappropriate knowledge. This scenario was common when farmers who had no knowledge of biological principles were first given information. A second conclusion from the model concerns the acquisition of a suitable minimum set of cognitive elements designed to help farmers interpret subsequent information and thus acquire appropriate knowledge. The farmers' understanding of natural phenomena generated a positive attitude towards new information. It was possible to change attitudes if the individual had access to new and meaningful information and transformed it into knowledge.

In short, interpretation follows the correct path if pre-existing and/or new knowledge allows individuals to process and understand information. Understanding generates new and appropriate knowledge that provides feedback to another interpretive cycle. Conversely, interpretation follows the wrong path if there is no appropriate pre existing or newly acquired knowledge that can be used to process new information, leading to incorrect interpretation and confusion.

The third group of factors that influence information interpretation consists of continuous inputs of external information which differs from the new information given to the farmers. It is important to remember that scientific information and farmers' knowledge does not interact in isolation, but rather within the dynamics of knowledge evolution in the farming community and that there is sometimes conflict and competition between them.

The Challenges

Indigenous Knowledge is constantly evolving. Its enrichment will depend on whether it can interact with new types of information and can be used to solve emerging problems. However, the process by which pre-existing knowledge and new information interact needs to be explored so that research and extension approaches can be designed to facilitate the acquisition of knowledge by farmers and counterpart the erosion of their prior knowledge.

Identifying forms of interactions and the factors that influence the process of information interpretation may be useful in evaluating extension approaches. Where necessary they can be adjusted with a view to creat a synthesis between indigenous and scientific knowledge that will be of benefit to decision makers. To facilitate knowledge formation and the modification and reinforcement of indigenous knowledge, scientific information should reach farmers in an appropriate form. This means that research and extension projects are needed that enhance the farmers' capacity to deal with the farming problems (Fig. 1).

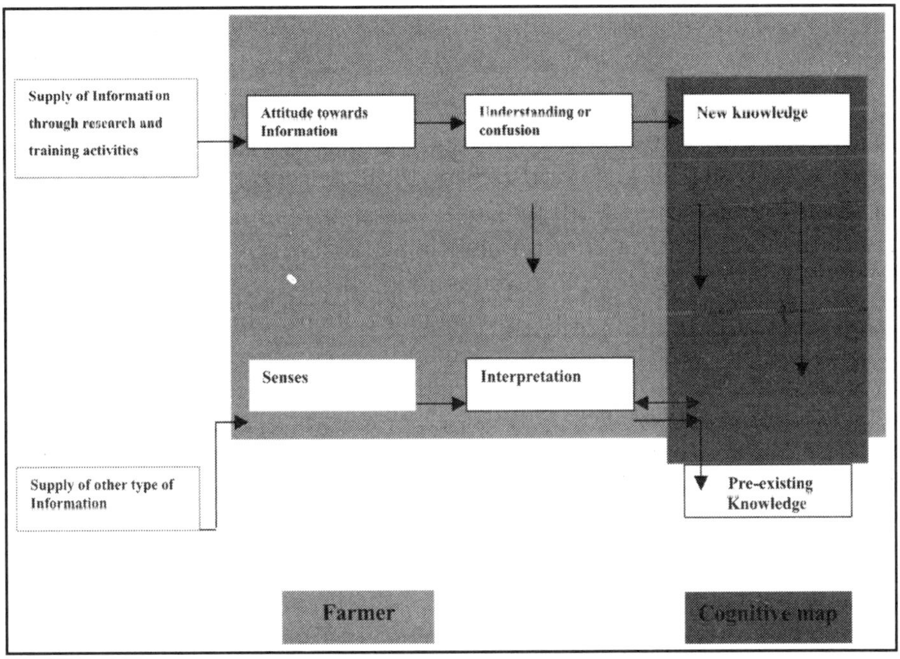

Fig. 1. Cognitive factors that influence the interpretation and acquisition of new Information (Adapted from Ortiz, 1997)

Today younger generations spend less time on the land and are less dependent on its resources; as a result they have less opportunity to learn from their elders and less incentives to do so. The loss of indigenous languages compounds the problem, as specialized term disappears and a literal language barrier makes it difficult for elders and young people to talk to each other. While perpetuating the use and transmission of IK depends on the members of indigenous communities themselves they can be helped or hindered by official altitude and actions. If it is regarded as an important source of accurate information and a cornerstone of community development, indigenous communities will continue to regard it as a vital resource well worth preserving for the benefit of future generations.

New Information and Communication Technology (ICTs)

In the present scenario of Indian agriculture, the public extension possibly can not provide additional qualified manpower to adequately address the complex demand of the farmers by reaching the millions of farmers. The growth and spread of new information and communication technology in rural India in recent years provide a viable alternative to overcome the physical barriers of face to face inter-personal communication. With economic liberalization and unprecedented development of new information and communication technology and educational infrastructure to create large pool of technical manpower in ICT (Information and communication technology), with wide range of information and communication technology option are being available even to remote and difficult areas. Appropriate choice and adaptation of new ICTs in different rural areas to communicate and educate millions of farm families will make all the difference in addressing challenges of agriculture in this millennium.

Up till now in India among various media, radio, television, literature and newspaper are certainly most utilized by the extension workers to transfer agricultural technology to the huge illiterate and literate segments of the rural area. We are observing a great transformation in agricultural extension approach in dissemination of knowledge. This revolution is due to more expansion in farming system fairly than the earlier accent of yield improvement. Even at previous juncture of agriculture extension growth, the style was *information supply mode* than *farmers' demand driven mode.*

Internet services, in conjunction with existing and more widely used communication media such as radio, television will enable the broadest enhancement of information and communication resources for rural people. National or regional agricultural market information systems hosted on the internet can be excellent information source for the staff of the radio stations throughout the region or the nation. Using information on current market prices including national variations and international figures broadcast by rural radio stations, farmers can negotiate better prices from the local buyers. Establishing rural internet access sites and facilities in concert with efforts to enhance horizontal communication networks among the agencies involved in rural and agricultural development is the essence of integrated approach. The web is surely the most promising way of extension services to reach more farmers with better services (Schmitz, 2001). The distance education for agriculture and rural development has been described as '*the third wave*' (Cook, 1998).

The internet is finding its way into the lives of people living in rural India. It does seem that the vast digital divide that has existed between rural and urban India might in its way to being bridged. The internet has finally reached in rural India in a small but significant way. The penetration of internet devices in India has been given in Table 1.

The emerging shift in Government strategy towards knowledge intensive services has created a climate more conducive to address enterprises, domestic infrastructure, education and the use of ICT to meet developmental needs. The Government has exempted the ICT industry from corporate income tax for 5 years. Such reforms have helped India to become increasingly integrated into the global economy through growth in the export of software and skill intensive software services such as call centres. The establishment of telecommunication regulatory authority of India (TRAI), opening of mobile phone to private operators in 1992 and finally opening of National long distance operations to private competition in 1994 are some of the strategic reforms. With the tele-density, 3.5 per cent in India, 3 million mobile users growing at 100 per cent per annum, 1.5 million internet accounts growing at 50 per cent per annum, increased access to telephones in Indian villages through the introduction of public call office (PCO) run by local shop keepers with 60 per cent of villagers having at least one phone, a large English speaking population and inexpensive ICT labour are important favorable factors. Public private partnership catalyzed by the IT Ministry, an IT act of 2000 aiming to set the overall strategy for the IT sector have further created an enabling environment for IT growth in India. With 20,000 web sites in Indian languages, 1,30,000 web sites in English and 1,6,70.3 MB per second international Bandwidth (Manzar, 2004), India provides an excellent opportunity for use of ICT for agricultural development for its 72 per cent rural population of 1.037 billion. The highest number of information kiosks implemented across rural sectors, 45% of the world ICT projects are implemented in India. Some of the biggest e-government implementations including digital voting system and proposal for rural information kiosks, one for each of the 6 lakh villages of India (Table 1).

Table 1: Penetration of Internet Devices in India

Information and Technology devices	March, 2000 (million)	March, 2008 (million) (Projected at present rate)
Installed base of PCs	4.3	30
No. of internet subscribers	1.0	35
No. of internet users	3.7	100
Fixed phones	26.0	125
TV sets	75.0	225
Cable TV	37.0	70

Experiences of rural internet users of many developing and developed countries indicate that the internet provides them with the very convenient method for quickly accessing large volume of information without being impeded by geographical barriers in the form of new ideas, discussion, expert advise, continuing education resources, increased global understanding and cultural awareness, and information that helps to make them better and more informed citizens. In addition to this, social benefits including new opportunities to overcome geographic isolation, increased social interaction, opportunities to organize and advocate for social change, equalization of urban rural disparities and new links between urban and rural communities were experienced. Agri-business users emphasize the internet value in enabling them to expand their markets to global audience and to establish national and global business networks and also alliances that would otherwise be inaccessible.

In our country, services provided by the Government are inadequate both in terms of infrastructure, technology and in empowering people with information. To provide information to local rural people through satellite based communication technology according to their needs and demands, information centres at village level can be established by providing local people in choosing the actual location of centre providing rent free space and electricity and recruiting volunteers. Each centre should contain data on agriculture, health related information, especially for rural farmers, women and children.

A directory of Government schemes should be made available to rural families on local prices of agriculture input and produce, cultural/ public events in locality, local transport/traffic detail including timing, grain prices, general and crop insurance schemes, hospital and medical practitioners, as well as information about integrated pest management in various crops. These databases should be in local language also. In addition, interactive CD ROMs on various issues can be made available. Information demand is different in each village hence great care should be taken to address the need for location specific information. It is stressed that internet initiatives for rural and agricultural development must consider the fact that different regions, organization, communities have different applications and technical needs. In some areas it is possible to have farmers and rural residents as direct internet users, while in other areas the capacity of intermediatory organizations need to be developed or assistance given in the establishment and promotion of community information centres linked to the internet.

There are cases of application of information and communication technologies in extension that have made a difference in the delivery of extension services in rural India. Some of these include the Warana Wired Village Project in Maharashtra, Milk collection in Dairy Co-operatives (NDDB), Information Village Project (M.S. Swaminathan Research Foundation), Information Technology Application for Indian Rural Postal System (CMC Ltd. Hyderabad), Knowledge Network for Grass root Innovations (IIM, Ahmadabad), Application of Satellite Communication for Training Field Workers and Extension Workers in Rural Areas (ISRO), Computerization of Mandal Revenue Offices (MRO's) and Computer Aided Administration of Revenue Department in Andhra Pradesh.

The Challenges

The IT based agricultural information on net is building up slowly but surely. There is still a lot of work to be done by policy makers to understand and act on this issue because the satellite based extension through the use of internet facility still has to go a long way to go to user friendly for rural people. People who work in the field of satellite based extension should have prior understanding that extension through satellite based extension in rural areas is not impossible but it require unique administrative, participatory and organizational efforts. Many organizations are supporting a number of SAUs to digitalize their technical information on the web. At the same time, research and training organizations both in public and voluntary sectors are being supported in building their capacity to digitalize the rural and agricultural information and make it available on the WWW. Efforts are being made to connect various districts of our country on the internet so that the farmers can get all the technical information on the web sites. If cyber connectivity will be made available to all India villages to take benefits of satellite based information and communication technology, the positive change in the face of rural development will definitely be possible.

People Participation in Research and Development

Farming system research evolved in the post-green revolution era with the growing perception of the failure of main stream agricultural research and extension institutions to generate and disseminate technologies widely adopted by small scale resource poor farmers. Technologies even sound by scientific standards is of limited value if it is not adopted. Conventionally agricultural researches had employed fundamentally *'top down approach'* to technology development which lack sufficient understanding of the management conditions under which small farmers in third world countries operate. The technologies developed through conventional research system and promoted for adoption by extension had problems of accessibility, technical feasibility, social acceptability, economic viability and compatibility with farmers' value system. Most of the station researches and extension programmes paid attention to technical aspects of the farmers' production problems. Farmers' production problems are by no mean merely technical problems but also socio cultural, economic, managerial and so on. This sort of thinking envisaged the need for participation and it was felt more in extension work because of the involvement of social scientists that are more tuned to behavioural sciences. There is growing realization among the researcher and extension workers that conducting meaningful research and extension calls for active partnership of the people for whom these are being done. Participation generally means involvement of the people in a process. It encompasses cooperation, coordination and contribution of people. Thus participation is a dynamic concept that changes from one situation to other situation. Contrary to general belief participatory research is useful to the researcher also and not only to the farmers. This process provides the traditional wisdom of the participants to become a part of the research and extension process.

The more advocated and discussed, the popular PRA (Participatory Rural Appraisal) offers a basketful of choices and best combinations based on objectives and purposes, resources, money and manpower. A few are being tried out by non-government organisations. Brief information regarding concept and purpose is discussed below with respect of the various approaches:

Social Map

It is the map of the village drawn without the help of scale to understand and simplify locations and structures of houses and other social facilities. The map is made by villagers based on their understanding indicating social setting on the map.

Purpose: Once a map is made by the villagers it becomes easy for them to describe locations within the village in terms of characteristics and conditions of the household the ownership of houses and other key features. In addition, it provides information regarding the dependency rations, beneficiaries under state/ central programme, health characteristics, and literacy. It is an useful basis for identifying problems in different households.

Resource Map

Like social map, resource map is a simple drawing indicating resource position of the village.

Purpose: To explore and understand the resource base of the village, village people's perception and experience with regard to utilization of resources, access to public resources, constraints and opportunities.

Time Line

Time line approach is a method of knowing the history of major recollected events in a community. It indicates a casual link between past and present. This method is also known as historical profile. In this technique a historical account is given by villagers of how different aspects of village life have changed and made impact on the social, economic, and agro-ecological bases in the village under consideration.

Purpose: It is useful for sensitizing the research and development agent to the importance of the past for understanding the present. Historical profile reveals important information for understanding the present situation in a community. It gives detailed account of the past of how things have changed over time, relationship and trends, the history of the technology development process and like. It facilitates discussion of the experiments already carried out by the villagers, crop histories, infrastructure facilities, livestock breed histories, fisheries, tree species, change in land tenure, outbreak of epidemics, drought and famines, change in Panchyat Raj administration and organisation, education change, population change, major political events, ecological degradation, folklore etc. precisely, it provides a summary overview of the key historical events in a community in relation to their socio-economic and political relevance for the present situation.

Time Trends

It is a simple technique of depicting quantitative changes over time in different aspects of rural life such as yield, price, population, livestock population, area under cultivation.

Purpose: To investigate into the changes taken place over time say for last ten years in quantitative items which are easily amendable to line diagram or bar diagram. These aspects could be yields, area under cultivation, price, rainfall, population, livestock population, interest rates, migration, birth and death rate, malnutrition rate, area under HYVs and any such directly quantifiable factor/variable.

Venn Diagram

Venn diagram is known after the man who created it. It is a diagram showing the key institutions and individuals in a village community and their relationship and importance for decision-making.

Purpose: As indicated, a Venn diagram explores the relationships between and among the key individuals, rural institutions, informal associations and such individuals and organizations including private and non-government which matter in the village development and upliftment.

Transect

A transect is a systematic exercise conducted by a multi-disciplinary team. It is a diagram of main land use zones. It is also called as farm walk.

Purpose: The purpose of a transect or a farm walk is to explore and study main land use zones. It compares the main features, resources, utilization pattern, problems and opportunities in each identified zone.

Mobility Map

It is both a data collection and an analysis tool. It is a tool to know the contacts of an individual or group of individuals of a particular village society with reasons thereof. It is also used to indicate the interactions within communities.

Purpose: Spatial mobility is used as an indicator for a person's contact, and knowledge of, the outside society. The mobility-maps of rural people based on socio-economic aspects, gender, and age groups, enable us to record, compare and analyse the interaction within communities and with other communities in a village, and to know the degree of freedom, empowerment, decision making, education, cosmopoliteness and consciousness dimensions of a given society. Precisely, it helps us to know the degree of mobility of rural groups and communities and importance of different areas.

Seasonal Calendar and Seasonal Analysis

Seasonal calendar is an exercise of showing the main activities, problems and opportunities through the annual cycle in simple diagrammatic form. It is a series of diagrams depicted on a single sheet of paper.

Purpose: To explore and identify the days, months of relatively great difficulty and vulnerability. It helps identifying significant variance which are likely to have an impact on the village people's life.

Livelihood Analysis

Livelihood analysis is an analytical technique exploring and obtaining information on behaviours, decisions and coping strategies in an individual households.

Purpose: To explore and interpret the behaviour, decisions and coping strategies of households with different socio-economic background. Factors such as household size and composition, number of labour migrants in the household, proportion of income by source, expenditure, livestock and land ownership, seasonality, relative income, credit and debit etc. could be included in the analysis.

Sustainability and Analysis

It is high order analytical exercise useful for periodic review of development activities.

Purpose : the purpose of sustainability analysis is to generate a high level of participation in electing the responses pertaining to sustainability issues like losses, deterioration in land, water, vegetation and other elements of eco-system due to biotic and abiotic factors. The sustainability matrix sharpens the group's analytical skills about identification of determinants of sustainability and helps bringing out important issues. It encourages the PRA group to ask a number of key questions which in turn lead to the decisions regarding the continuation and modification of the activities. Group discusses each determinant and makes an action plan for follow up using technological information available.

Wealth Ranking

Wealth Ranking is an analytical exercise carried out by local key informants based on their own criteria such as income, assets, employment status and other local measures of well being.

Purpose: The purpose of wealth ranking is to explore the villages people's local wisdom, their criteria used in classifying the entire households of a village wealth ranking is based on the assumption that community members have a good sense of who among them is more or less well off, poor or below poverty line. The purpose is to investigate the perceptions of wealth differences and inequalities in community to discover local indicators and criteria of wealth and well being, and establish the relative position of households in a community. Finally to know the poorest of the poor and their relative poverty.

Ranking

Ranking means placing items or objects or activity in order following single or joint criteria. Ranking method include preference ranking, direct matrix ranking and pair-wise ranking.

Purpose: To explore preferences of individual community members, their ranking criteria and priorities. It enables to understand people's decision making process according to usefulness and need. Ranking done by women enables to understand their priorities in development process and provides better understanding of the agro-ecosystem.

Agricultural Technology Management Agency: A Stepping Stone Towards Agricultural Extension Reforms

As a focal point for integrating research and extension activities and decentralizing day to day management of public Agricultural

Technology System (ATS). Agricultural Technology Management Agency (ATMA) is a registered society of key stakeholders involved in agriculture and research activities for sustainable agricultural development in the district. The concept of ATMA has been introduced as autonomous organization with an objective of integrating research, extension and all other stakeholders at the district level to support the farmers need and interests through an integrated approach of strategic plan.

As a registered society, it is responsible for technology dissemination at the district level. Involvement of farmers can be achieved at the village level through Farmers Interest Groups (FIGs), at block level as a member of Farmer Advisory Committee (FACs) and at the district level as a member of ATMA governing board.

The concept of ATMA envisages a paradigm shift from "top down" to "bottom up" approach in planning and implementation of agricultural development programmes. In order to strengthen Research-Extension-Farmer linkages and to provide an effective mechanism for coordination and management of activities of different agencies involved in generating location specific technology as well as dissemination and adoption of appropriate technology at district level and below, ATMA has been introduced in 24 districts of 6 states (Andhra Pradesh, Bihar, Orrisa, Punjab, Himachal Pradesh and Maharashtra) and the draft for implementation of ATMA in the state of J&K is almost ready. Being a registered society, it can get project funds, spend them and enter into contracts, arrange for facilities and keep revolving accounts.

Functions of ATMA

1. To decentralize decision making at district level.
2. To increase responsibility of stakeholder/farmer input in programme planning and resource allocation particularly at the block level.
3. To increase linkages with all the line departments, research organizations, NGOs and all the agencies associated with agricultural development activities in the district.
4. To accelerate programme coordination and integration so as to ensure effective implementation of programme issues like farming system innovations, farming organizations, technology gaps, natural resource management etc.
5. Responsible for technology dissemination at district level.

Objectives of ATMA

1. To identify location specific needs of farming community for farming system based agricultural development.
2. To set up priorities for sustainable agricultural development with a farming system approach.
3. To draw plans for production based system activities to be undertaken by farmers/ultimate users.
4. To execute plans through line departments, training institutions, NGOs, farmers organizations and allied institutions.
5. To coordinate efforts being made by various line departments, NGOs, farmer organizations and allied institutions.
6. To strengthen Research-Extension -Farmer linkage in the district and to promote collaboration and coordination between various state funded technical departments.
7. To facilitate empowerment of the farmers/producers through assistance for their increased participation in planning, marketing, technology dissemination and agro-processing etc.
8. To facilitate market intervention for value addition to farm produce.

Women Empowerment

Emancipation of women is an essential pre-requisite for economic development and social progress of the nation. Improving the economic status of women has now been recognized as an important component for any development programme. Indian women especially ruralites are handicapped by social customs, traditions, social evils and above all illiteracy, which needs special attention so that their potential and vigour for the growth and development of society and the nation can be fully utilized. To improve their status, it is pertinent to understand the flaws in the process of their development and consequently of their society.

Women in rural India play a major role in shaping the country's economy, since women constitute the biggest traditional group of unskilled worker in agriculture. At present, the women work force in agriculture and allied sectors is estimated at about 61 million which amounts to about 30% of the total rural workers in the country. They participate in different crop production and food processing operations including storage, packing, transport and marketing. Most of the work related to the management of cattle and other farm animals is done by women besides burden of household management. Studies have shown that the Indian rural women work for about 14-16 hours a day to carry out the most arduous activities on farm and at home. Rural women perform more than sixty percent of farm activities, still no voice of theirs' has ever emerged on any platform headed by them. There are various constraints and challenges faced by rural women as such their empowerment is a distant dream. Although the modernization of agriculture is taking place at a rapid pace, but the jobs attended by women remained more or less same. Considerable work has been done to develop agriculture with major emphasis on technical and economic achievements. Very little attention has been paid to socio-cultural developments especially the gender issues. Important gender issues involved are often taken for granted or totally ignored. As a result, technology development and transfer programmes have generally been carried out on the assumption that the technologies are either gender neutral or that the men are main user and decision makers (Castillo, 1985). This is often incorrect because women have quite different technological needs than men due to their different level of education, experience, skills, physique, stamina etc. Therefore, technology design will not be relevant to the user's needs, and transfer programmes will be ineffective unless these differences are recognized and acted upon.

Women farmers are unable to assimilate the changes due to lack of education and literacy. Cultural and mobility constraints also inhibit any commercial initiative on the part of women. The process of modernization has adversely affected the employment of women in industry as industry and agriculture they are the main victim of the process of economic transformation. Several studies have shown that the women worker irrespective of the job they perform are equal in performance and efficiency to a male worker in identical situation and in certain aspects they are more efficient than male like reliability, promptness, punctuality etc. In recent era, network of promotion agencies and industries have been set up at the national and state level to promote women entrepreneurship through support measures in the form of policy, finance, infrastructure, training, raw material and marketing facilities etc. However, in spite of their best efforts these agencies are unable to bring about the desired change and impact. Promotion of different organizations at different levels among women seems to be most viable and potent method to rope in the rural women to the national mainstream for their over all socio-economic development, development of the society and nation as a whole.

The realm of women participation in agriculture and allied activities in a remote cold arid area of the country where the temperature falls up to 35°C below zero and even a single blade of grass is not available in the peak winter may be quite encouraging for the policy planners, the women of the zone possess amazing entrepreneurial quality but there are also some grey areas of gender participation in economic transformation through agriculture and allied activities in Ladakh which still stand unattended. Time has come to focus our all attention for the improvement of overall deprived conditions of these rural women through achieving and accomplishing their economic independence and self reliance and self sustenance. Our wisdom, resources and administration coupled with technological breakthrough are to be channelized to overcome their deprivation, poverty and drudgery. It is to be well understood that without economic liberalization of women, no emancipation of women can truly occur. It is only when the women is allowed to participate as an equal partner in the most vital process of development of the nation and it is considered respectable for women to work shoulder to shoulder with man in society that the chain that binds and subjugate women to men can be broken. Also the role of NGOs in the women empowerment could not be over sighted. Presently there are as many as 484 NGOs working in the state. By and large the professional focus of these NGOs, is to the larger extent on rural development and poverty alleviation. However, women a potential component of the society may not be over sighted and as such functioning of some of the NGOs is gender specific and exclusively for the uplift of the women folk. These NGOs generally frame rural women SHGs in which women are educated and prepared for establishing independent income and employment generating enterprises. With the concerted efforts, the women are encouraged for their self-sustenance, socio-economic and cultural development. There is urgent need that some NGOs should come up addressing the gender issues on the aspects of giving credit, rewards and recognition, entrepreneurship development and their promotion and their attachment with suitable credit organisations.

Conclusion

An important distinction must be drawn between possible increase in production from the present era of cultivation and herd of present size and those due to expending the area or increasing the herd size. The most important single measure to support an increasing population is the increase in the area under cultivation. For this improvement of irrigation facilities especially in the areas where lack of irrigation facilities are already restricting the area cultivated. Besides, growing more grain, increased irrigation could permit the growing of alfa alfa both for conservation and feeding which could make a big contribution to the standards of livestock nutrition and productivity. There is great opportunity for introducing a wider range of secondary crops, vegetables and trees. Genetic improvement of the cereals from outside source will be difficult because of severe restrictions on length of season. Thus, the researches need to be done in the zone itself by crossing the local strains with the imported one so as to suit climatically. The low milk yield of local cow naturally promotes recommendations to raise this by impartation and use of pedigree bulls. Improved husbandry and faster growth should however make it possible to raise productivity by existing size of herd. the prospects of livestock improvement are complementary but the immediate attractions of increasing milk yield by importing high grade bulls are unlikely to be successful unless measures can also be taken to increase and improve the fodder. Hardy breeds of European cattle such as Scottish highland, Galloway and Welsh black should be tried as sires of cow to be crossed with yak. There seems little point in mechanization at present but undoubtedly the moderate use of fertilizers could raise production to a higher plane but research is needed on the natural sources of Nitrogen. If these are indeed *Azactobactor, Rhizobium, Blue-green algae* then it is important to know whether artificial nitrogen fertilizers has a depressing effect on them. All such work must be based on a full understanding of and sympathy with the local agriculture system and its restrictions both environmental and social. Indeed it is essential; to start with clarifying the reasons why the improvement, expansion or development are thought necessary at all and considering whether technological solutions to such demands are in the best interest of the people, to what extent they are being persuaded because of externally imposed political or economic pressures or because of professional preconceptions learnt in a different situation.

References

Castillo, G.T. (1985). Women in rice farming system (WIRFS) Research programme proposal. Prepared in IRRI. November.

Cook, Jonathan F. (1998). Distance Education for Agriculture and Rural Development. The Third Wave. Background paper presented at the workshop on Distance Education for Agriculture and Rural Development, 17-18 February, 1998. AERDD, The University of Reading.

Cruikshak, J. (1998). The social life of stories: narrative and knowledge in the Yukon Territory, University of Nebraska Press, Lincoln.

De Kruijf H.; Von Liebenstein,G. and Tick, A. (1995). Indigenous knowledge system for plant protection: impressions from symposium; Indigenous Knowledge and Development Monitor 3(2):14-16.

De, Walt, B. (1994). Using indigenous knowledge to improve agriculture and natural resource management. Human Organisation 53(2):123-131.

Korten, D.C. (1980). Community organizations and rural development: A learning process approach; Public Administration Review. Sept.-Oct. 480-503.

Malim, T. (1994). Cognitive process; attention, perception, memory, thinking and language. Mac Millan, London.

Manzar, Osama (2004). Adversity to success the world's best e-content and e-creativity experience, The country Paper India, Global ICT summit 2004, Hong Kong, Digital Empowerment Foundation.

Ortiz, O. (1997). The information system for IPM in subsistence potato production in Peru. Experience of introducing innovative information in Cajamara Province. Ph.D. Thesis. Agricultural Extension and Rural Development Department, University of Reading, UK.

Rolling, N. (1990). The agricultural research technology transfer interface. A knowledge system perspective in Making the link. Agricultural research and technology transfer in developing countries. Kaimovits, D. (ed.). West view Press, USA. pp.1-41.

Salas, M. (1994).The technicians only believe in science and can not read the sky: the cultural dimensions of the knowledge conflict in the Andes. In: Beyond farmers first. Rural people's knowledge, agricultural research and extension practice. Scoones, I. and Thompson, J. (eds.). Intermediate Technology Publications, London. pp. 41-52.

Schmitz, J.G. (2001). Agricultural Information Systems. Proceedings of the India Soy Forum. Bhatnagar, P. (ed.) SOPA: Indore.

Van, Vlaenderen H. (1999). Problem Solving: A process of reaching common understanding and consensus. South African Journal of Psychology 29(4): 166-77.

Young, Pang (1995). A successful model of combating poverty in China: The application, recreation and practices of farmer led approaches in agriculture extension. Paper presented in International Workshop on Farmer led approaches to agricultural extension. July 17-22, 1995, IIRR, Silang, Cavite, Philippines.

□□□

Van, Vlaenderer H. (1999). Problem Solving: A process of reaching common understanding and consensus. South African Journal of Psychology 29(4) 166-77

Young-Fang (1995). A successful model of combating poverty in China: The application recreation and practices of farmer led approaches in agricultural extension. Paper presented in International Workshop on Farmer led approaches in agricultural extension, July 17-22, 1995, IIRR, Silang, Cavite, Philippines.

21

Conservation of Genetic Diversity in Trans-Himalayas

SANDHYA GUPTA AND SANJAI K. DWIVEDI*
Tissue Culture and Cryopreservation Unit,
National Bureau of Plant Genetic Resources, New Delhi-110 012
*CEPTAM (DRDO), Metcalfe House, Delhi-110 054

The Trans-Himalayan region, which lies north in the rain-shadow of the main Himalayas, is usually described as "high altitude cold desert". The landscape is characterized by a distinct lack of natural forests showing a high degree of endemism (only found locally). Along the river banks and valleys in Ladakh, some greenery does exist with willows, poplars, wild roses and many herbaceous plants and shrubs having colonized the land, and even extended up the mountain slopes. Most of the vegetation is, however, stunted.

The plant species that manage to exist in this region are well adapted to the harsh, extreme environmental conditions. Compared to the main Himalayas, this region is floristically impoverished and has affinities with the main Himalayas and diverse regions such as Afghanistan, Tibet, the Mediterranean region and the Indian desert areas. About 750 species of flowering plants have been reported from this region (Chundawat and Rawat, 1994). Unique floristic features of this region include a large number of endemics, a high proportion of monocots and the absence of large tree species. Predominant taxonomic groups include Asteraceae (Compositae) and Brassicaceae (Cruciferae).

Animal Genetic Resources

Distribution

The cold desert (Greater and Trans-Himalayan tracts) regions have severe seasonal and resource scarce environments. Covering the bulk of Indian Himalayas, they are a rich repository of biodiversity values and ecosystem services. In the eastern extremity of the Trans-Himalayas, the Zanskar and the Ladakh ranges merge into the southern margins of the Tibetan plateau. The huge, brackish marsh and lake ecosystem houses water bodies like Tso Moriri, Hanle and Chushul, the habitat of some very threatened fauna, the best known being the black-necked crane which breeds in these bogs and marshes. The northern and western parts of the zone in the Indian region lie mainly above the snowline, and this area includes Siachen, an 1,180 sq km glacier, said to be the largest outside the polar regions.

Though, the environment here appears harsh and inhospitable, a wealth of animal life exists. The fauna in this region is surprisingly diverse and has elements from the Himalayan as well as Mediterranean and Tibetan regions. Birdlife is diverse and the high altitude wetlands are breeding grounds for large number of migrants, especially water birds. Muscicapidae (flycatchers, thrushes, redstarts and chats), Sylviidae (warblers and whitethroats), Fringillidae (finches and buntings), Accipiteridae (hawks), Scolopacidae (sandpipers, snipes and curlews), Passeridae (pipits, wagtails, accentors and sparrows) and Anatidae (ducks and geese) are well represented. Birds found in this region that are of high conservation value include the black-necked crane (*Grus nigricollis*), Tibetan sandgrouse (*Syrrhaptes tibetanus*), Tibetan snowcock (*Tetrogallus tibetanus*), chukar (*Alectoris chukar*), golden eagle (*Aquila chrysaetos*) and the bar-headed goose (*Anser indicus*).

Two characteristic species of Tibetan plateau have very limited distributions in the central mountains; (i) Kiang or Tibetan wild ass (*Equus kiang*) is found from the high plains of eastern Ladakh into the Karnak area of the upper Chang Chu, and single animal sporadically wanders farther west into nullahs of the upper Markha catchment;

(ii) Tibetan argali (*Ovis ammon hodgsoni*), much scarcer than kiang, have established small outlying populations in the central mountains on at least three occasions during the 19th and 20th centuries. One of these was reported in the 1880s at an unnamed locality south of the Indus, and another in the 1930s. The latest of these appeared in the late 1970s in the upper Rumbak valley and is still thriving (Fox *et al.*, 1991).

Habitat separation between shapu on the one hand and bharal and ibex on the other is a typical example of that between wild sheep and goats. The distribution of bharal and ibex is of more interest, with bharal occurring principally in the eastern parts of Ladakh and ibex in the west. The habitats they occupy are usually indistinguishable on a number of observed factors, such as geology, ruggedness, vegetation, altitudinal range, yet the geographical separation between the two species is quite clear and there is little overlap range (perhaps 3-4% in Ladakh) along the contact zone. A few mixed groups of ibex and bharal have been reported in the Shun-Shadi area of Zanskar, but such occurrences are rare and yet hybridization has not been reported. This pattern of largely exclusive ranges, with a long contact zone is repeated elsewhere when the two species' ranges meet. Discussing the separate distributions of ibex and bharal in the Karakoram, Schaller (1977) observed that "Competition for resources would be inevitable and coexistence impossible without a geographic partitioning of ranges. Sympatric survival is not possible in a simple habitat for caprids of a similar size". They have apparently partitioned the habitat geographically, reflecting dispersal from different directions-ibex from the west and bharal from the east. A similar zoogeographical pattern is repeated by other species pairs in Ladakh; marmots (*Marmota himalayensis* and *M. caudata*); hares (*Lepus oiostolus* and *L. capensis*); and snowcock (*Tetraogallus tibetanus* and *T. himalayensis*).

The two sheep species, shapu and argali, have generally separate distributions in Ladakh that just come into contact at the eastern edge of the mountains in the Gya-Miru area, and also in Rumbak valley, a tributary of the Indus. Hybrids between the two very occasionally result, with 3-4 specimens reportedly obtained during the last *ca.* 140 years, one of them from the Rumbak area, from Miru and in the 1870s from an unspecified locality (Ward, 1924; Mallon, 1998).

The mountain ungulate community in Ladakh shows a typical Trans-Himalayan pattern. In Spiti, which lies southeast of Ladakh, bharal and ibex occur in separate ranges which just meet in the Kibber area, and argali appears sporadically via valley-head passes on the Tibetan border (Burrard, 1925; CADARI, 1996). The mountain ungulate fauna of Hunza in Pakistan occupied Kashmir, northwest of Ladakh, also matches that of Ladakh. Shapu are restricted to a narrow band along valley bottoms, ibex are widespread and occur at higher elevations up to the snow line, but have a separate range from bharal. Kiang occurs in northeastern Hunza and argali (in this case *O. ammon poloi*) occurs on the fringes of the area around the Khunjerab and Kilik passes.

Diversity

The largest number of wild sheep and goats in the world, are found here. The ungulate fauna of the Himalayas include species such as the chital, sambar, wild pig, Asian elephant, species that are found in other parts of the country too. There are however various cervids, moschids, bovines and equines unique to the Himalayas or limited to the high mountain chains of Central Asia. The distinct species and/or sub-species of ungulates include the nayan or great Tibetan sheep (*Ovis ammon hodgsoni*), the urial or shapu (*Ovis orientalis*), the bharal or blue sheep (*Pseudois nayaur*) and the ibex (*Capra ibex*). It is noteworthy that this assemblage of species/subspecies constitutes *ca.* 50% of India's ungulate fauna (Schaller, 1977, 1998; Shackleton, 1997). Many of these species are wild relatives of sheep, goat, horse/donkey and yak, thus adding value to this assemblage as an important genetic pool. Found on the plateau of the Trans-Himalayas, and in a small part of eastern Ladakh, are one species each of an antelope and a gazelle. The Tibetan antelope (*Pantholops hodgsonii*), which is better known as the chiru, is a little animal that lives at attitudes of over 5,000 m and is known to cross over from the Tibetan plateau into the Chang Chen Mo valley of northeastern Ladakh. The Tibetan gazelle (*Procapra picticaudata*) whose stronghold is the sprawling Tibetan plateau is occasionally sighted in northeastern Ladakh. Smaller animals of the region include pikas, marmots and Tibetan hares.

Very little is known about the other faunal groups. Only a few amphibians and reptiles (approximately 12 species) are adapted to the region. Amphibians are represented by the toads (Bufonidae), reptiles include agamids (Agamidae-three species of toad agamas are found in this region, even in the Chang Thang), skinks (Sinicidae), geckos (Geckonidae) and the snake family Colubridae (three species of colubrid snakes). Endemic fish diversity is believed to be considerable, however, much has not been published on fish diversity in the region. A number of species belonging to the leech family (Homaopteridae) and a carp species (Cyprinidae) have been reported from the region. Invertebrate diversity in the region is largely undocumented and still remains to be explored in detail. A number of surveys were conducted in the Trans-Himalayan regions during the seventies (Mani, 1962) and a few recent studies include Maheshwari (2000) and also surveys by the Wildlife Institute of India. Most of these species evolved to inhabit the niches produced by spectacular mountain building during the Cretaceous and Tertiary that created new, usually cold and bleak landscapes, and are well adapted to these harsh environments. An understanding of their ecology can thus help in better management of the entire region (http://www.wii.gov.in).

There is recent evidence suggesting that domestic livestock deplete the density and diversity of wild herbivores in the cold deserts of the Himalayas by imposing resource limitations. To ascertain the degree and nature of threats faced by Himalayan ibex (*Capra sibirica*), the nature of interaction between them and livestock in Spiti region of Indian Trans-Himalaya were studied. Resource use patterns (over space, habitat and food dimensions) of ibex and livestock were assessed and the observed patterns were compared against a simulated competition-free environment (by null models for community analyses) to assess the impact, livestock may have on ibex. Analyses involved Monte Carlo simulation algorithms, Nonmetric Multidimensional Scaling and Classification Trees. Results indicated that domestic horses and migratory goat and sheep can impose resource limitation on ibex and hence are a potential threat. But since the local agro-pastoral community is experiencing major socio-economic changes, there is a gradual decline in per-capita house holdings against an increase in cash crop acreage. This opens up certain possibilities for long-term conservation of ibex and other wildlife in the region, which supports the greatest diversity of mountain mammals on earth (Bhatnagar *et al.*, 2002).

Conservation Status

Within India high altitude cold desert covers approximately 186,000 sq km which is also one of the least (0.4%) protected regions under the protected area network in India. The region has many endangered species listed in schedules I & II of the Indian Wildlife (Protection) Act, 1972, Government of India, and also in IUCN's Red Data Book. The Trans-Himalayas experience extremes of cold and arid climatic conditions which account for its low productivity. Animals occur in low densities and require larger areas to maintain a viable population. The Trans-Himalayan region is an intrinsically fragile ecosystem and its overuse has pushed it close to ruin at many locations. This area has remained unstudied by ecologists and the paucity of information on the Trans-Himalayas is recognized by managers and conservationists. Snow leopard *(Panthera uncia)* an endangered large predator of the region was considered a flagship species for the overall conservation of this fragile ecosystem on the lines of "Project Tiger." This called for urgent conservation measures which required scientific information especially on the ecology of snow leopard and its associated species. Conservation status of animal species in Trans-Himalayan region is presented in (Table 1).

A number of mammal species found in the region are endangered and are listed under Schedule I of the Wildlife (Protection) Act, 1972. Examples include the snow leopard (*Uncia uncia*), the Ladakh urial or shapu (*Ovis vignei vignei*), the Tibetan argali or nayan (*Ovis ammon hodgsoni*) and the Himalayan ibex (*Capra ibex sibirica*). Species such as the wild yak (*Bos grunniens*), the Tibetan gazelle or goa (*Procapra hodgsoni*) and the Tibetan antelope or chiru (*Pantholops hodgsoni*) found in the Chang Thang region of the Trans-Himalaya are similarly under threat.

It is basic to the understanding of natural systems that extreme climates create habitats of great fragility. The Trans-Himalayas is no exception. Of the estimated 186,000 sq km land area of the Trans-Himalayas in India, the Protected Area (PA) network extends to only 0.4 per cent, making it the lowest for any of the 10 zones in the Indian subcontinent. The largest contiguous protected area is the Hemis High Altitude National Park in Ladakh which covers about 600 sq km Another seven 'wildlife reserves' are distributed over a total area of 860 sq km, also in Ladakh, where only hunting, and not the land use for traditional grazing and other related activities are regulated.

A number of biodiversity conservation planning documents have focused on the Trans-Himalayas. The National Biodiversity Strategy and Action Plan (NBSAP) Project initiative focused on districts within the Trans-Himalayas and has identified region specific issues. Research on important umbrella species such as the snow leopard and its prey species have contributed to biodiversity research in general. Currently, as a result of a number of such initiatives, there seems to be a good coverage of research topics, species and ecosystems.

Historically, the region has supported very low population densities and even now population densities are low in most parts of the Trans-Himalayas. This makes PA management easier and in most places no active management is required. This gives PA managers room to concentrate on areas of heavy grazing (livestock numbers have increased tremendously), tourist influx and other problems. Though human-wildlife conflicts have existed even in the past (at a much smaller scale), solutions that combine traditional responses and ingenious insurance and incentive schemes seem to be effective.

Table 1: Conservation status of selected animal species in Trans-Himalayan region

Name	Zoological name	Vernacular name	Taxonomy	Conservation status
Siberian ibex	*Capra sibirica* Pallas	Kin (Spiti), Skin (Ladakh), Tangrol (Kullu, Lahul)	Family: Bovidae Sub family: Caprinae Tribe: Caprini	IUCN: Low risk CAMP: Vulnerable IWPA: Schedule I CITES: Not listed US ESA: Not listed
Flare horned or kashmir markhor	*Capra falconeri* Wagner	markhor (Punjab and Kashmir), raphoche (male) rawache (female)	Family: Bovidae Sub family: Caprinae Tribe: Caprini	IUCN: Endangered CAMP: Critical IWPA: Schedule I CITES: Appendix I US ESA: Endangered
Bharal, Himalayan blue sheep	*Pseudois nayaur* Hodgson	Knabo (Ladakhi, Tibetan), nabo (Spiti), bharal (Hindi, Pahari)	Family:Bovidae Sub family: Caprinae Tribe: Caprini	IUCN: Low risk CAMP: Low risk, least concern IWPA: Schedule I CITES: Not listed US ESA: Not listed
Ladakh urial	*Ovis vignei vigne* Blyth	Shapo(male), Shammo, Shanmar (female) (Ladakhi)	Family: Bovidae Sub family: Caprinae Tribe: Caprini	IUCN: Endangered CAMP: Endangered IWPA: Schedule I CITES: Appendix I US ESA: Endangered
Tibetan argali	*Ovis ammon hodgsoni* Blyth	Nayan, nyan (Ladakh, Tibetan, Spiti)	Family: Bovidae Sub family: Caprinae Tribe: Caprini	IUCN: Vulnerable CAMP: Critically endangered IWPA: Schedule I CITES: Appendix I US ESA: Endangered
Tibetan antelope	*Pantholops hodgsonii* Abel	Tstosh (Ladakhi, Tibetan), chiru (parts of Tibet)	Family: Bovidae Sub family: Panthalopinae Tribe: Saigini	IUCN: Endangered CAMP: Critical IWPA: Schedule I CITES: Appendix I US ESA: Not listed
Tibetan gazelle	*Procapra picticaudata* Hodgson	Gowa (Ladakhi, Tibetan)	Family: Bovidae Sub family: Antilopinae Tribe: Antilopini	IUCN: Low risk CAMP: Critical IWPA: Schedule I CITES: Not listed US ESA: Not listed
Yak	*Bos grunniens* Linnaeus	Dong (Ladakhi, Tibetan)	Family: Bovidae Sub family: - Tribe: Bovini	IUCN: Vulnerable CAMP: Critical IWPA: Schedule I CITES: Not listed US ESA: Endangered

Source: Bhatnagar (2002)

Conservation Issues

Though the Trans-Himalayas is a sparsely populated region, the problems for this unique ecozone are compounded by the near-complete absence of natural forests which to an extent might have been able to mitigate the human and livestock pressure on the sparse vegetation and water resources. As determined by resource availability, the population densities of the region's wild animals are very low and widespread. All in all, this is an extremely fragile ecosystem. The past few years have witnessed a rising invasion of what were once considered inaccessible areas. There are a growing number of nomadic pastoralists with their increasing herds of livestock, and a simultaneous rise in military and other developmental activities. Tourism, which on the face of it appears so benign, has a dark side which brings severe ecological degradation with it. Tourists, for instance, among other things, need hot water and heating. So woody plants, which are critical to the survival of several species, quickly succumb to the 'urgent' need for fuel. Added to such problems is the fact that many mammals of the Trans-Himalayas are under intense hunting pressure for their fur. At one time traditional weapons, traps etc. kept the hunting in check, but today automatic weapons, powerful binoculars and a thriving trade have combined to take a devastating toll of wild animals (Rawat and Sathyakumar, 2002).

Several factors have led to low abundance and poor conservation status of ungulates in the Himalayan region. Firstly, the Himalayan ecosystem is relatively young, fragile and low in primary productivity. With the increase in human population the area has undergone rapid degradation, fragmentation and loss of wildlife habitat. Poaching and trade of animal parts, competitive exclusion by the domestic livestock, faulty land use practices, and human-wildlife conflicts are other factors affecting wild ungulates and other faunal communities (Gaston *et al.*, 1983; Green, 1985; Vinod and Sathyakumar, 1999).

Habitat Loss

Excessive degradation and fragmentation eventually leads to habitat loss. Systematic studies documenting loss of ungulate habitats in the Himalayan region are lacking. Hence, it is difficult to point out clear cases of this phenomenon. However, in most of the sectors, there are plenty of evidences indicating the shrinkage of wildlife/ungulate habitat. Green (1986) reported that over 70% of potential musk deer habitat has already been lost due to habitat loss and habitat degradation in the southern side of the Greater Himalayas. Displacement of human populations for the developmental projects, construction of roads along the sensitive habitats, encroachment of forests for agriculture and heavy infestation of exotic weeds are other causes of habitat loss in the region.

Competition with Domestic Livestock

Owing to high seasonality and low primary productivity, the Himalayan region supports relatively low ungulate/herbivore biomass. It is therefore, obvious that with the increase in the biomass of domestic livestock in many areas, wild ungulates have suffered competitive exclusion. Although animal husbandry is one of the main stay of livelihood in the Himalayan region, the management of livestock especially disease surveillance, rotational grazing and pasture management have been neglected leading to conflicts with wildlife as well as PA managers.

Poaching

Mountain ungulate densities in Ladakh are not notably high, in comparison with other areas, and their significance lies much more in their extensive and unfragmented ranges and the fact that overall, populations for most species are not declining. This is in large part due to the lack of a hunting tradition among local people and traditional Buddhist disapproval of killing animals. The species that has suffered most through illegal hunting is the shapu, whose habitat, lying along the main Srinagar-Leh highway, is easily accessible. The narrow-linear pattern of distribution, with its high ratio of periphery to area also makes it susceptible to fragmentation because it is accessible at many points, and there is no remote central core to act as a population reservoir, unlike the case with bharal and ibex (Mallon, 2002).

All mountain ungulates of the Greater Himalayas are seriously threatened due to poaching for meat, skin/hide, and for products such as the 'musk' from Musk deer. Poaching for meat (bush meat hunting) is common in many parts of the Greater Himalayas, particularly in the Eastern Himalayas. Species such as the goral, tahr, serow, takin are poached for meat/hide by local villagers and indigenous people throughout their distribution range. There are instances of wild mountain ungulate meat served in local restaurants in towns or villages adjoining wilderness areas. Although, the extent of poaching by local villagers or indigenous people is not known, there are evidences of the consequences of high poaching levels in many areas where mountain ungulates have become either locally extinct or occur in very low densities. Poaching for sport and meat by the personnel of the security forces in the international border areas, also have led to serious impacts on mountain ungulate populations. Poaching of musk deer for 'musk' is rampant through out the Greater Himalayas for its high commercial value (about US $65,000/kg in 1985) in the international markets (Green, 1986; Sathyakumar, 1993). This has led to local extinctions of this species in many parts of the Greater Himalayas. As a result, the once continuous distribution of musk deer is now confined to some isolated pockets and in most of these areas they occur in very low densities. Similarly, poaching of Himalayan tahr and goral have either resulted in local extinctions or very low densities in many areas.

Strategies

To save the last remaining threatened wildlife of this zone, several proposals have been mooted. If enacted, these will serve to increase the region's protected areas to well over 10 per cent of the total zonal area, of which 6.8 per cent would be set aside for National Parks and Sanctuaries, thereby facilitating major inputs in terms of infrastructure, finance and manpower. One of the best ways of monitoring wild habitats is to hone in on 'indicator' species-species whose presence 'indicates' the health of the habitat and whose absence suggests decay. The snow leopard is one such species for the high mountains. At the International Snow Leopard Symposium (1986), the establishment of Project Snow Leopard was recommended, and several snow leopard reserves have since been created in Hemis, Siachen Shyok and Rupshu. Rodgers and Panwar (1988) have suggested that the entire Trans-Himalayas (in India) be accorded protection under UNESCO's Man and Biosphere Programme. Further, the Hemis National Park with its adjacent monasteries is a prime candidate for inclusion into the World Heritage Site network on account of its natural and cultural value.

Infrastructure and Logistics

The Trans-Himalayas still remains one of the remotest biogeographic zones in the country. Infrastructure and logistics are still poor for the region as a whole. Field stations and equipment, long-term monitoring stations, multi-institutional setups (like Wildlife Institute of India-International Snow Leopard Trust model) need to be tested. Advanced facilities such as GIS labs are also lacking in this region. Transport and communication facilities need to be improved for the Forest Department as well as research personnel. Forest guards in particular need to be outfitted with appropriate equipment. Data stations need to be set up to monitor climate and gather data on the effects of climate change. Wetlands should be given priority as they are directly affected.

Research

There are still many faunal groups that are largely unstudied with respect to taxonomy as well as status, distribution and ecology. Since most of the ecosystems are simple and host fewer species (than the main Himalayas), systematic surveys could be carried out for basic documentation and taxonomy. Spatial information on the distribution and status of medicinal plants could be given priority. Additional research and monitoring is required on species that are currently under threat or being commercially exploited. Human-wildlife conflict and its potential impacts should be studied in more detail. More databases and clearing-house mechanisms are needed for the region. There is a requirement for a database centre for the Trans-Himalayas.

Much of the Trans-Himalayas is still largely undisturbed and has low human population densities. In many such areas no additional management interventions are required. At the same time, it is important to identify critical areas requiring immediate management intervention. For example, ecotourism facilities in the vicinity of Tso-Moriri and Korzok have been known to cause a variety of problems. Such sites need to be given special attention. Socio-economic data on the region needs to be compiled and made available to biodiversity practitioners. Land use zones and patterns need to be studied. Socio-economic evaluations and alternate livelihood practices also remain to be researched.

Protected Area Management

The region has a large protected area (PA) network which has not been completely effective in conserving these unique values. The human population densities are much lower (usually < 1 per sq km) than in most other parts of the country (over 300 to a sq km). However, even such small populations can come into conflict with strict PA laws that demand large inviolate areas, which can mainly be achieved through relocation of the scattered settlements. However, according to Bhatnagar (2008) in this landscape, relocation is not a tenable strategy for conservation due to a variety of reasons.

Officers at the departmental level should be given orientation to multi-disciplinary exercises and need to be encouraged to develop projects of that nature. Local forest and wildlife officials should be identified who can administer such projects effectively. Community based management of human-wildlife conflicts should also be explored. Development of management plans incorporating the various aspects of biodiversity conservation specific to the region should be undertaken.

Most of the protected areas in the Himalayan region lack adequate man power and funds for proper management. In addition, several PAs have ill defined boundaries leading to conflicts between the local communities and PA management. There is an urgent need to strengthen the management of most of the PAs allocating more well trained and motivated staff, budget and infra-structure.

Several PAs and Reserved Forests in the Himalayan region need to be brought under community reserves where local people could be made partners in conservation and management. Through various programmes such as community based eco-tourism and participatory management of natural resources the illegal activities such as trade of animal parts and poaching could be gradually brought down.

Sustainable Use and Livelihoods

Sustainable use and alternate livelihoods projects need to be initiated. Sufficient incentives need to be provided to local communities to conserve biodiversity effectively. Currently there seems to be a point of view that there are more incentives to poach and trade than conserve biodiversity. This needs to be given special attention since there is trade in species (especially large mammals such as snow leopard and chiru). A number of community based incentive programmes such as livestock insurance and incentives in dealing with grazing and snow-leopard depredation (initiated by NGOs such as the NCF and ISLT and their allies) are gaining ground. However, for large-scale success, a larger number of such projects need to be initiated by governments also. Projects that examine how the private sector can contribute meaningfully to biodiversity conservation in the region should be encouraged. Existing sustainable use and exploitation models especially relating to plant species need to be reviewed. Of primary significance is the exploitation and sale of seabuckthorn given its widespread presence in the region and the recent attempts by various organizations to develop beverages and other products. Medicinal plant conservation and utilization also needs to be brought under the sustainable use and benefit-sharing frameworks, so that economic and intellectual rights go to local communities. Tourism contributes to the economy of the region, especially to urban centers such as Leh and along the major trek routes. The concept of home stays is also beginning to get established. Since the area is ecologically fragile, more assessments are required to look at sustainable ecotourism. There is an overall need to strengthen local institutions for biodiversity conservation and to enhance rural livelihoods.

Care should be taken to avoid establishing large-scale plantations of exotic poplars (*Populus* spp.) and willows (*Salix* spp.), which may turn out to be detrimental to local biodiversity conservation. Conservation in the region is largely based on umbrella/ flagship species approaches and revolves around the conservation of species such as the snow leopard, ibex, *bharal* and *kiang* and species associated with them.

Policy and Advocacy

Inter-disciplinary policy/advisory groups need to be formed. In the current scenario, where emerging challenges are largely livelihood based, it is unreasonable to expect any one department to single handedly manage these issues. There are no institutional arrangements for collaboration among NGOs, communities and the government. A number of networking projects need to be funded and initiated. Policy related work especially with sustainable development and livelihoods needs to be distinct from that of the main Himalayas, since the Trans-Himalayan zone is distinct. Sustainable policies for specific issues such as intellectual property rights, medicinal plant conservation and biotechnology for the region need to be prepared.

There is an urgent need to increase trans-boundary co-operation between India-Nepal, India-China and India-Myanmar to control the illegal trade of wildlife products. Measures should be taken to register all the arms (licensed guns etc.) with the Wildlife warden or the Divisional Forest/Wildlife Officer.

Currently, there is no organised system of harvesting wild medicinal and aromatic plants and uncontrolled harvest often results in the degradation of habitat. There is a need to evolve policies related to rotational harvest of medicinal plants for the benefit of communities and there is a need to control excessive pressure on the land. As part of eco-development measure cultivation of medicinal plants needs to be promoted in the buffer zones of various PAs where human pressure for these commodities is excessive.

Plant Genetic Resources (PGR)

The vegetation of the region is generally dry alpine steppe or dry alpine scrub (Champion and Seth, 1968). Several ecological factors such as altitude, temperature, rainfall, and slope played an important role in determining the vegetation type, structure, and composition. In all, six vegetation types have been identified within on physiognomic basis. The upper limit of vegetation growth was 5,000 m to 5,200 m, beyond which bare rocks and permanent ice predominated (http://www.snowleopardnetwork.org). In this region, there are a number of species of great medicinal and economic importance and as a result have great conservation importance. Wild relatives of numerous species including onion (*Alium* spp.), rhubarb (*Rheum*) and *Horedum* (barley) are found in the Trans-Himalaya. It is believed that over 50 per cent of plants in the region have recorded medicinal value (Kala, 2001). Common medicinal and aromatic plants include species such as *Physoclaina praelta* and a number of species belonging to the genera *Artemesia, Tanacetum, Delphinium* etc. Trees and shrubs, particularly *Juniperus* and *Betula utilis* which have aromatic as well as medicinal properties seem to have been over-harvested. The major fruits that are cultivated in the area include apricot, apple, walnut, mulberry and seabuckthorn have great nutritional value. The Physio-chemical properties of major apple varieties, seabuckthorn, mulberry and apricot varieties which are grown in Ladakh have been studied (Dwivedi *et al.*, 2008). The cultivated vegetable crops include local turnip, onion, and leafy vegetable, *viz., Ustakh, Dums Khala, Sangtso* etc. The forest trees like Juniper, *Elagnus, Ulmus*, poplar, willow etc. also grow successfully in the region. These crops have been successfully grown in the region since time immemorial and have been well adapted to the stressful condition of the region.

The plant diversity available in Trans-Himalayan region may play an important role in the crop improvement and breeding programme for the development of abiotic stress tolerant varieties. Survey, collection and evaluation of some of these crops have been taken up by various organizations like FRL, SKUAST, HPKVV, GBPUAT, ICAR etc. However, limited attention has been paid to the proper conservation of the plant genetic diversity of the region. Over the years,

India has developed sound scientific management regimes for *ex situ* conservation and access to its genetic resources. National Plant Genetic Resources (NBPGR) as the nodal agency is addressing the task and coordinating the PGR management activities of several institutions, scientific societies and non-governmental organisations. It aims at efficient management of plant genetic resources by providing convenience of access to the various crop improvement programmes. It also encompasses the National Genebank Network.

PGR Management

The sustainable management of natural resources is a primary global concern, as increasing population and rapid technological advances are putting tremendous pressure on these resources. During the last three decades of this century rapid progress in the field of agriculture was witnessed, along with the collection, conservation and sustainable utilization of PGR. Conservation of PGR has now become one of the most widely discussed issues in the field of agriculture and environment. Three international negotiations impacted on this rise to prominence. These are the Convention of Biological Diversity, the agreement on Trade-Related Aspects of Intellectual Property Rights as part of the Agreement establishing the World Trade Organization and the International Treaty on Plant Genetic Resources for Food and Agriculture. These negotiations, and the rapid advancements in the fields of molecular biology, biotechnology and bioinformatics, have led to the emergence of new legal, political and technological regimes that dictate PGR management. The issues of ownership and utilization of germplasm and benefits there from, and related discussions resulted in a paradigm shift from PGR being a *'common heritage of mankind'* to *'sovereign rights of nations'*. The current barriers of free flow of genetic resources, increasing application of intellectual property rights and vast potential of biological wealth yet to be tapped through bio-prospecting and genetic engineering, has placed greater demands on nations to adjust to the changing scenario of management.

Ex Situ Conservation of PGR

There is need to conserve this valuable wealth using *ex situ* and *in situ* conservation approaches. Most of the fruit and agroforestry trees growing in the cold regions are perennials and are orthodox seeded species. The orthodox seed may be conserved in the form of seeds in the permafrost genebank and/or at -20°C in the seed genebank.

However, these trees are cross pollinated and produce heterozygous progeny. Thus, conservation through seed is not useful as far as conservation of true-to-type genotype is concerned. Some of these species produce recalcitrant or intermediate seeds which cannot be dried to the low moisture level required for proper seed storage and thus, are not amenable to storage in seed genebank. Conventionally, fruit crops are conserved in field genebank (orchard, clonal repository, glass houses etc.). Similarly, the medicinal plants and vegetable crops growing in the cold arid regions are clonally propagated and can be conserved in the field genebank. However, field genebank, besides being vulnerable to natural calamities, insects-pests and diseases; require more space, labour and other resources. Tissue culture technology offer great potential for conserving the germplasm as an alternative or a complement to the field genebank. There are two major *in vitro* conservation strategies: (i) slow growth for short- to medium term-conservation and (ii) cryopreservation for long-term conservation of plant material.

Slow growth techniques allow clonal plant material to be stored under *in vitro* conditions with extended subculture intervals. Slow growth methods aim for reducing the growth rate of plant material so as to prolong the subculture interval without causing any adverse effect on the plant tissue. Various methods may be used alone or in combination to get effective slow growth storage of various crops. Cryopreservation is relatively a new conservation method and protocols have been developed for conservation of various crops for long-term storage at -150 to -196° C under liquid nitrogen or vapour phase. Cryopreservation techniques may be applied for the long-term storage of the plant material that may include shoot apex, somatic embryo, seed, excised zygotic embryo, dormant vegetative bud and pollen.

Climate Change

The Himalayas is the highest but the youngest mountain belt (20 to 60 million years B.P.) of the earth running in arc shape for about 2,500 km. It has more than 90 peaks above 6,000 m and contains about 50% of all glaciers outside of the polar environments (Bahadur, 1993). All glaciers in this region are in general recession since last 150 years (Mayewski and Jeschke, 1979). Gangotri, Siachen, Bara Shigri and Patsio are major glaciers in this region which are showing retreat with different rates and their respective tributary glaciers are completely disconnected from main body of glaciers (Kumar *et al.*, 2009). A comparatively longer climatic reconstruction covering major part of last glacial period from Ladakh suggests that this site, at least since last 35,000 yr B.P., has been experiencing a predominantly semiarid climatic condition, punctuated by short episodes of increased moisture evidenced by the expansions of *Betula* and other local tree within the steppe.

Projects and Initiatives

Biodiversity conservation in resource poor areas and fragile ecosystems need a different set of protection measures. This is especially so in the light of exploitation of valuable natural resources, especially agricultural, medicinal and aromatic plant species. In the Trans-Himalayas there are evidences of unsustainable practices and an absence of regulation of commercialization. Existing exploitation practices, commercialization, benefit-sharing and IPR aspects of seabuckthorn as well as of a variety of medicinal plants (many of which are used in traditional Tibetan medicine) need to be reviewed and appropriate institutional capacities need to be developed for this purpose. It should also be examined as to how the private sector can contribute meaningfully to such enterprises.

As is the case with the Himalayas, the Trans-Himalayas is also rich in medicinal and aromatic plant resources and indigenous agro-pastoral diversity. Therefore, under the framework of proper policy and guidelines, these resources can benefit local people. Some such initiatives have already begun and need to be monitored for sustainability like the utilization and commercialization of seabuckthorn (*Hippophae rhamnoides*). As an alternate livelihood option, ecotourism has gained ground and has received considerable attention from researchers.

As a result of increased human and livestock numbers in the past few decades, there has been an escalation in human wildlife conflict. Depredation by snow-leopards and wolves is common and local people lose considerable revenue from livestock that are predated on by these carnivores. Domestic livestock also compete for food resources of wild ungulates. Currently, a few organizations have been working at the grass-roots level to bring about socio-economic change within local communities and provide incentives for the protection of these large mammals. Incentives for keeping livestock out of certain pastures and insurance and compensation programmes for loss of livestock are being tried out. Most of these projects have been initiated by NGOs such as the Nature Conservation Foundation, the International Snow Leopard Trust and the Snow Leopard Conservancy. A number of positive developments have been reported in this context and it would be worthwhile to investigate if such schemes can be initiated at a larger level. The initial components of such a project could be aimed at testing the efficacy of such initiatives.

A project dealing to address current problems relating to protected area management, PA delineation, infrastructure and human resources development is necessary for the Trans-Himalayas. Currently, as outlined in the preceding section, there are problems associated with the legal status of some PAs (some are entirely classified as inviolate core areas). Such classification is impractical and can cause problems further on. The Forest Department in Ladakh and Lahaul and Spiti district has very few staff members. Therefore at the current strength, there is very little patrolling and monitoring that can be done in these areas. A number of experts have also pointed out that PA manager in the Trans-Himalayas need to have a specialized set of skills different from those required for the rest of the country. Therefore capacity building of the PA management system is necessary. Infrastructure, transport and communication facilities are also lacking. Although this has helped in insulating the region from unregulated development, some critical infrastructure needs have to be met. These include GIS labs, field stations, library and documentation centers.

Special Action Plan for Trans-Himalayas

Most policies, strategies and action plans do not account for the distinctness of the Trans-Himalayas. Conservation and development in this region needs to be specifically suited to the environmental fragility and unique cultural aspects of this region. The Trans- Himalaya is widely different form all the other bio-geographic zones including the main Himalayas on the other side of the divide, and adequate expertise needs to be developed in terms of research and human resources to develop specific strategies for this region. These could be with respect to key aspects such as the management of human-wildlife conflict, ecotourism and heritage based conservation, sustainable use, etc. or for the management of species and ecosystems that are unique to the region. There should also be separate database and clearing-house mechanisms for the Trans-Himalayas. Similarly, at the moment, inter-institutional collaboration is lacking in the region. Therefore, it would also be useful to develop a dynamic network of institutions involved in this region.

Future Thrust

1. Individual capacity needs such as training in taxonomic capacities and PA management need to be addressed for the region. There is a need for increasing the number of trained PA management staff. Since the needs of this region are specific, the Forest Department staff should be given special training on aspects such as the potential impacts of climate change, planting of exotic species, minimizing human-wildlife conflict, wildlife trade etc. Networking needs to be facilitated between individuals involved in biodiversity conservation and personnel from other departments and organizations, especially the army. Research and training opportunities need to be improved for almost all aspects of biodiversity conservation.

2. There are hardly any institutions dealing with the Trans-Himalayas as a thematic mandate. This is particularly so with regard to government institutions. Other than the Forest Department, a large central institution for the region is absent. Existing institutional capacities are especially lacking in terms of scientific infrastructure such as advanced equipment, and GIS and molecular labs. Field stations and communication facilities need to be upgraded and there needs to be networking with the armed forces for providing logistic facilities in remote border areas. Library and documentation centers and a clearing-house facility need to be established. In this region a number of institutions such as the Armed Forces are involved in biodiversity related issues (e.g. The Field Research Laboratory, Leh). Awareness generation among these institutions about biodiversity needs to be brought into the mandate of the primary organizations. For many of the institutions, current missions are towards the protection of certain species (chiru, snow leopard) or ecosystems. Although flagship/ charismatic, large-vertebrate centric approaches are needed and still pertinent to a large extent, there needs to be more focus towards conservation of overall diversity. Currently, the onus is on these scattered NGOs to develop sustainable use models as well as in developing incentives to mitigate human wildlife conflicts.

3. The Trans-Himalayan zone is yet to have an institution dedicated to holistic biodiversity conservation of this fragile zone. The same applies to an action plan for the area. A plan for this region would need to incorporate three major aspects: a sustainable use policy, guidelines and regulations for appropriate use of biotic resources and ecotourism; environmental impact assessment (EIA), and mitigation of human-wildlife conflict also need to be addressed at the system level.
4. To conserve the valuable plant genetic resources of the cold arid regions, there is need to develop reserve areas for *in situ* conservation of the biotic and abiotic resources. The *ex situ* conservation approaches would certainly of much help where the elite germplasm may be conserved in the field genebanks or *in vitro* repositories under various organizations. The elite germplasm of vegetatively propagated species may also be conserved in the National Genebank using tissue culture technology for short-to medium-term storage, and under cryopreservation for long-term storage.

Conclusions

There is no doubt that Ladakh has a key role in the conservation of Trans-Himalayan biodiversity. The factors most favourable to wildlife conservation are demographic patterns and traditional attitudes to hunting. The low human population is not distributed evenly, but clumped in major valleys at sites where cultivable land and water for irrigation coincide. The current conservation situation is relatively favourable when compared to many other parts of the Himalayan region. Firstly, Ladakh does not have the acute hunting problem faced by many parts of the Himalayan region. Secondly, the Trans-Himalayan environment is not susceptible to the sudden degradation following deforestation that has devastated wide tracts of the monsoon Himalayas. However, it is still fragile and no-one knows to what extent increased exploitation of pastures and hill slopes can take place before irreversible damage occurs, with serious long-term consequences for ungulates and animal husbandry. A clear priority is an inter-disciplinary study on the interaction of land use and available natural resources, to assess the impact of current rates of grazing and shrub removal on the vegetation and on wild animal populations. An adequate legal framework exists and a good protected area network has been notified: the success of future conservation will depend on successful translation into effective action on the ground.

References

Bahadur, J. (1993). The Himalayas: A Third Polar Region, Snow and Glacier Hydrology. In: Proceedings of the Kathmandu Symposium, November 1992. IAHS Publ.no. 218.

Bhatnagar, Y.V. (2002). Species of the Trans-Himalaya and other Arid Tracts. In: Mountain Ungulates of the Indian Subcontinent, ENVIS, Wildlife Institute of India, Dehradun. (http://wiienvis.nic.in/ungulates/pagechapter2.htm).

Bhatnagar, Y.V. (2008). Relocation from wildlife reserves in the Greater and Trans-Himalayas: Is it necessary? Conservation and Society 6(3): 263-270.

Bhatnagar, Y.V.; Mishra, C.; McCarthy, T. and Bagchi, S. (2002). Wildlife conservation in the Trans-Himalayas: An insight into ecology and economics of the pastoral way of life. (http://www.wii.gov.in/ars/2002/sumanta.htm).

Burrard, G. (1925). Big Game Hunting in the Himalayas and Tibet. Herbert Jenkins, London.

CADARI (1996). Biophysical and socio-economic study of the Kibber area, Himachal Pradesh, 1993 & 1994. Unpublished report. CADARI Spiti Expeditions, Canterbury, UK.

Champion, H.G. and Seth, S.K. (1968). A Revised Classification of Forest Types of India. Manager of Publications. New Delhi.

Chundawat, R.S and Rawat, G.S. (1994). Indian Cold Deserts: A Status Report on Biodiversity. Wildlife Institute of India, Dehradun.

Dwivedi S.K.; Stobdan, T.; Yadav, A. and Singh, S.B. (2008). Fruit Production in cold arid Regions of Indus-Challenges and opportunities. In: Advances in Agriculture, Environment & Helath, eds. S.B. Singh *et al.* Satish Serial Publishing House, Delhi. pp. 61-81.

Fox, J.L.; Nurbu, C. and Chundawat, R.S. (1991). Tibetan argali (*Ovis ammon hodgsoni*) establish a new population. Mammalia 55: 448-552.

Gaston, A.J.; Garson, P.J. and Hunter, M.L. (1983). The status and conservation of forest wildlife in Himachal Pradesh, Western Himalaya. Biol. Conserv. 27: 291-314.

Green, M.J.B. (1985). Aspect of the Ecology of Himalayan Musk deer. Ph.D. Thesis. Cambridge University, Cambridge.

Green, M.J.B. (1986). The distribution, status and conservation of the Himalayan musk deer *Moschus chrysogaster*. Biol. Conserv. 35: 347-375.

http://www.gov.in.

http://www.snowlepordnetwork.org.

Kala, C.P. (2001). A Study of Traditional Health *Care & Medicinal Plants.* In: Conserving Bio-Diversity in the Trans-Himalaya: New Initiatives of Field Conservation in Ladakh. First annual Technical Report (1999-2000), Wildlife Institute of India, International Snow Leopard Trust and US Fish and Wildlife Service.

Kumar, V.; Singh, S.K. and Venkataraman, G. (2009). Climate change effects on Glacier recession in Himalayas using Multitemporal SAR data and Automatic Weather Station observations. Geophysical Research Abstracts, Vol. 11, EGU2009-7116-1.

Maheshwari, G. (2000). A survey of bio-indicator community of Chironomidae (Diptera) of North-West Himalayan lakes. National consultation on conservation of high altitude wetlands inLadakh. WWF-India at Leh, July 2000.

Mani, M.S. (1962). Introduction to High Altitude Entomology. Methuen and Co., Ltd., London.

Mallon, D.P. (1998). Ecology and Conservation of Mountain Ungulates in Ladakh. Ph.D. Thesis. Manchester Metropolitan University.

Mallon, D.P. (2002) My Memoir of the Wildlife Surveys in Ladakh - In: Mountain Ungulates of the Indian Subcontinent, ENVIS, Wildlife Institute of India, Dehradun. (http://wiienvis.nic.in/ungulates/pagechapter14.htm).

Mayewski, Paul A. and Jeschke, Peter A. (1979). Himalayan and Trans-Himalayan Glacier Fluctuations Since AD 1812, Arctic and Alpine Research, 11(3): 267-287.

Rodgers, W.A. and Panwar, H.S. (1988). Planning a Protected Area Network in India. Vol. 1. A report. Wildlife Institute of India, Dehradun. 341p.

Rawat, G.S. and Sathyakumar, S. (2002). Conservation issues in the Himalayan region of India –In: Mountain Ungulates of the Indian Subcontinent, ENVIS, Wildlife Institute of India, Dehradun. http://wiienvis.nic.in/ungulates/pagechapter9.htm.

Sathyakumar, S. (1993). Musk Deer. Sanctuary, Asia. Vol.12 (5): 52-57.

Shackleton, D.M. (ed.) (1997). Wild Sheep & Goats & their Relatives. Status Survey and Conservation Action Plan for Caprinae. IUCN, Gland, Switzerland and Cambridge, UK. 390 p.

Schaller, G.B. (1977). Mountain Monarchs: Wild Sheep and Goats of the Himalaya. Chicago University Press. Chicago. 425 p.

Schaller, G.B. (1998). Wildlife of the Tibetan Steppe. University of Chicago Press, Chicago.

Vinod, T.R. and Sathyakumar, S. (1999). Ecology and conservation of ungulates in Great Himalayan National Park, Himachal Pradesh. In: An Ecological Study of the Conservation of Biodiversity and Biotic Pressures in the Great Himalayan National Park Conservation Area - An Ecodevelopment Approach. Forestry Research Education and Extension Project - Great Himalayan National Park (FREE-GHNP), Final Project Report, Wildlife Institute of India, Dehra Dun. Vol. 1- 6.

Ward, A.E. (1924). The mammals and birds of Kashmir and the adjacent hill provinces. J. Bombay Natural History Society 30: 253-259.

❑❑❑

22

Ethnobotanical Uses of Plants by Tribal Communities

BASANT BALLABH AND O.P. CHAURASIA
Defence Institute of High Altitude Research (DIHAR), C/o , 56 APO

The high barren mountains, large valleys, long rivers, uneven land of Ladakh geographically divide the region into five main valleys - Leh, Nubra, Changthang, Zanskar and Suru (Chaurasia and Singh, 1996-2001). Due to its harsh climate and remoteness the Ladakh region has been brought under scheduled tribe status by the Government of India. It is least populated area of the country where population density is about 2-3 person per sq km. There are many tribal communities found in Ladakh such as Boto's, Balti's, Changpa, Drukpa, Gadhis, Gujjars and Bakarwals etc., but it is mainly dominated by Boto (the Buddhists) tribal community. The traditional culture, costum, language and clothings among Buddhists and Mohamadans are more or less similar but they perform different religious practices (Jina, 1996). The Ladakh has very short cropping season from April to September thus vegetative growth starts at the beginning of summer season and the flora reaches to full bloom during August and starts vanishing by the end of September. Propagation through vegetative means such as roots, suckers, runners, bulbs, bulbils, rhizomes and tubers is most remarkable feature of high altitude plants, most of them are perennial in nature, and multiply through vegetative methods. The cold desert flora is highly important for tribal communities of Ladakh because they utilize various parts of plants, *viz.*, roots, leaves, stems, resin, flowers and fruits etc. in daily life for different purposes such as food, shelter, medicine, fodder, forest and fuel.

Keeping in view the rich ethnobotanical wealth of cold arid Ladakh, Field Research laboratory, (Defence Research and Development Organisation), Leh, has surveyed this region since the last decade of 20th Century and collected numerous species of specific ethnobotanical importance with relevant information. The collected species were elaborate identified with the help of various floras, monograms and herbaria of J&K and DD.

Description of Ethnobotanical Species

The prolonged and snow, bound winter prevailing in Ladakh region forced the local inhabitants to fulfil their daily needs from their surroundings. The tribal communities of Ladakh has very significance ethnobotanical lore. The important ethnobotanical species of Ladakh region are being described here according to the classification system of Benthum and Hooker along with family name, botanical names with citation, vernacular name (E=English, H=Hindi, L=Local name etc.) and ethnobotanical uses as follows :

Angiosperms

Dycotyledons

Ranunculaceae

1.*Aconitum heterophyllum* Wall. ex Royle. Ill. Bot. Himal. 36, 1835; Vern: Attees (H), Aconite (E), Bona-karpo (L).

Uses: An important plant of all medical systems. The roots are used against tooth-ache, high fever, stomach complaint, gastric trouble and headache.

2.*A . violaceum* Jacq. ex Stapf. Ann. Roy. Bot. Gard. Calc. 10 (2), 144, 1905. Vern: Pattees (H), Aconite (E), Bona-nagpo (L).

Uses: The roots are commonly used as a substitute of *A. heterophyllum*. Useful against gout, arthritis, disorder of heart, amnesia, loss of body heat, leprosy and paralysis.

3.*A quilegia fragrans* Benth. In Manud Botanist 4: t. 181, 1840; *A. vulgaris* var. *grandiflora*, sensu Aitchison; *A. vulgaris* Linn. var. *pyrenica* (DC.) Hk. f. *et* T.; Vern: Fragrans Columbine (E).

Uses: The plant has aromatic flowers which are commonly used for ornamental purpose.

4.*D elphinium brunonianum* Royle. Ill. Bot. Himal. 56. 1834; Vern: Himalayan Larkspur (E), Lunde-kaown or Bya-rgod-spos or Poldo or Laddar Temosa (L).

Uses: The infusion of leaves, stem, flower and fruits is used for herbal medicines of intestinal pain and insecticide. The seeds and vegetative parts of the plant contain an alkaloid "Delphinine" which is used as an insecticide. Flowers are used during worship and religious ceremonies.

Berberidaceae

5.*B erberis lyceum* Royle. Ill. 64. Trans. Linn. Soc. XVII: 94. Vern: Berberry (E) Daruhaldi (H) (Annonymus, 1948).

Uses: The root extract is used in medicine for preparing tonic, cough, chest, throat troubles and eye problems in Ladakh. Fruits are edible.

6.*B . ulicina* Hk.f. and T. In FBI, I: 111, 1872; Vern; Berberry (E), Kiraring or Chhelpa or Chhiner or Chhinisrangama or Khizer (L).

Uses: Fruit yield a purplish-violet dye, which is used in medicines. Root and bark are sometimes used for preparing tonic in Ladakh. Fruits are edible.

7.*P odophyllum hexandrum* Royle. Ill. Bot. Himal. 64. 1834 *P. emodi* Wall ex Royle.; *P. emodi* var. *hexandrum* (Royle) Chatt. *et* Muker.; *P. hexandrum* var. *jaeschkei* (Chatt. *et* Muker.) Browich.; Vern: Himalayan Mayapple (E), Demokusu or Papra or Ol-mose (L).

Uses: The roots are used against skin problems, hardening of skin and skin diseases by local Amchis in Ladakh. The young and ripe fruits are used in high altitude mountain sickness. The Fruits are edible.

Papaveraceae

8. *Meconopsis aculeata* Royle. Ill. Bot. Himal. t. 15. 1833; FBI, I: 118, 1875; Debnath *et al.* Nayar.; Vern: Himalayan blue poppy (E), Landrementok or Ut-pal sngon-po or Achatsarmum (L).

Uses: The entire plant is used in medicine of ulcers, disorders of lungs, liver and inflammation, and is beneficial in treating fever associated with bone and deformed spongy cavernous bones and hot disorders. The flowers are used as ornament.

9. *Hypecoum leptocarpum* Hk. f. *et* T.; F. B.I, I: 120. 1872; Vern: Little Poppy (E), Parpata (L).

Uses: A bitter taste and qualities are cool, blunt and light.

Fumariaceae

10. *Corydalis govaniana* Wall. Tent. Fl. Nip. 55; Cat. 1431; Vern: Ralchhat nagpo or Stong-Zil (L).

Uses: The roots are used as antipyretic, diuretic, gastric against pain and muscular pain.

Brassicaceae

11. *Brassica campestris* Linn. Syn. *B. rapa* (Linn.) Hartm Hanlb. Skand.; *B. rapa* Linn.; Vern: Pili Sarson (H), Field mustard (E), Nuskar (L).

Uses: Local people in Ladakh mainly use oil extracted from seeds for edible and cooking purposes. It is also used as green vegetable in summer.

12. *B. caulorapa* Pasq. Syn. *B. oleracea* var. *gongylodes* Linn.; WI, II: 292-293, 1985; Vern: Ganth Gobi (H), Knol-khol (E), Kushutam (L).

Uses: Knol-khol is the most important and widely cultivated vegetable of all the valleys by local people of Ladakh.

13. *B. oleracea* Linn. var. *botrytis* Linn., sub var. *cauliflora* D.C.; WI, II: 277, 1985; Vern: Phool Gobhi (HL), Cauliflower (E).

Uses: The curd and leave stalks are used for vegetable.

14. *B. oleracea* Linn. var. *capitata* Linn. f.; WI, II: 284-290, 1985; Vern: Bandh Gobi (H), Cabbage (E), Bandh or patta Gobhi (L).

Uses: The head and leave stalks are used for vegetable.

15. *B. rapa* Linn. Syn. *B. compestris* Linn. var. *rapifera* (Metzg.) Sinsk; FBI, I: 156, 1872; Vern: Shaljum (H), Turnip (E), Niuma (L).

    Uses: The turnip is popular vegetable of local tribes of Ladakh. It is also used as an excellent fodder for cattle and winter feed for livestock. It is also used to cure swelling of foot due to cold injuries.

16. *Capsella bursa-pastoris* (Linn.) Medik. Pflanzen-Gatt. 85, 1792; *Thlaspi brusa-pastoris* Linn.; Vern: Shepherd's purse (E), Shamso or Sog-ka (L) (Kachroo *et al.*, 1977).

    Uses: The herb is cooked as vegetable by local tribes. It is also used against swelling, pus and pain, stop vomiting, beneficial in treating kidneys, lungs, nerve disorders and fluid retention in the body by Amchis (Chaurasia *et al.*, 1996).

17. *Chorispora sabulosa* Camb. In Jacq. Voy. Bot. 15. t. 15. 1844; *elegans* Camb.; *C. elegans* var. *sabulosa* (Camb) Schulz *et* var. Schulz.; *C. sabulosa* var. *eqlandulosa* Narayanswamy ex Naithani & Uniyal.; Vern: Imbuk (L).

    Uses: Young leaves and shoots are cooked as vegetable by locals. The scented flowers are used as ornamental purposes.

18. *Christolea crassifolia* Camb. In Jacq. Voy. Bot. 4: 17 t. 17. 1844; Vern: Kharla or Sanak (L).

    Uses: The tender shoots are consumed as vegetable.

19. *Lepidium latifolium* Linn. Sp. Pl. 644. 1753; Vern: Dittander (E) & Sauson or Seoji (L).

    Uses: Fresh leaves are boiled and washed thoroughly with water and cooked as vegetable like mustard.

20. *Raphanus sativus* Linn. Sp. Pl. (1753), 669; Vern: Muli (H), Radish (E), Labook (L).

    Uses: Local varieties are well adapted and used for vegetable and salad. The root is used in remedy of kidney stone, jaundice and scorbutic conditions.

21. *Thlaspi alpestre* Linn. ex Hk. f. & T. In Linn. Soc. V: 177, 1853; Vern: Bumbuk or Bre-ga (L).

    Uses: The fruits are useful in treating cough, kidney problems of hot nature and vaginal discharge.

Capparidaceae

22. *Capparis spinosa* Linn. var. *himalayensis* (Jafri) Jacobs. in Blumea 12 (1): 419. 1965. *C. himalayensis* Jafri.; *C. leucophylla* Collett.; Vern: Ber (H), Kaper bush (E), Kabra (L).

    Uses: Young and delicate leaves and flower buds are used as vegetable. The tender leaves are boiled, washed thoroughly for 5-8 hours in water and fried with onion and other spices, which gives a delicious vegetable dish. Fruits are edible.

Caryophyllaceae

23. *Arenaria bryophylla* Fern; *A. musciformis* Wall. ex Edgew. & Hk. f. T. in FBI, I: 237; 1974; Vern: Lekhum (L).

    Uses: The plant is dried in shade made into powder and used against kidney and urinary troubles in Ladakh. It gives good fuel material for winter use.

24. *Stellaria media* (Linn.) Vill. Syn. *Alsine media* Linn.; D.E.P., VI (3), 358; Vern: Strawort or Stitch wort or Chickweed (E), Khalta (L).

    Uses: The tender leaves and stalks are eaten as vegetable, raw or boiled.

25. *S. palustris* Ehrh. ex Retz. Fl. Scand. Prod. Ed. 2: 106. 1795; Majumdar.; *S. glauca* With.; Vern: Khalta (L).

    Uses: The tender shoots and leaves are boiled and eaten as potherb in far-flung areas of Ladakh. It is also used as fodder.

26. *Thylacospermum caespitosum* (Camb.) Schischk. Herb. Fl. Ross. 9: 90. 1932; *Periandra caespitosa* Camb.; *T.rupifragum* (Kar. & Kir.) Schrenk.; Vern: Tegarcan or spang (L).

    Uses: It is used for beautiful decoration and as ornamental plant. It is also used as fuel during scarcity in winters.

Tamaricaceae

27. *Myricaria germanica* (Linn.) Desv. ssp. *alopecuroides* (Schrenk). Kitamura, Fl. Afgan. 227. 1960; *M. alopecuroides* Schrenk; *M. germanica* auct. Non. (L.) Desv.; Vern: Umbo (L).

    Uses: The whole plant is used as medicine for rheumatism. It is used for forestry, fuel and fodder.

28. *Tamarix gallica* Pallas. II. (Desv.) Thist. Dyer. Syn. *T. troupii* Hole. FBI, In part, non Linn. D.E.P., 5: (3). 411 in part; C.P., 116; FBI, I: 248, 1975; Vern: Tamarisk (E), Agrelta (L).

    Uses: The plant is used as forestry plant, wood as fuel, branches for making baskets and leaves as fodder.

Hypericaceae

29. *Hypericum perforatum* Hk. f. & T. F.B.I, I : 253. 1874; Vern: Common St. John's Wort (E), Meerang (L).

    Uses: A paste of leaves is used in treatment of wounds and cuts. Similarly it is used in septic wounds of animals. A drug hyperforin, isolated from it is active against gram-positive bacteria.

Malvaceae

30. *Althaea rosea* (Linn.) Cav. Diss. II (1786) 91, t. 29, f. 3. Syn., *Alcaea rosea* Linn.; Vern: Hollyhock (E), Hallo or mDog-ldan. pho-lcam., Mo-tog-ha-lo. (L).

    Uses: The roots and flowers of this cultivated ornamental plant are used in treating hot disorders of the kidneys, womb and also in treating all kind of vaginal discharge with burning sensation.

31. *Lavatera kashmiriana* Camb. In Jacq. Voy. Bot. 29, t. 32; Vern: Sazamool (H), Wild Hollyhock (E), Sazposh (L).

    Uses: Tender shoots and leaves are washed, boiled and again washed then cooed as vegetable in Ladakh.

32. *Malva verticillata* Linn. Sp. Pl. 689. 1753; *M. parviflora* Linn.; Vern: Chinese Mallow (E), lCam-pa ma-ning (L).

    Uses: The fruits are used to treat dropsy, thirsts, diarrhoea, infected sores and dilate all vessels in the body. Young leaves and tender shoots are boiled and cooked as green vegetable.

Zygophyllaceae

33. *Peganum harmala* Linn. Sp. Pl. 444. 1753; Vern: Wild Rue (E), Sepan (L).

    Uses: This plant is supposed to be used for treatments of more than 70 kinds of different ailments in Ladakh. The seeds are narcotic and used for child bath, fever, stomach complaint, anthelmintic, diuretic, tonic, appetizer and joint pains, aphrodisiac, urinogenital complaints, lactation, menstrual complaints smoke of seeds is inhaled for exhalation.

34. *Tribulus terrestris* Linn. D.E.P. 6 (4): 18; Vern: Chhotigokhru (H), Land caltrop or Puncture vine (E), gZe-ma, Rasha, Kokulla (L).

    Uses: Fruits are considered useful in liver complaints, leucoderma, impotency, cold, cough, urinary disorders and weakness; infusion is useful in gout (Kirtikar and Basu, 1935).

Geraniaceae

35. *Biebersteinia odora* Stephan ex Fischer.; *B. emodi* Jaub. & Spach.; Vern: Chhekerpa or darkspos or Khardung (L).

    Uses: Paste of whole plant is used against cure of cuts, ulcers, skin sores and wounds.

36. *Geranium pretense* Linn. Sp. Pl. 681. 1753; FBI, I: 429, 1875; Vern: Meadow Cranesbill (E), Gugchuk or Ga-dur (L).

    Uses: The leaf extract is used as febrifuge, analgesic, used against fever from influenza and infectious disease, to check dysentery and diarrhoea in Amchis system of medicine.

Vitaceae

37. *Vitis vinifera* Linn. Sp. Pl., 202, 1753; Vern: Angoor (H), Grape (E), Gun (L).

    Uses: It is rare fruit in Ladakh mainly grown in Turtuk, Kargil, Khalsi and Nimmu areas for fruit on small scale.

Fabaceae

38. *Astragalus cicerifolius* Royle ex Fisch. In Bull Soc. Nat. Mosc. 24 (2): 404, 1853; Vern: Yakpo (L).

    Uses: The whole plant is relished by sheep and goats.

39. *A. rhizanthus* Royle ex Benth. In Royle. Ill. Bot. Himal. 200. 1835; Vern: Dukchal or Sralchun or krelsed or Sarma (L).

    Uses: Leaves are grazed by sheep and goats. A decoction of plant is useful in skin infection (Ballabh, 2002).

40. *A. zanskarensis* Benth ex Bunge. Mon. 1: 43, 1868; Vern: Deanko or Nanchu (L).

    Uses: The leaves are commonly grazed by sheep and goats.

41. *Cicer microphyllum* Benth. In Royle, Ill. Bot. Himal. 200. 1835; *C. soongaricum* Auct. Non. Stepf. ex DC.; Vern: Wild Gram (E), Seri or Srad-Kar (L).

    Uses: It is considered as the best fodder plant and used for animals, grazed by sheep and goats in the pastures of Changthang valley. Unripe seeds are eaten by shepherded.

42. *Lathyrus sativus* Linn. Sp. Pl. (1753), 730; Vern: Khesar (H), Local pea (E), Sanma or Sandil (L).

    Uses: Seeds are commonly used as pulse, fruits as vegetable and whole herb is used as fodder in Ladakh. It is also used for making flour when mixed with wheat, which is considered the best source of flour for thupka preparation.

43. *Medicago falcata* Linn. DC. Prodr. II: 172; Vern: Yellow flowered Alfalfa (E), Ole (L).

    Uses: It is found throughout Ladakh and one of the most common fodder crops of the region. It is used green and as hay during winter.

44. *M. lupulina* Linn. Sp. Pl. 779. 1753; Vern: Ole or Bu-su-hang pho (L).

    Uses: The leaves, stem, flower and fruits are used to treat hot disorders of lungs, coughs and heals wounds. The plant is used as nutritive fodder in Ladakh.

45. *M. media* Linn.; CDP, I: 78, 1996; Vern: Lucerne (E), Ole (L).

    Uses: It is found throughout Ladakh and one of the most common fodder crops of the region.

46. *M. sativa* Linn. D.E.P., V, 199; Vern: Alfalfa or Lucerne (E), Ole (L).

    Uses: It is grown throughout Ladakh and only common fodder crop of the region used as fresh and as hay.

47. *Melilotus alba* Medik ex Desr. Sp. Pl. 779. 1753; Vern: White sweetclover (E), Jellek or Ole (L).

    Uses: It is cultivated fodder crop of the region and considered best quality fodder.

48. *M. officinalis* Linn. Pall. Reise 537. 1776; Vern: Yellow sweet clover (E), Jellek or Ole or rGya-spos (L).

Uses: The decoction of entire plant is used as an antidote for fever caused due to poisoning, asthma, lotions, enemas and other infectious diseases. The whole herb is used for fodder.

49. *Oxytropis microphylla* (Pallas) DC. Prodr. II: 279; *O. chiliophylla* Royle.; *Phaca microphylla* Pallas.; Vern: sTag-sha nagpo (L).

Uses: The leaves are used in treating infectious fever, inflammation, subsides swelling, healing wounds and all kind of poisoning.

50. *Pisum sativum* Linn. Syn *P. arvense* Linn. D.E.P., VI(1), 276; Vern: Matar (H), Pea (E), Sremma (L).

Uses: It is most important leguminous crop of Ladakh, cultivated for pulse and vegetable.

51. *Trifolium pratense* Linn. sp. Pl. 768. 1753; Vern: Tiripatra (H), Purple or red clover (E)

Uses: The whole plant is used as fodder and also grazed.

52. *T. repens* Linn. Sp. Pl. 767. 1753; Vern: White or Dutch clover (E)

Uses: The herb is used for good quality of fodder. It is also grazed by horses and animals. The leaves contains 17.56 per cent crude protein which is best for milking animals.

53. *Trigonella foenum-graecum* Linn. Sp. Pl. (1753), 777; Vern: Methi (H), Fenugreek (E)

Uses: Local farmers are now adapting its cultivation with the help of suitable varieties and greenhouses, it is cultivated for green.

Rosaceae

54. *Malus pumila* Mill. Syn. *M. communis* DC.; *M. sylvestris* Hort. Non. Mill.; *M. domestica* Borkh.; *Pyrus malus* Linn.; Vern: Seb (H), Apple (E), Kushu (L).

Uses: Apples are used for fruits and cultivated at lower altitudes in all the valleys. Apples are sold fresh in local market during August and September.

55. *Potentilla anserina* Linn. Sp. Pl. 495. 1753; Vern: Silver weed (E), Penma or Troma (L).

    Uses: A decoction of leaves is used in diarrhoea and dysentery. It is also useful in case of kidney stones and arthritis in Ladakh.

56. *Prunus armeniaca* Linn. Sp Pl, (1753), 474; Vern: Khubani (H), Apricot (E); Chuli (L).

    Uses: Apricot is the only natural fruit crop grown in the entire Ladakh. It is used fresh or dehydrated for prolonged winter.

57. *Rosa ecae* Aitchinson. Wall. ex Dcne. In Jacq. Voy. Bot. 4: 64. t. 75. 1844; Vern: Jangali Gulab (H), Afghan rose (E), Siah (L).

    Uses: The whole plant is used as ornamental, fodder and fuel in Ladakh.

58. *R. webbiana* Wall. ex Royle. Ill. Bot. Himal. t. 42. f. 2. 1834 *et* 208. 1835; Vern: Jangali Gulab (H), Wild Rose (E), Siah or Madpo (L).

    Uses: Fruits are edible. Flowers are ornamental.

Saxifragaceae

59. *Bergenia stracheyi* (Hk. f. & T.) Engl. In Bot. Zeit. 26: 842. 1868; *Saxifraga stracheyi* Hk. f. & T.; *S. ligulata* Sensu Aitchison.; Vern: Pashanbhed (H), Tiang or gatilas (L).

    Uses: In Zanskar areas, roots of the plant are used against stomach complaints. The roots are also used as febrifuge, swelling of limbs etc.

60. *Ribes alpestre* Wall. ex Decne. In Jacq. Voy. Bot. 4: 64. t. 75. 1844; *R. grossularia* auct. non. Linn.; Vern: Asian gooseberry (E), Askuta or Zasoot (L).

    Uses: The ripe fruits are eaten. The leaves are used as fodder, stem is used for fuel in Ladakh.

61. *R. orientale* Desf. Hist. Arb. 2: 88. 1809; *R. villosum* Wall.; Vern: Asian gooseberry (E), Askuta or Zasoot (L).

    Uses: The berries are edible, mainly eaten by children. The whole plant is used as fuel in Ladakh.

62. *Saxifraga cernua* Linn. Sp. Pl. 403. 1753; FBI, II: 390, 1879; Vern: Bri-rta sa-'dzin dman-pa (L).

    Uses: The entire plant is used to reduce pus, accumulated in the lungs, chest and is useful in the treatment of inflammation of the nerves.

63. *S. stenophylla* (Royle) Hulten in Sv. Bot. Tidskr. 58. (1): 94. 1964; *S. stenophylla* Royle.; *S. flgellaris* var. *stenophylla* (Royle). Clarke.; Vern: Saxifrage (E), Titsarzing or Yenahu (Dri-rta-sa-Mzin) Man-pa (L).

Uses: The whole plant is used against spreading of putrid blood and pus in lungs and inflammation in veins.

Crassulaceae

64. *Rhodiola heterodonta* (Hk. f. & T.) A. Boiss. In Komarov, Fl. USSR. 9. 32. 1939. *Sedum heterodontum* Hk. f. & T.; Vern: Rose root or Stone crops (E), Shro-lo dkar-po (L).

Uses: The young leaves and tender shoots are edible. This is then mixed with curd by which a local delicious dish "Tantur" is prepared in Ladakh.

65. *Rhodiola imbricata* Edgew. In Trans. Linn. Soc. 20: 47. 1846; *Sedum imbricata* (Edgfew.) Walp.; *S. imbricatum.* Hk. f. & T.; *S. rhodiola* auct. non. DC. Clarke.; Vern: Rose root or Stone crop (E), Shrolo marpo (L).

Uses: The roots are used in medicine against lung problems, cold, cough, fever, loss of energy and pulmonary complaints. It is also used in preparation of health tonic and it restores memory.

Onagraceae

66. *Epilobium angustifolium* Linn. Sp. Pl. 347. 1753; FBI, II: 582, 1879; *Chamaenerion angustifolium.* (Linn.) Scop.; Vern: Rose way willow herb (E), Utpalwenabo (L).

Uses: The whole herb including flower is used against stomach disorders. The plant is used for ornamental purposes.

Apiaceae

67. *Bunium persicum* (Boiss.) Fedtsch. Rastit. Turkestan 612. 1915; *Carum persicum* Boiss.; *C. bulbocastianum* Auct.; Vern: Sahi Zeera or Kala Zeera (H), Caraway black (E), Zeur (L).

Uses: Seeds are highly fragrant and used for flavouring local dishes. The bulbs/tubers are also eaten as vegetable after removing toxicity (Singh *et al.*, 2001).

68. *Carum carvi* Linn. var. *gracile* Sp. Pl. 263. 17539; Vern: Kala Jeera (E) Umbu, Kosnyot or Go-nyod or Go- sNyod (L).

    Uses: The seeds and green leaves are used as condiment for flavouring local dishes or eaten as pot herb. The roots are cooked as vegetable.

69. *Chaerophyllum reflexum* Lindl. In Royle, Ill, 232; FBI, II : 691. 1879; AFKH, 210, 1983; WI, III: 455-56, 1992; Vern: Lcha-wa or Young (L).

    Uses: The whole plant is used as fodder for animals in Ladakh. The roots are used to promot heat; tonic is used in anemia, swelling of limbs, stomach problem.

70. *Coriandrum sativum* Linn. Sp. Pl. (1753), 256.; Vern: Dhaniya (H), Coriander (E), Oshu (L).

    Uses: Coriander is cultivated in many places of Ladakh during summers for green leaves as well as for spice.

71. *Daucus carota* Linn. Sp. Pl. (1753), 242; Vern: Gajar (H), Carrot (E), Sarakhturam (L).

    Uses: It is now commonly grown in Ladakh for vegetable.

72. *Ferula jaeschkeana* Vatke. in Ind. Sem Hort. Berol. Append. 2. 1876; Vern: Sampharu or Chuklam (L).

    Uses: The roots are used in rheumatism and popularly as poultice. Leaves are used as fodder.

73. *Heracleum pinnatum* C.B. Clarke. FBI, II : 712. 1879; Vern: sPru-ma (L).

    Uses: The roots are sometimes used to treat inflammation and pain caused by vulnerable fever, intestinal worms.

74. *Prangos pabularia* Lindl. D.E.P. VI (I), 334; Vern: Prangos (E); Prangs or Palano (L).

    Uses: Prangos is the most important fodder plants of Dras. Roots are used as diuretic. Seeds are considered carminative. Roots and seeds are also used in energetic drugs.

Rubiaceae

75. *Rubia cordifolia* Linn. Syst. Nat. ed. 12, III: 1768. 229. *R. munjista* Roxb.; Vern: Manjith (H), Indian Madder (E), bTsod (L).

Uses: The stem and roots are effective against blood diseases, chest complaints, discoloration of skin and leucoderma, inflammation, menstrual disorders etc.

Dipsacaceae

76. *Morina longifolia* Wall. FBI, III: 216, 1882; Vern: Kim (L).

Uses: The root extract is said to be useful against worm infestation and boils.

Asteraceae

77. *Achillea millefolium* Linn. Boiss. Fl. Orient. III, 255; Vern: Gandana (H), Milfoli (E), Chuang (L) (Stewart, 1916-17).

Uses: Amchis used the plant extract as insecticide and also against skin infections. The leaves are used for stomach complaints. Flowers are used in tonic.

78. *Anaphalis triplinervis* (Sims) Clarke var. *intermedia* (DC) Airy Shaw *A. nepalensis.* (Spreng) Hand. Mazz. DC.; *Elichrysum nepalensis* Spreng.; Vern: Yaktso (L).

Uses: The flower buds are used as spice and eaten raw as salad.

79. *A. triplinervis* (Sims) C.B. Clarke var. *monocephala* (DC) Airy Shaw (*A. nubigena* DC.); Vern: Phulumentock or sPra-rgod (L).

Uses: The paste of flower head is used against wounds. The stem, leaves, flowers and fruits are beneficial in epidemic diseases, toxicity from poisons.

80. *Artemisia annua* Linn. Sp. Pl. 847, 1753; Vern: Artemisia (E)

Uses: It is cultivated for an essential oil, which contains a drug named 'Artemisinin' is used in malaria, jaundice and as diuretic.

81. *A. brevifolia* Wall. ex DC. Prodr. 6: 103. 1838. *A. 1maritima* Linn.; *A. maritima* Linn. var. *thomsoniana* Clarke, Comp.; Vern: Kirmala (H), Worm seed (E), Khamchu or Phur-nag (L).

Uses: Leaf extract is also used against stomach problems and internal worms of intestine mainly in children by Amchis.

82. *A. dracunculus* Linn. Sp. Pl. 849. 1753; *A. glauca* Pallas ex Willd; Vern: Little Dragon Mugwort (E), tsa-chheh or Shersing or Burtse (L).

    Uses: It is observed that the leaf or flower extract is mainly used against toothache, urinary problems and sometimes against intestinal problems. Its oil is edible.

83. *A. gmelinii* Web. ex Stechm. Dissert. de Artem 30. 1775; *A. sacrorum* Ledeb.; Vern: Nurcha or Burtse or Khamchu or Burmak or Khampa shridi (L).

    Uses: Leaves and flowers extract is recommended for headache, cold, cough and in abdominal upsets.

84. *Aster diplostephioides* Benth. (DC.) Cl. Comp. Ind. 45, 1876. *Heterochaeta diplostephioides* DC; Vern: Kiamentok or Me-tog Lug-mig rigs-dang-po (L).

    Uses: The handsome flowers are used for ornamental purposes in Ladakh. The flowers are useful in influenza, contagious fever, bleeding, cold and poisoning.

85. *Bidens pilosa* Linn. in FBI. III: 309, 1882; Vern: Local or Gurgur Chai (L).

    Uses: The leaves are used for making most common local tea called *Gurgur Tea* with butter, milk and sugar.

86. *Callistephus chinensis* R. Br. Linn, WI - Raw Materials- II, 15; Vern: Aster (E), Skalzangmentok (L).

    Uses: It is now plenty seen in Leh and Nubra valleys and also cultivated for flowers near Gompas and residence of locals in other parts of Ladakh.

87. *Centaurea depressa* M Bieb. Fl. Taur. Cauc. 2: 346, 1808; Vern: Vashakha (L).

    Uses: The decoction of whole plant is useful in chest pain, headache, fever, cold and cough in Amchi system of medicine.

88. *Chrysanthemum pyrethroides* (Kar. & Kir.) B. Fedtsch. Rast. Turk. 737. 1915; *Richteria pyrethroides* Kar. & Kir.; *C. richteria* Benth.; Vern: Serpan or A-byag-gzer-joms (L).

    Uses: The flowers are used against rheumatism, fever and healing of external injury of boils, wounds and excess serous fluid.

89. *Cremanthodium ellisii* (Hk. f.) Kitam. var. *C. plantagineum* Maxim. forma. *maximowiczii* (Franch.). R. Good.; Vern: Lukumentok or Rekompa or Ming-chen-ser-po (L).

Uses: The whole plant is analgesic, useful in fever from contagious disease, swelling in the throat, reduce pain in upper body regions.

90. *Helianthus annuus* Linn. Sp. Pl. (1753), 904; Vern: Surajmukhi (H), Sunflower (E), Kanglanyesher (L).

Uses: Sunflower is now grown in all the valleys by local people for flower and beautification in Ladakh. Seeds are used for oil. Seeds are diuretic, expectorant, used in treatment of cough and cold.

91. *Inula racemosa* Hk. f. In FBI, III: 292, 1882; Kitamura; *I. helenium* auct. non. Linn. Aitch.; Vern: Puskarmool (H), Manu (L).

Uses: Roots of this plant are considered to be useful in bronchial asthma when mixed half of the amount with Kuth. Harmonise wind disease and blood, digestive, analgesic.

92. *Lactuca dolichophylla* Kitam. in Hara. Linn. *L. logifolia* DC. Prodr. 7: 135. 1838; *L. wallichiana* Tuisl.; Vern: Khala (L).

Uses: Young leaves, flower buds and shoots are boiled, washed with fresh water and cooked as vegetable. The protein and mineral quantity is found 27.1% and 17.5%, respectively.

93. *L. sativa* Linn. Sp. Pl. II: 1, 119, Syn. *L. scariola* Linn. var. *sativa* (Linn.) Clarke.; Vern: Salad (H), Lettuce (E).

Uses: The young leaves are boiled, washed thoroughly and cooked with "Thukpa" and it is commonly used as salad by the tribes during summer.

94. *Saussurea bracteata* Decne. In Jacq. Voy. Bot. 94. t. 102. 1844; Vern: Phansi or Spengsetobo or Span-rtsa-Do-bo or Jar-bag (L).

Uses: Flowering buds are used to cure cut, boils, against influenza, fever, intestinal pain, contagious disease, poisoning and arthritis.

95. *S. lappa* (Decne.) Sch.-Bip. In Linnaea, 19: 331, 1846; *S. costus* (Falc.) Lipsch.; Vern: Costus (E), Kuth (L).

Uses: The roots are used in bronchitis alongwith Manu in equal quantity.

96. *Sonchus oleraceus* Linn. Sp. Pl. 794. 1753; Vern: Milk Thistle (E), Khala (L).

Uses: The green leaves are eaten raw as salad or cooked as vegetable.

97. *Tanacetum gracile* Hk. f. & T.; *T. gracilis* D. Don.; *T. gracile* Hk. f.; Vern: Khamchu or Thorchokma (L).

Uses: The leaves and flowers are used as febrifuge and anti-worm.

98. *Taraxacum officinale* Wigg. Prim. Fl. Hols. 56. 1780; *Leontodon taraxacum* sensu. Aitchison.; Vern: Common Dandelion (E), Han or Khur-mang or Sanma, Yamngi (L).

Uses: The leaves are boiled then mixed with curd and used for making delicious local dish called "Tantur". Roots are also eaten raw.

99. *Waldheimia glabra* (Decne.) Regel. in Acta. Horti. Petrop. 6: 310. 1879; *Allardia glabra* Decne.; *W.tridactylites* sensu Aitchison.; Vern: Solonakpo or Sasomantso or Palu (L).

Uses: The paste of whole plant is used in septic wounds. The flowers are used as ornamental purposes.

100. *W. tomentosa* (Decne.). Regel. in Acta. Horti. Petrop. 6: 308. 1879; *Allardia tomentosa* Decne.; Vern: Makungla or Solo-karpo (L).

Uses: A paste of leaves and flowers is useful against sprains, rheumatism and wounds.

Campanulaceae

101. *Codonopsis clematidea* (Schrenk) Clarke. In Hk. f. FBI, III: 433. 1880; *Wahlenbergia clematidea* Shrenk.; Vern: Bruktung or Mokhting (L).

Uses: The roots are consumed by the local people either raw or cooked in Ladakh. The plant is considered useful in mild stomach ache, in-digestion and stimulant.

102. *Codonopsis ovata* Benth. in Royle, Ill. Bot. Himal. 235. t. 69. f. 3. 1835; Vern: Ludut or Chamelia (L).

Uses: The flowers are also used as ornamental plant. The whole plant is used in medicine same as *C. clematidea.*

Plumbaginaceae

103. *Acantholimon lycopodioides* (Girard) Boissa. In DC. Prodr. XII, 632, 1848. *Statice lycopodioides* Girard.; Vern: sNyalo or Longze (L).

Uses: Plant ash is believed to be given with milk in case of heart weekness and cardiac disorders. It is the best source of fuel during winter in Ladakh.

Primulaceae

104. *Primula macrophylla* D. Don. Prodr. El. Nep. 80. 1825 (excl. syn. *P. moorcroftiana*). *P. stuartii* Wall. var. *perpı rea* (Royle) Hk. f.; *P. purpurea* Royle.; *P. nivalis* Pall. var. *macrophylla* (D. Don.) Pax. & Knuth.; Vern: Sulumentok or Kangla or Nakla or Kelche-Karpa (L).

Uses: Amchi use the leaves and rootstock in cold and cough. It is used as beautiful ornamental plant used in religious ceremonies.

105. *Primula rosea* Royle. Ill. Bot. Himal. t. 76. f. 1. 1834, *et* 311. 1836; (encl. var. *elegans* Duby. Hk. f.); Pax. & Knuth.; Vern: Sulumentok or Brab-lcam (L).

Uses: It is also used for ornamental purposes by local tribes.

Gentianaceae

106. *Gentiana algida* Pallas. var. *nubigena* (Edgew) Kush. In Acta. Hort. Petrop. 15(2): 266. 1898; *G. nubigena* Edgew.; Vern: Tiktas or sPang-rgyan-dkar-po (L).

Uses: The flowers are used against inflammation of pharynx, bronchitis, cough and epidemic fever. Taste is bitter and potency cooling (Singh and Chaurasia, 2000).

107. *G. tianschanica* Ruper. In Mem Acad. Sci. Petesb. 7: ser. 19, 61. 1869; *G. decumbens* auct. Non. Linn.; *G. dahurica* auct. Non. Fiesh.: Clarke.; *G. olivieri* sensu Chowdhery & Wadhwa.; Vern: Wanglo or Kyi-lce nag-po (L).

Uses: The flowers are used against swelling of the joints due to gout, arthritis. The whole plant is edible and it is used as salad and vegetables.

108. *Gentianella moorcroftiana* (Wall. ex Griseb.) Airy Shaw; *Gentiana moorcroftiana*. Wall.; Vern: Spang-Gain-karpo (L).

Uses: The whole plant is used as purify blood, antitoxin, febrifuge. Useful in cold, cough, fever, headache and pulmonary track and fever due to poison etc.

109. *Lomatogonium rotatum* (Linn.) Fries ex Nym Cons. Fl. Europ. 500. 1881; *Swertia rotata* Linn. ex Hk. f.; Vern: Tsemrang (L).

Uses: Roots are crushed and made into a decoction, which is considered useful against fever and body pains.

110. *Swertia petiolata* Royle ex D. Don. In Trans-Linn. Soc. 18: 519. 1837; *S. petiolata* var. *lahulensis* (Kerner). Clarke.; Vern: Heising or Zadtik or Tikta (L).

Uses: The decoction of whole plant is used in fever body ache, gall disorder, headache and as a health tonic for liver and general weakness.

Boraginaceae

111. *Arnebia euchroma* (Royle) John. I. M Johnston. Contr. Gary. Herb. 73: 49. 1924; *Lithospermum euchromon* Royle; *Macrotomia perennis* (Schrenk) Bioss; *M. grandis* Bornm Oesterr.; *M. euchroma* (Royle) Pailsen.; *Arnebia euchroma* var. *grandis* (Bornm) Kazmi.; Vern: Ratanjot (H); Demok. or Dre-mog (L).

Uses: The mature roots yield a fat soluble red edible dye, which is used in Buddhist Gompas for dying god and goddess statue made of barley (*Hordeum vulgare*) for worship; also colouring of milk products and food items.

112. *A. guttata* Bunge. Index Sem Horti. Dorpat. 7. 1840; *A. tibetana* Kurz.; Vern: Ratanjot (H), Demok or Dre-mog (L).

Uses: The mature roots yield a fat soluble edible dye of inferior quality, which is also used as a substitute of *A. euchroma.*

113. *Lindelofia stylosa* (Kar. & Kir.) Brand in Englm Pflanzner. Ht. 78: 85. 1921; *Cynoglossum stylosum* Kar. & Kir.; *Lindelofia benthamii* Hk. f.; Vern: Makpen (L).

Uses: This plant looks beautiful and sometimes used as ornamental plant, best for flower beds and fodder.

Convolvulaceae

114. *Convolvulus arvensis* Linn. Sp. Pl. 153. 1753; Vern: Hirankhuri (H), Field bindweed (E), Tiktikma (L).

Uses: The roots are reported to be used as purgative. It is a common fodder plant of Ladakh.

Cuscutaceae

115. *Cuscuta reflexa* Roxb. Cor. Pl. II. 3. t. 104; Vern: Amarlata (H), Dodder (E).

Uses: The paste of climber is used against warts and sores in Ladakh. The whole palnt is used against burns, cuts and cough.

Solanaceae

116. *Capsicum annuum* Linn. syn. *frutescens* C. B. Clarke (FBI) in part, non Linn.; FBI, IV: 239, 1885; *C. purureum* Roxb; *C. minimum* Roxb; C.B. Clarke.; Vern: Mirch (H), Chilli (E)

Uses: It is used as condiment and pickle in Ladakh. It is also used as megico-religious plant in case of removing evil sprit and bad eyes affect (Jain, 1991).

117. *Datura stramonium* Linn. Bernh. in. Linnaea, 1833; Vern: Datura (H), Thornapple (E)

Uses: The drug is narcotic and anodyne. It affects sympathetic nervous system, but does not affect the motor or the sensory nerves.

118. *Hyoscyamus niger* Linn. Sp. Pl. 179. 1753; Vern: Khursani Ajwain (H), Henbane (E), Gyelamtag or Rhang-phrom lang-thang-rtse or Phagun (L).

Uses: Seeds are used to subsides pain due to pathogenic diseases, tooth ache, pharyngitis and used in treating malignant ulcers.

119. *Lycopersicon esculentum* Mill. D.E.P. V, 100; *L. lycopersicum* (Linn.) Karst.; Vern: Tamatar (H), Tomato (E), Tamatar (L).

Uses: Tomato is well adapted and common vegetable and salad.

120. *Physalis alkekengi* Linn. Bailey. 1947. 3. 2608; Vern: Chinese Lantern or Bladder Cherry or Winter Cherry (E), Shoklo (L).

Uses: The fruits are used in the treatment of urinary disorders, gout and rheumatism. The fruits are applied in high fever.

121. *Physochlaina praealta* (Decne.) Miers. In Ann. Meg. Nat. Hist. Ser. 2.5: 473.1850; *Hyoscyamus praealtus* Walp.; Vern: Langthang (L).

Uses: Seeds are used as vermifuge and leaves are narcotic and used in ulcer and eye diseases. Seeds are also used as febrifuge, vermifuge, heal wound, tumors, anthelmintic and analgesic.

122. *Solanum nigrum* Linn. Sp. Pl. 186. 1753; Vern: Black nightshade (E), Makoi (H), Tsigma (L).

Uses: The tender shoot and leaves are edible. The leaves and fruits are used against skin diseases.

123. *S. tuberosum* Linn. D.E.P., VI (3), 265, C.P. 1028; Vern: Alu (H), Potato (E), Alu (L).

Uses: Potato is top ranked cultivated vegetable crop of Ladakh. Potatoes are cheaply available throughout the season in local markets.

Scrophulariaceae

124. *Euphrasia officinalis* Linn.; Benth. In DC. Prodr. X: 552; Vern: Eyebright (E), Kangchuk (L).

Uses: The whole plant is used as tonic and fresh juice is subjected to relieve eye problems, as tonic and jaundice.

125. *Lagotis kunawurensis* (Royle ex Benth.) Rupr. Ion Mem Acad. Sci. St. Petesberg. Ser. 7, 14 (4): 64. 1869; *Gymnandra kunawurensis* Royle ex Benth. *G. cashmeriana* Royle ex Benth.; *Lagotis glauca* Gaertn. *cashmeriana* (Royle ex Benth.) Hk. f.; *L. glauca* auct. non. Gaertn.; Vern: Tsermo or Hong-len (L).

Uses: The entire herb is used against hot disorders of the liver, lungs small intestine, blood, disturbed fever and helps purification of the blood.

126. *Lancea tibetica* Hk. f. & T. In Hk. f. Kew. 9: 244. t. 7. 1857; Vern: Raikse or Chagna or sPa-yang rtsa-ba (L).

Uses: The extract of whole plants is to be recommended by the Amchis as a tonic. Strengthen lungs, pulmonary function, dries blood and pus. Leaves are useful for healing wounds.

127. *Pedicularis longiflora* Rudolph. spp. *tubiformis* (Klotz) Pennell.; *P. tubiformis* Klotzsch.; *P. tubiflora* Hk. f.; Vern: Lousewort (E), Phakchang or Lugru-ser-po (L).

Uses: The flowers are useful in treating hot disorders and inflammation of the liver and gall bladder, excessive seminal discharge, oedema and hangover.

128. *Pedicularis mollis* Wall ex Benth. Scroph. Ind. 52. 1835; Vern: Lugru-mar-po (L).

Uses: The tender leaves are used as salad.

Orobanchaceae

129. *Orobanche hansii* Ker. Nov. Pl. Sp. 2: 15. 1870; *O. cernua* Loefl. var. *hansii* (Ker). Beck, Monogr.; Vern: sGro-shang-rtse (L).

Uses: The leaves, stem, flowers and fruits are used to restores kidney heat, promots the functioning of kidney and checks haemorrhage.

Lamiaceae

130. *Dracocephalum heterophyllum* Benth. Lab. Gen. *et* 738, 1835; Vern: Zinkzer or Zimthigle (L).

Uses: Aromatic flowers and leaves are used against cold, cough and headache (Jain and Rao, 1983).

131. *Elsholtzia densa* Benth. In Lab. Gen. *et* Sp. 714. 1835; Vern: Tstase or Philongtso or Bye-rug nag-po (L).

Uses: The stem leaves, flowers and fruits are used against all kind of pathogenic diseases in the anus, uterus, skin, gastro-entrites region and rashes.

132. *Elsholtzia eriostachya* (Benth.) Benth. Lab. Gen. *et* Sp. Pl. 163. 1833; Incl. var. *pusilla* (Benth.). Hk. f.; *Aphanochilus eriostachya* Benth.; Vern: Tsatsa or Betso or Bye-rug ser-po (L).

Uses: The stem leaves, flowers & fruits are used against all kind of pathogenic diseases in the anus, uterus, skin, gastroenteritis and rashes.

133. *Mentha longifolia* Linn. Hudson. Fl. Angel. 221. 1762; *M. spicata* Linn. var. *ongifolia* Linn.; *M. sylvestris* Linn. (incl. var. *royleana* Wall. ex Benth.) Hk. f. (incl. var. *royleana*); *M. royleana* Wall. ex Benth.; *M. longifolia* var. *royleana* (Benth.) Rech. f.; *M. longifolia* var. *royleana* (Benth.) Raizada & Saxena; Vern: Jangali Pudina (H), Horse Mint (E), Phololing (L).

Uses: It is commonly used as chatani and as flavouring agent to local dishes. Its leaves are eaten with garlic in case of stomach problems and as pain reliever.

134. *Nepeta floccosa* Benth. Lab. Gen. *et* Sp. 736. 1835; Vern: Shamalolo or Songmongo or Shangukaram or Jimthegala (L).

Uses: The smelling flowers are used as flavouring agent by locals. A decoction of leaves is used against malaria, fever and cough.

135. *N. longibracteata* Benth. Lab. Gen. *et* Sp. 737. 1835; Vern: Prianku or ongmongo (L).

Uses: The flowers and leaves are used as flavouring agent. An extract of whole plant is useful in inflammation of stomach and liver.

136. *Perovskia abrotanoides* Kiril. in Bull Mosq. 1841, 15. t. 1; Vern: Iskiling or Tarubu (L).

Uses: The fresh leaves are used as flavouring agent. It contains a lavender type oil and smell.

Plantaginaceae

137. *Plantago depressa* Willd. Enum Pl. Hort. Berol. Suppl. 8. 1813; *P. tibetica* Hk. f. & T. ex Hk. f.; Vern: Tha-ram or Rambusuk (L).

Uses: An infusion of whole plant is useful in stomach pains and dysentery.

Amaranthaceae

138. *Amaranthus spinosus* Linn. Sp. Pl. 991; 1753; Vern: Chaulai (H) Amaranthus (E), Neu (L).

Uses: It is commonly used as green vegetable fresh or dehydrated both in Ladakh.

Chenopodiaceae

139. *Beta vulgaris* Linn. Sp. Pl. 222, 1753; *B. vulgaris* var. *benghalensis* Roxb. Fl. Ind. II: 59; Vern: Palak (H), Indian spinach beet (E), Mangol palak (L) (Aswal and Mehrotra, 1987).

Uses: Mangol palak is one of the widely grown vegetables of Ladakh. It is most common in market round the year fresh as well as solar dehydrated.

140. *Chenopodium album* Linn. Sp. Pl. 219. 1753; Vern: Bathua (H), Wild Spinach (E), Neung or Emn (L).

Uses: Whole herb is used as vegetable either fresh or dehydrated for prolonged winter use.

141. *C. botrys* Linn. Sp. Pl. 219. 1753; Vern: Sagani or Shanik (L).

Uses: Whole herb is used for green vegetable or sometimes dried for winter use by local tribes. The leaves contain 29.5% minerals, 2.07% calcium, 525 mg/100 g phosphorus and 109 mg/100 g irons with valuable source of energy upto 246 kcl. The aromatic flavour is added to the cheese and soup. The plant is used in catarrh and humoral asthma, also used as anthelmintic.

142. *Spinacea oleracea* Linn. Sp. Pl. 1027, 1753; Vern: Palak (H), Spinach (E), Vilayati palak (L).

Uses: It is common vegetable crop of Ladakh during early summer.

Polygonaceae

143. *Aconogonum tortuosum* (D. Don.) Hara. Flow. East. Himl. 632. 1966. (*Polygonum tortuosum* D. Don.) Prodr.; Vern: Serpalulu. or Itimeng (L).

Uses: It is considered the best fodder for milking animals.

144. *Fagopyrum esculentum* Moench. Meth. 290. 1794; FBI, V: 55, 1888; *Polygonum fagopyrum* Linn.; Vern: Buckwheat (E), Dayat (L).

Uses: Leaves and young shoots are washed thoroughly with water and boiled then cooked as vegetable as well as dried for prolonged winter use. The nutritive value showed that it is a good source of protein 25.1%, minerals 21.3%, phosphorus 574 mg/100 g and rich in energy. Seeds are also used for making flour to prepare different dishes.

145. *F. tataricum* (Linn.) Gaertn. Fruct. 2: 182. t. 119. f. 6. 1790; *Polygonum tataricum* Linn. Don. Prodr.; Vern: Duckwheat (E), Dayat (L).

Uses: The young herb is first boiled in hot water and then used for vegetable, its seeds are used for making chapattis after grinding. The plant is used as fodder.

146. *Oxyria digyna* (Linn.) Hill. In Hort. Kew. 158. 1769; *Rumex digyna* Linn. *Oxyria reniformis* Hooker; Vern: Mountain Sorrel (E), Lamanchu or Chu-Lchum (L).

Uses: The plant contains vitamin 'C', which is eaten either raw as salad or cooked for vegetable in Ladakh. It is also used in chatani. According to nutritive value it is rich in fibre 10.8%, iron 120 mg/100 g and mineral 29.8% contents.

147. *Rheum webbianum* Royle. Ill. Bot. Himal. 318. t. 99. f. 1. 1836.; Vern: Himalayan Rhubarb (E), Lachhu or Chu-rtsa (L).

Uses: The tender leaves are used in vegetable; leaf stalk are rich source of vitamin 'C' they are used in salad, chatani and in custard. Leaf-stalks are cut, outer membrane removed and made into small pieces and then mixed in salad or custard.

148. *Rumex patientia* Linn. spp. *orientalis* Benth. ex Schult. f.; *R. orientalis* Benth. ex schult. f.; *R. angulatus* Rech. f.; spp. *tibeticus* Rech. f.; Vern: Patience dock (E), Soma or Shommena (L).

Uses: The tender plant is washed thoroughly, boiled and used for vegetables in Ladakh during early summer, the stem is also eaten. It has 29.4% protein, 12.3% minerals, 9.4% fibre, 35.1% carbohydrate, 509 mg/100 g phosphorus and richest sources of energy 316 kcl. The roots are sweet and also eaten raw.

Elaeagnaceae

149. *Elaeagnus angustifolius* Linn. Sp. Pl. 121, 1753; *E. hortensis*, M Bieb.; Vern: Chhiolik (H), Oleaster (E), Shersing (L).

Uses: Fragrant flowers of this plant are used as a religious symbol and kept in the houses and gompas. The fruits are edible.

150. *Hippophae rhamnoides* Linn. spp. *turkestanica* Rousi. In Ann. Bot. Fenn. 8: 208. 1971; *H. rhamnoides* sensu Hk. f. *H. tibetana* Schlecht.; Vern: Seabuckthorn (E), Tsermang or Sastalulu or Cherker or sTar-bu (L).

Uses: The whole plant is economical. Fruits are rich source of vitamin 'C' and used for making juice, jam, chutney and squash. Roots are useful to improve soil fertility by nitrogen fixing ability due to presence of *Frankia*. Leaves are good source of fodder for sheep, goat and donkey. The leaves contains 22.72% crude protein which is best for milking animals. The fruits, leaves and seeds are considered as an anti-aging, anti-cold, restores memory and energetic.

151. *H. tibetana* Schlect. In Linnaea 32: 296 1863; Vern: Seabuckthorn (E) Tsermang or Sastalulu (L).

Uses: Leaves are good source of fodder for sheep, goat and donkey. Fruits are rich source of vitamins 'C' and used for making juice and squash.

Urticaceae

152. *Humulus lupulus* Linn. D.E.P., IV, 302; C.P., 759; Vern: Hops (E).

Uses: The flowers are used for extraction of essential oil, in preparation of beer and other drinks. It is tonic, nervine and diuretic.

153. *Urtica hyperborea* Jacq. ex Wedd. In Ann. Sci. Nat. Ser. 4, 1: 180. 1854; Vern: Dzatsutt or Zazut (L).

Uses: The leaves are used for preparing local delicious dish called as "Thupka" in Ladakh, which is considered warmth. The protein value of this plant is 28.3%, minerals 19.5%, carbohydrate 30.2% and it is good source of energy about 268 kcl.

Ulmaceae

154. *Ulmus wallichiana* Palchon. In Ann. Sc. Nat. Ser. 3: 10. 227, 1848; Vern: Elm (E), Umbuk (L).

Uses: The bark is used as herbal shampoo for washing hair, bone fracture and dislocation of bones; leaf and wood is used as fodder and fuel.

Moraceae

155. *Morus alba* Linn. D.E.P., V : 279; Vern: Sahtoot (H), Mulberry (E), Oshe (L).

Uses: Fruits are edible. Leaves are used as fodder, wood is used as fuel and for making baskets in Ladakh. Mulberry leaves are used for rearing silkworms.

Juglandaceae

156. *Juglans regia* Linn. Sp. Pl. (1753), 997; Vern: Akhrot (H), Walnut (E), sTarga (L).

Uses: Seeds are eaten and wood is used for making agricultural implements, tools and fuel. The bark is sometimes used for making dye (Singh and Arora, 1978).

Salicaceae

157. *Populus ciliata* Wall. ex Royle. Ill. Bot. Himal. 346. t. 98. f. 1. 1839; Vern: Pak burta (H), Himalayan Poplar (E), Yarpa (L).

Uses: Most important woody tree of Ladakh, which is commonly used for timber, furniture, fuel forestry and fodder etc. The branches are used for making supporting wood for baskets.

158. *P. nigra* Linn. var. *italica* Koehne syn. *P. italica* Moench.; *P. nigra* var. *pyramidalis* Spach.; Vern: Black Poplar (E), Yulat (L).

Uses: One of the most important woody trees of Ladakh, which is commonly used for timber, tools, furniture and fuel.

159. *Salix alba* Linn. Sp. Pl. 1021. 1753; Vern: European or White Willow (E), Mulchang (L).

Uses: The hard wood is used as timber, furniture, stem wood as fuel, green leaves as fodder. A decoction of leaves is used in case of dysentery.

160. *S. babylonica* Linn. D.E.P. 6(2). 388; FBI, V: 629, 1888; WI, IX: 171-172, 1972; DIFME, 158, 1991; CDP, I: 110, 1996; IMP, III: 2365, 1999; Vern: Weeping Willow (E).

Uses: The wood is used as timber, fodder, fuel, forestry and ornamental plant. The branches are used for making baskets.

Monocotyledons

Orchidaceae

161. *Dactylorhiza hatagirea* (D. Don.) Soo. in Ann. Univ. Sci. Buds. Sec. Biol. 3 341. 1960; *Orchis hatagirea* D. Don.; *O. latifolia* auct. Non. Linn.; *O. maculata* sensu Aitchison.; Vern: Salam Panja (H), Spotted Heart Orchid (E), Ambolakpa (L).

Uses: An extract of roots is also used against urine and kidney problems. The tubers are considered energetic, and health improving, which are recommended for weak people also (Chaurasia *et al.*, 2001).

162. *Epipactis helleborine* (Linn.) Crantz. Stip. Austr. 2. ed. 2. 467. 1769; Seidenfadan & Arora.; *Serapisas helleborine* Linn.; *E. latifolia,* (Linn.) All.; Vern: Penginlo (L).

Uses: Amchis used the infusion of plant as blood purifier.

Iridaceae

163. *Iris lactea* Pallas. var. *I. ensata* Thunb. In Trans Linn. Soc. II: 328, 1794; FBI, VI: 272, 1892; Vern: Sosan (H), Iris (E), Tesmamentok or Tais-ma or dRis-ma (L).

Uses: It is used for fodder and ornamental plant, which is fed to cattle to increase quantity of milk. The leaves are used for making baskets.

Liliaceae

164. *Colchicum luteum* Baker. in Gard. Chron. 1874, 33; Vern: Hirantutiya (H), Golden collyrium (E), Tukapa (L).

Uses: A decoction of corms and seeds are used against stomach complaints and paste of corm with mustard oil is used against gout and joint pains (Irshad *et al.*, 1989).

165. *Polygonatum verticillatum* (Linn.) All. Fl. Pedem 1: 131. 1785; *Convallaria verticillata* Linn.; Vern: Mitha dudia or Salam dana (H), Whorled Soloman's seal (E), Ra-mnye (L).

Uses: The roots restore bodily strength, and useful in the treatment of pain in the kidneys and wind diseases in Ladakh.

166. *Tulipa stellata* Hk. f. var. *chrysantha* Boiss.; FBI, VII: 355, 1897; Vern: Wild Tulip (E), Kapichog (L).

Uses: Bulbs are consumed by the locals as raw or cooked as vegetable or separate during early summer in Ladakh. Wild tulip is used for ornamental purposes due to presence of beautiful flowers.

Alliaceae

167. *Allium carolinianum* DC.; *A. blelandum* Wall.; DIFME, 17, 1991; Vern: Koshuk or Sukosh (L).

Uses: Fresh and tender leaves are edible; dried leaves used as flavouring agent (Singh *et al.*, 2001).

168. *A. cepa* Linn.; Roxb. *l. c.* 142; Wall. Cal. 5072; Vern: Piyaz (H), Onion (E), tSong (L).

Uses: Onion is used for edible purpose fresh as well as dehydrated during winters by local tribes. The bulbs are anthelmintic, used in asthma, nose bleeding, blisters, boils, bronchitis, ear complaints, eye complaints, expectorant, sedative, skin diseases, stomach disorders, vomiting, wounds and as cooling.

169. *A. przewalskianum* Regel. Allior. Monogr. 164, 1875. *A. jacquemontii* Regel, *A. jacquemontii* Kunth. *A. rubellum* auct. Non Biob.: Hk. f.; Vern: Jangali Lahsun (H), Wild garlic (E), Skotche (L).

Uses: Leaves are edible; used as condiment, vegetable and chatanee. The leaves are also used against dysentery and stomach troubles.

170. *A.sativum* Linn.; Roxb. *l. c.* 142; Wall. Cal. 5072; Vern: Lahsun (H), Garlic (E), Sgogpa (L).

Uses: Garlic is used for flavouring food as condiment fresh as well as in prolonged winters. The bulbs are also used against joints pain, gout, stomach ache, back ache and as appetizer.

Poaceae

171. *Agropyron repens* (Linn.) P. Beauv. Agrost. 102; Vern: Zamak (L) (Hooker, 1872-1897).

Uses: This is the useful wild fodder grass.

172. *Avena sativa* Linn. Sp. Pl. 79, 1753; Vern: Jai (H), Common Oat (E), Yupo (L).

Uses: It is used for easy digestion, fever and its tincture is useful in nerve and uterine tonic. It is cultivated mainly for food, feed and fodder.

173. *Hordeum vulgare* Linn. Sp. Pl. (1753), 84; Vern: Jau (H), Barley (E), Swah, sowa, Sherok or Grim (L).

Uses: Barley is very common crop of the area, it is widely cultivated in Changthang valleys of Ladakh. Used as food, fodder and medicine etc. The grains are used for preparation of local beer called "Chhang".

174. *Phragmites australis* (Cav.) ex Steud. Nom Bot. Ed. 2, 2: 324. 1841; *Arundo australis* Cav.; *Phragmites communis* Trin. Fund. Agrost. 134. 1820 (based on *Arundo phraqmites* Linn.); *P. communis* var. *stenophylla* Boiss.; Vern: Reed (E), Piachnig (L).

Uses: The whole plant is used as fodder, fuel, thatching, brooms.

175. *Poa sikkimensis* (Stapf) Bor. Kew Bull. 7: 130. 1952; *P. annua* Linn. var. *sikkimensis* Stapf.

Uses: The grass is useful both as green and dried fodder.

176. *Sorghum vulgare* Pers. Syn. I, (1805), 101. *Andropogon sorghum* Brot.; Vern: Jawar (H), Sorghum (E), Toue (L).

Uses: It is rarely cultivated in lower altitudes and used for food and fodder both grain and grass. Sorghum grain is an important article of food over vast dry areas of the country.

177. *Stipa orientalis* Trin. In Ledeb. Fl. Alt. 1: 83. 1829; Vern: Pilli (L).

Uses: It is a common grass mainly grazed by wild animals, sheep, goats and yak also collected for fodder as hay.

178. *Triticum aestivum* Linn. Sp. Pl. (1753), 86, *T. aestivum* Lam.; Vern: Gaihoon (H), Wheat (E), Tow (L).

Uses: It is one of the major cultivated cereals crop and commonly grown in all the valleys and mainly used for food as pava, roti, chapatti, biscuits, local breads and tandoori roti etc. Wheat is also used in preparaion of local drink called "Chhang".

Gymnosperm

Ephedraceae

179. *Ephedra gerardiana* Wall. ex Stapf. In Akad. Wiss. Wien, Math. Naturwiss. Kl. Denks 56: 75. t. 18. 1889; *E. vulgaris* auct. Non. Rich.; Vern: Asmania (H), Ephedra (E), Chhapat. or mTshe-ldum Lug-mtshe (L).

Uses: The roots, stems and branches are used as a tonic, against fever, stop bleeding, blood purification, snuffing and bronchial asthma. It is also used as an important fuel and fodder during winters (Polunin and Stainton, 1989).

Cupressaceae

180. *Juniperus communis* Linn. var. *saxatilis* Pallas, Fl. Ross. 1 (2): 12. t. 4. f. A. 1788; *communis* non. Linn. Hk. f.; Vern: Common Juniper (E), Shukpa (L).

Uses: The plant is considered religious and pious, cultivated near habitations and the leaves of the plant are burnt on coal or cow dung cakes for purification of house in the morning hours. It is suitable as ornamental shrub.

Conclusion

Ethnobotany plays an important role in day to day life of tribal communities since the time immemorial. The traditional knowledge of tribal communities of Ladakh is of ethnobotanical importance. They utilize numerous plants and their various parts, *viz.*, roots, leaves, stems, flowers and fruits for daily use. Out of the above 180 described plants they are using about 93 species for different medicinal purposes,

70 species for edible purpose, 35 species for fodder, 20 species for ornamental purpose, 12 species for fuel purpose, 10 species for forestry purpose, 6 species for megico-religious purpose, 5 species for different house hold purpose, 4 species for dye yielding purpose etc.

Now a days, due to modernization and availability of needful items in the nearest market the dependency on wild sources is decreasing rapidly. But in remote villages where there is no communication facilities available tribal population is still depedent on wild sources for their daily needs. Therefore, it is concluded that the plants of surroundings are always been beneficial for varius purposes such as food, shelter, medicine, fodder, forestry, decoration, colouring and other daily requirements of tribal communities. The scientific cultivation, conservation and sustainable utilization of above plants by involving tribal farmers would be highly beneficial to conserve the rare and endangered species and the traditional knowledge for coming generations.

References

Anonymous, (1948-76). The Wealth of India- Raw Materials., I-IX. C.S.I.R., New Delhi.

Aswal, B.S. and Mehrotra, B.N. (1987). Ethnobotanical Studies on the Flora of Lahaul Valley (North-West Himalayas). In: Recent Advances in Plant Sciences. Sharma, M.R. and Gupta, B.K. pp: 116-130.

Ballabh, B. (2002). Ethnobotany of Boto tribe of Ladakh Himalaya. Unpublished KUC, Nainital.

Chaurasia, O.P. and Singh, B. (1996-2001). Cold Desert Flora, (I-V). Field Research Laboratory, Leh Ladakh.

Chaurasia, O.P.; Singh, Brahma and Ballabh, Basant (2001). Potential Aromatic Flora of Himalayan Cold Desert : Ladakh & Lahul - Spiti. J. Econ. Taxon. Bot. 25(1): 91-97.

Hooker, J.D. (1872-1897). A Sketch to the Flora of British India, Oxford.

Irshad, A.; Navchoo, A. and Buth, G.M. (1989). Medicinal System of Ladakh, India. J. Ethnopharmacology 26: 137-146.

Jain, S.K. and Rao, R.R. (1983). Ethnobotany in India - An overview. Bot. Surv. Ind. Howrah.

Jain, S.K. (1991). Dictionary of Indian Folk- Medicine and Ethnobotany. Deep Publications, New Delhi.

Jina, P.S. (1996). Ladakh the Land & the People. Indus Pub. Co., New Delhi.

Kachroo, P.; Sapru, B.L. and Dhar, U. (1977). Flora of Ladakh. Bishen Singh Mahendra Pal Singh, Dehra Dun.

Kritikar, K.R. and Basu, B.D. (1935). Indian Medicinal Plants. Lalit Mohan Basu, Allahabad, 2nd Edn.

Polunin, Oleg and Stainton, Adam (1989). Flowers of the Himalaya. Oxford University Press. New Delhi.

Singh, B. and Chaurasia, O.P. (2000). Medicinal Flora of Indian Cold Desert. Acta Hort. 523(13): 65-72.

Singh, B.; Chaurasia, O.P. and Ballabh, Basant (2001). Edible wild plants of Trans-Himalayan Cold Desert. In : Plant Diversity of the Himalaya. Pande, P.C. and Samant, S.S. Ynodaya Prakashan, Nainital, pp. 483-512.

Singh, H.B. and Arora, R.K. (1978). Wild Edible Plants of India. I.C.A.R., New Delhi.

Stewart, Ralph Randles (1916-17). The Flora of Ladakh & Western Tibet. Bull Bot. Tor. Club. 43: 571-590.

□□□

23

Sowa-Rigpa : The Traditional Science of Healing

PADMA GURMET
Sowa-Rigpa (Amchi) Research Centre (CCRAS)
Ministry of Health and Family Welfare, Govt. of India, Leh, Ladakh, J&K

The Himalayas are always known for their rich herbal resources. Unlike other parts of Himalayas the Trans-Himalayan regions has a barren landscape and lifeless appearance. But, it possess very rich biodiversity with more than one thousand rare and valuable plant species supporting large number of wild and domestic animals, birds and insects. The life and cultural diversity of inhabitants of these Trans-Himalayan regions are largely influenced by these wild plants; basic commodities from food to shelter, medicine, fodder, and plants for socio-cultural needs were largely collected from these wild plants sources. Due to its unique geographical and environmental condition the plants of high altitudes have high medicinal and aromatic value. As per our study, out of around 1,000 plant species collected from Ladakh and Lahul-Spiti regions, 525 plants has medicinal uses in various traditional systems of medicines. The significance and use of Himalayan medicinal plants are mentioned in most traditional medical systems of Asia like Ayurveda, Chinese medicine, Unani and *Sowa-Rigpa* etc. Particularly *Sowa-Rigpa* (Tibetan Medicine/Amchi) system of medicine is specialized in use of High altitude medicinal plants.

"*Sowa-Rigpa*" (Science of healing) popularly known as Amchi or Tibetan medicine is among the oldest and well documented codified medical tradition of the world. It has been the traditional medicine of Tibet, Magnolia, Bhutan, some parts of China, Nepal, Himalayan regions of India and few parts of former Soviet Union etc. In India, *Sowa-Rigpa* is prevalent in Himalayan regions of Ladakh, Himachal Pradesh, Sikkim, Arunachal Pradesh, Darjeeling-Kalingpong and now in Tibetan settlements all over India. *Sowa-Rigpa* has been the only medical system prevalent in Ladakh and Trans-Himalayan regions until recent times. Public health has been managed very appropriately in these regions under the ancient socio-political system through *Sowa-Rigpa*. Every major village and hamlet had their own Amchi (practitioners of *Sowa-Rigpa*) to provide health facility since ages. Being originated from India, the majority of theory and practice of *Sowa-Rigpa* is similar to Indian medical system "Ayurveda" followed by few Chinese principals and then prevailing Tibetan folklore. The incorporation of Buddhist theories and principals and skillful use of Himalayan plants and mineral are among special feature of this system. *Sowa-Rigpa* being a holistic medical system with complete knowledge to achieve good physical, mental, social and spiritual health is a major resource of Trans-Himalayan Buddhist societies. The revalidation with modern scientific methods and tools, *Sowa-Rigpa* can add many new dimensions in the concept of health, management of disease and high altitude medical research.

Herbal Wealth

Since the dawn of history man has been in search of ways to find cure and relief from mental and physical illnesses. Nature and natural resources particularly plants have always been source of remedy for health and longevity. Therefore, the history of medical botany dates back to the origin of civilization. The Himalayas have been known for rich medicinal plants since the time immemorial and have been source for medicinal plants for most of Asian traditional medical systems like Ayurveda, Chinese medicine, *Sowa-Rigpa*, Siddha, Unani and many folk medicines. Particularly, the references of Himalayas and Himalayan plants is mentioned in India from Vedic period, the great Indian epics like Ramayana and Mahabharata have mentioned about medicinal plants of Himalayas. The most commonly practiced traditional medicine system of Trans-Himalayan regions *"Sowa-Rigpa"* has the maximum references and number of Trans-Himalayan medicinal plants mentioned in literatures (Table 1).

Table 1: Some important commercially traded medicinal plants of Trans-Himalayas

S. No.	Botanical name	Trade name
1.	*Aconitum heterophyllum*	Patis/ Atis/ Aconite
2.	*A. violaceum*	Meetha patis/ Aconite
3.	*Allium victorialis*	Wan pran
4.	*Arctium lappa*	Jangli kuth
5.	*Arnebia euchroma*	Rattanjot
6.	*Artemisia brevifolia*	Murin, Seski
7.	*Berberis jaeschkeana*	Rasaunt, Daruharidra
8.	*B. pseudumbellata*	Rasaunt, Daruharidra
9.	*Betula utilis*	Bhojpatra
10.	*Bunium persicum*	Kala zira, Shahi zira
11.	*Carum carvi* var. *gracile*	Vialaati zira, Konyoet
12.	*Chaerophyllum acuminatum*	Mithi patis
13.	*C. villosum*	Mithi patis
14.	*Codonopsis ovata*	Sardana
15.	*Dactylorhiza hatagirea*	Salam panja, Hath panja
16.	*Ephedra gerardiana*	Ephedra, Somakalpa, Somlata
17.	*Ferula jaeschkeana*	Kaindal
18.	*Gentiana kurroo*	Karu
19.	*Hippophae rhamnoides*	Chharma, seabuckthorn
20.	*Hyoscyamus niger*	Henbane, Khursani Ajwain
21.	*Hypericum perforatum*	Basant
22.	*Inula racemosa*	Manu/ Pushkarmool
23.	*Juniperus communis*	Hapusha, Hauber
24.	*J. macropoda*	Hapusha, Hauber
25.	*Malva sylvestris*	Sutsal
26.	*Meconopsis aculeata*	
27.	*Picrorhiza kurrooa*	Karoo/ Kutki
28.	*Podophyllum hexandrum*	Bankakri, Mayapple
29.	*Rheum speciformae*	Rewandchini, Rhubarb
30.	*R. webbianum*	Rewandchini, Rhubarb
31.	*Rhododendron anthopogon*	Talisapatra, Talishma
32.	*Rubia cordifolia*	Manjishtha
33.	*Sassurea lappa*	Kuth, Costus root, Kustha
34.	*Taraxacum officinale*	Dandelion, Hundh
35.	*Thymus serpyllum*	Ban ajwain
36.	*Tribulus terrestris*	Chota gokhru
37.	*Valeriana jatamansi*	Jatamansi

Sowa-Rigpa

Sowa-Rigpa is popularly known as Amchi medicine in most part of the Indian Himalayas. Derived from the Mangolian word of "*Am-Rjay*" means superior of all; the practitioner of this medicine are known as Amchis. Till early 1960s Amchi medicine used to be the only health care facility for the people of these regions and even after the introduction of allopathic medicine with all government support it could replace *Sowa-Rigpa* system in many parts of Himalayan Buddhist society due to its strong socio-cultural background. Amchis have not only social respect but also spiritual respect as the representatives of *Sangyas-Manla* (Medicine Buddha) and their services for ailing beings are priceless. Every major village and hamlet has been having an Amchi of their own since ages. Besides treating the patients as doctor of the village, Amchis are the most learned and resourceful persons of the village, therefore being an Amchi has been a matter of great dignity in the Trans-Himalayan societies. It takes several years to become a skillful Amchi, which requires hard theoretical and practical trainings. In most of Himalayan regions Amchis are trained through *Gudpa* (lineage) system in families, where knowledge is passed down from father to son in generations. After finishing their training the new Amchi has to give an exam, in front of entire community in presence of few expert Amchis in a ceremony, to enroll himself as an Amchi of the village. But for higher studies many Amchis used to go to Tibet. The relation between Amchi and patients has been always cordial due to strong religious background and social systems, where Amchis never used to ask for cost for their medicine and services, whatever the patient wish or afford they can offer and even most of the time it goes without any price. The villagers used to offer crops during harvesting time and free labour every year to the Amchi family for their services. With time and modernity things are rapidly changing and the ancient ways of practicing Amchi medicine is replacing with modern formal clinics and professional trainings (six years BTMS degree course *etc.*).

Historical Background

The history and origin of *Sowa-Rigpa* has been a matter of debate among the historians and scholars of *Sowa-Rigpa*, there is various conflicting account about the origin of this medical tradition, some scholars believes that it is originated from India, some says China and others consider it to be originated from Tibet itself. According to Tibetan history, before introduction of Buddhism during 7th-8th Century, *"Bon"* a kind of shamanistic religion was prevalent in Tibet and all the major art and sciences of Tibet was closely influenced by *Bon* traditions. Accordingly a *Bon* medical tradition (folklore) was also prevalent in Tibet. But introduction of Buddhism from India had tremendously changed the socio-political and cultural system of Tibet, the rich arts and science of India had created a revolution in Tibetan history. Thus the majority of theories and practices of *Sowa-Rigpa* are also similar to Indian medical system "*Ayurveda*" followed by few Chinese principals and then prevailing Tibetan folklore. The first *Ayurvedic* influence came to Tibet during 3rd Century AD but it became popular only after 7th Century with the approach of Buddhism to Tibet. There after this trend of exportation of Indian medical literature, along with Buddhism and other Indian art and sciences were continued till early 19th Century. India being the birth place of Buddha and Buddhism has always been favourite place for learning Buddhist art and culture for Tibetan students; lots of Indian scholars were also invited to Tibet for prorogation of Buddhism, Indian art and science. This long association with India had resulted in translation and preservation of thousands of Indian literature on various subjects like religion, science, art, culture and language in Tibetan language. Out of these around twenty-five text related to medicine are also preserved in both canonical and non-canonical forms of Tibetan literatures. Many of these were further enriched in Tibet with the knowledge and skills of neighbouring countries and their own ethnic knowledge. Present Tibetan medical system "*Sowa-Rigpa*" (Science of healing) is one of the classic examples of it. r*Gyud-Zi* (*Chatus Tantra*/ four tantra) the fundamental text book of this medicine was pioneered by Bhagwan Buddha 2,500 years ago and propagated by well known Indian scholars like Jewaka, Nagarjuna, Vaghbhatt and Chandarnandan etc. around 7th-8th Century it was translated into Tibetan by Verochana and Kashmiri Pandit Chandaranandan. But before introducing the *rGyud-Zi* into Tibet, it was amended and enriched by then Tibetan court physician Youthog Yontan Gonbo and other scholars by incorporating the elements of Tibetan folk-medicine, Chinese and Persian medicine to make it harmonious with the environment, cultural and geographical condition of Tibetan Plato. After Tibet the impact of *Sowa-Rigpa* along with Buddhism and other Tibetan art and sciences were spread in neighboring Himalayan regions.

Philosophical Background

The fundamental text book *rGyud-Zi* being taught by Buddha; it is closely linked with Buddhist philosophy and way of life. Wherein we believe that whether we are physically healthy or not, all of us are sick until we attain Nirvana even though disease might not be manifested, it is always present in dormant form until we give up root cause of sickness-ignorance. Ignorance gives rise to three basic roots of sickness *Nespa-gsum* (*Tri-dosha/* Three humors), i.e., *Dod-chags* (desire/lust) root cause of *rlung* (Air, Vata), *Zes dang* (anger) root cause of *mkris-pa* (*Pitta,* Bile) and *Ti-mug* (mental darkness) root cause of *pad-kan* (*Kapha,* Phlegm), (Table 2). Due to presence of 84,000 afflictive emotions in mind, it gives rise to same number of disease; thus Bhagwan Buddha preached 84,000 kinds of teachings. The animate and inanimate phenomena of the universe are composed of same material basis of *Jung-wa-lna* (five elements) due to the karma of all the living beings. The behavior and conduct of physician and patient are also largely based on Buddhist principles. A Buddhist tantra and mantra ritual plays a very important role in the treatment of patients.

The Basic Theory

Sowa-Rigpa is based on the principles of *Jung-wa-lna* (English-five elements; Sanskrit-*Panch-mhabhuta*) and *Nespa-Sum* (English-three humors; Sanskrit-*Tri-dosh*). All animate and inanimate phenomena of this universe are composed of *Jung-wa-lna;* namely *sa, Chu, Mai, rlung* and *Nam-mkha* (Roughly translated as earth, water, fire, air and space; Sanskrit-*Dharti, Jal, Agni, Vayu* and *Akash*). The science of physiology, pathology and pharmacology of this system are established on these theories. Which means our body is composed of these five cosmo-physical elements of *Jung-wa-lna;* when the ratio of these elements becomes imbalance in our body, disorders result. The medicine and diet used for the treatment of disorders are also composed of the same five basic elements. In the body these elements are present in the form of *Nespa-sum* (three humors) *Luszung-ldun* (English-seven physical constituents; Sanskrit- *Sapt dhatu*) and *Tema-sum* (English-three excretion; Sanskrit-*Tri-mala*). In drugs, diet and drinks they exist in the form of *Ro-tug* (six tastes), *Nuspa* (potencies), *Yontan* (quality) and *Zu-rjes* (post-digestive taste). It is in context of this theory that a physician would use his knowledge, skills and experience in treating a patient, using the theory of similarity and dissimilarity of five elements.

Table 2: Root cause of sickness, *Nyes-pa* (humors) and its relation with five elements

Cause of sickness (three poisons)	*Nyes-pas* (humors/ *Tri-dosa*)	The element base of *Nyes-pas*
Dod-chags (desire/lust)	*rlung* (*Air, Vata*)	Air
Zes dang (anger)	*mkris-pa* (Bile, *Pitta*)	Fire
Ti-mug (mental darkness)	*Pad-kan* (*Kapha*, Phlegm).	Earth and water

Healthy and Unhealthy Body

A healthy body is typified as a state of balance of three humors (*Nespa-sum*), seven physical constituents (*Luszungs-ldun*) and three excretions (*Trima-sum*) in the body with complete harmony of five aggregates (five senses). The three humors are air (*rlung*), bile (*mkris-pa*) and phlegm (*pad-kan*) which is further divided into fifteen type of humors. Seven physical constituents are: nutritional essence, blood, muscle tissue, fat tissue, bone tissue, bone marrow and regenerative fluid; the three excretions are: stool, urine and sweat. Humors are the biological representative of five elements, physical constituents are the basic tissue elements of the body and excretions are the waste product of the body whose proper elimination is essential for good health. Healthy and unhealthy state of body is largely dependent on the balance and imbalance of these twenty-five elements due to proper diet and behavioral patterns. The subject of human body is finely explained in *Sowa-Rigpa* with different chapters on embryology, anatomy, physiology and neurology etc. are explained in an appropriate manner with functions of each and every parts of the body.

Outline of Disease

It has already been mentioned that the science of *Sowa-Rigpa* is closely linked with Buddhist philosophy, where it is believed that all the breathing creatures on the earth are sick until they don't give up the root cause ignorance. Due to ignorance three mental poisons, anger, desire and mental darkness, are being born with us, who act as basic origin of illness and give birth to three humors. When the ratio of three humor air, bile and phlegm, misbalances due to circumstantial conditions such as diet, lifestyle, seasonal and mental conditions gives rise to different kinds of disorders. Diseases can be classified in terms of different methods such as location in the body, type etc. Due to presence of 84,000 different types of afflictive emotions in the mind the number of disorders also goes up to 84,000, these can be condensed into 1,016 types, further 404 types, 101 types, three humors and finely into two types of hot and cold disorders. The outline of disorder is explained in detail with complete note on entrances of disease, locations, pathways, time of arising, fatal effect side effects etc.

Diagnosis

A physician of *Sowa-Rigpa* employs three main tools for diagnosing a patient, i.e., visual diagnosis, diagnosis by touch and diagnosed by questioning. Visual diagnosis comprises of two main methods of tongue examination and urine alysis. Various diseases can be diagnosed by checking the nature, color and smoothness of the tongue; urine can be analyzed on three different stages: fresh urine, while it is cooling down and when it is cold by looking at its color, speed of discoloring, vapor, odor, bubbles, sediments etc. (Table 3). Diagnosis by touch is represented by the advanced technique of pulse examination followed by touching the body for temperature and smoothness etc. Pulse is the most important and major diagnostic tool in *Sowa-Rigpa,* which is explained under thirteen, general topics like preparatory conduct, proper time for examination, place, pressure of physicians finger, technique, constitutional pulse, astonishing pulse, general and specific pulse, death pulse etc. (Table 3). Questioning is another mode of diagnosing a patient; history of case, present condition, family background and changes in body are main questions to be asked.

Table 3: Some major tools of diagnosis

Nyes-pas **(humors/ *Tridosa*)**	**Tongue texture**	**Urine examination**	**Pulse qualities**
rlung (Air, *Vata*)	Reddish, dry and boorish	Whitish blue like water with large bubbles when stirred	Empty, floating with irregular beats
mkris-pa (Bile, *Pitta*)	Thick yellowish coating	Yellowish-red with profuse steam, fetid and fast disappearing small bubbles appears when stirred	Fast, overflowing with taut beats
Pad-kan (*Kapha*, Phlegm)	Whitish sticky coating with pale, smooth and moist texture	White with distinct odour, saliva like bubbles appears when stirred	Sunken, slow with very weak beats

Methods of Treatment

The methods of treatments have four major sections of diet, behaviour, medicine and accessory/external therapies. Right administration of these four sections is very important for treating a patient in an appropriate manner. Minor ailments can be treated with proper regulation of diet and behaviour only, a patient with medium state of condition can be treated with medicine like decoction, powder, pills, purgatives, emetics etc. and in advanced stages disease can be cured by application of accessory/ external therapies like moxabustion, venesection, fomentation, massage, surgery etc. Along with curative medicine, special emphasis is given in *Sowa-Rigpa* for preventive measures like proper regulation of diet and behaviour in different seasons and circumstances. All the disorder can be condensed into three humors, accordingly the treatment of disease are also possible under three humor basis. Taste and potency is among the major tools for treatment and formulating medicine, which is guided by the five elements (Table 4). Buddhist rituals and mantra also plays a very important role in the treatment of disease and in pharmaceutical process etc. The role of a physician is very important in treatment of a patient; therefore special emphasis is given on the qualities, motives, approach and conduct etc. of a physician with strict guidelines.

Table 4: Dominant elements of six tastes and their effects on *Nespa-sum* (three humours)

The dominant element	Taste	Agrevates*Nespa*	Pacifies*Nespa*
Earth + Water	Sweet	*Pad-kan* (Phlegm, *Kapha*)	*rLung* (Air, *Vata*) & *mkris-pa* (Bile, *Pitta*)
Earth + Fire	Sour	*mkris-pa* (Bile, *Pitta*)	*rLung* (Air, *Vata*) & *Pad-kan* (Phlegm, *Kapha*)
Water + Fire	Salty	*mkris-pa* (Bile, *Pitta*)	*rLung* (Air, *Vata*) & *Pad-kan* (Phlegm, *Kapha*)
Water + Air	Bitter	*rLung* (Air, *Vata*) &*Pad-kan* (Phlegm, *Kapha*)	*mkris-pa* (Bile, *Pitta*)
Fire + Air	Pungent	*mkris-pa* (*Pitta*, Bile)	*Pad-kan* (Phlegm, *Kapha*)
Earth + Air	Astringent	*rLung* (Bile, *Pitta*) & *Pad-kan* (Phlegm, *Kapha*)	*mkris-pa* (Bile, *Pitta*)

Pharmacology

Based on the theory of *Jung-wa-lna*, pharmacology of *Sowa-Rigpa* believes that every substance on the Earth has medicinal value and therapeutic efficacy. Medicinal substance are classified into eight major categories, i.e., *Rinpoche sman* (priced metal and stone), *sa sman* (drugs from mud and earth), *rdo sman* (stones), *shing sman* (drugs from trees), *rtsi sman* (essence and exudates medicines), *thang sman* (plant ingredient for decoction/plateau medicine/shrub), *sno sman* (herb) and *srog chags sman* (animal parts) (Table 5). The ingredients of medicine are used in both single and compound forms, compound medicine is based on two major permutation and combination that is *ro* (taste) based combination and *nus pa* (potencies) based combination. There are about seventeen different categories of medicinal preparation out of which some important preparations are decoction, powder, pills, lincutus, medicated ghee, Bhasmas, medicated wine, paste, medicated bath etc.

Table 5: Classification of medical substances

S.No.	Classification	Some examples
1.	*Rinpoche* (precious metals & gemstones)	Gold, silver, copper, iron, turquoise, quartz, pearl, conch, coral, lapis lazuli etc.
2.	*rDo* (stone/minerals)	Limestone, alabaster, tin ore, marcasite, hematite, lead monoxide, quartz, asbestos , fossil stone etc.
3.	*Sa* (soils/earth)	Sulphur, gold sand, yellow ochre, saltpetre, chalcanthite, ochre, natron etc.
4.	*Shing* (trees)	Sandalwood, eaglewood, *Bombax* sp.,*Cassia fistula*, seabuckthorn, juniper etc.
5.	*rTsis* (mucilaginous/ nectarous medicine)	Bezoar, bamboo pith, saffron, cardamom, musk, clove, nutmeg etc.
6.	*Thang* (shrubs / plateau medicine)	*Berberis* sp., *Justicia adhatoda* L., *Rubus ellipticus*, *Rubus* sp., *Zanthoxylum nepalense* etc.
7.	*sNo* (herbs)	*Sassurea lappa, Rubia cordifolia, Tribulus terrestris, Picrorhiza kurrooa, Dactylorhiza hatagirea* etc.
8.	*sRog-chags* (animal product)	*Piebald rhinoceros*, elk horn, musk, bear bile, sheep brain, goat brain, cow-dung, cow urine etc.

The element composition of a drug has to be determined on the basis of the six major tastes, i.e., sweet, sour, saline, pungent, astringent and bitter. Out of which, element composition of sweet taste is predominate by *sa* (earth) and *chu* (water) elements; sour taste is predominate by *mai* (fire) and *sa* (earth); saline is predominate by *chhu* (water) and *mai* (fire); bitter is predominated by *chhu* (water) and *rlung* (air); pungent taste is predominated by *mai* (fire) and *rlung* (air); astringent taste is dominated by *sa* (earth) and *rlung* (air) element. Sweet, sour, saline and pungent taste alleviates *rlung nespa* (air humour), which is dominated by *rlung* (air) element (Table 4). Bitter sweet and astringent alleviate *nespa mkris pa* (bile humour), which is dominated by *mai* (fire) elements. Pungent sour and saline taste alleviate *nespa pad kan* (phlegm humour) which is dominated by *sa* (earth) and *chu* (water) elements. Other important factor for understanding drugs composition and action are *Zu zes sum* (three post-digestive tastes) *nuspa-rgyad* (eight potency) and no *bo* (specific action) of particular drugs etc. The therapeutic efficacy of drugs is also dependent on the above factors.

Sowa-Rigpa (Amchi) Research Centre, Leh

Recognizing the importance of *Sowa-Rigpa* for public health in Ladakh and its research potentials, a *Sowa-Rigpa* (Amchi) Research Centre, Leh, an institution dedicated for research and development of *Sowa-Rigpa* and Trans-Himalayan Plants, has been set up by Govt. of India as a peripheral Unit of Central Council for Research in Ayurveda and Siddha, Department of AYUSH, Ministry of Health and Family Welfare Govt. of India. It was established in 1976 in Leh-Ladakh with the mandate of Research and Development of *Sowa-Rigpa*. The main aim and objective of the Institute is to conduct fundamental and clinical research on various formularies of *Sowa-Rigpa* and to explore and document herbal and mineral resources of Ladakh Himalaya.

Activities and Achievements

(i) Clinical Research

Under this project, clinical research has been conducted on various diseases with Amchi formularies and parameters. In most cases Amchi medicine proved to be very effective. Some major cases taken up for study are:

- *Padkan-sMugpo* (Peptic ulcer)
- *Tek-dum* (Arthritis)
- *Pags-nad* (Skin diseases)
- *Thaglung-stod-tsangs* (Hypertension)

(ii) Herbal and Mineral Exploration

Under this head, extensive medico-botanical survey of Ladakh and Lahul-Spiti has been undertaken by this unit. Total around 1,000 plant species were collected, out of which 500 plant species were found to be used in various Indian systems of medicines. Medicinal plants used by Amchis are documented in the form of research papers and a comprehensive check-list book on Trans-Himalayan medicinal plants has been published. Around fifty different mineral stones were collected from different places of Ladakh.

(iii) Literary Research Project

Sowa-Rigpa has a rich literary wealth in Tibetan language, which includes large number of translated Indian classical medical literatures and its commentaries written by Indian and Tibetan scholars. A list of source and number of ancient medical literature in Ladakh has been identified. A Bibliography of Tibetan manuscript translated from Indian medical literature was also prepared.

(iv) Extension Activities

Various reorientation trainings, seminars and workshops have been organized by this Centre to develop and strengthen Amchi practice in Ladakh. The research findings of the Unit and new developments in the field of traditional medicine and medicinal plants are being conveyed to Amchis through these activities.

(v) Cultivation and Conservation of Medicinal Plants

Availability of medicinal plants is important for survival of all the herb-based medical system of the world. Due to current increasing demand for herb-based products, ecological changes, unscientific and over exploitation etc. are causing great threat for survival of many plant species. To prevent such situation in Himalayas *Sowa-Rigpa* Research Centre, Leh has been engaged in cultivation and conservation of some important and threaten medicinal plants of Ladakh. Few rural Amchis (Traditional healers) have been adopted by this Centre for cultivation of medicinal plants. A project for establishing Trans-Himalayan medicinal plants garden have been also initiated by this Centre for conservation and display of cold desert plants.

Conclusions

The Trans-Himalayan regions have very rich medicinal plants resources with complete knowledge system *"Sowa-Rigpa"* to utilize them systematically for various disorders. *Sowa-Rigpa* is a rich accumulation of science, art and philosophy with history of continuous practice and progress for more than 2,500 years. It is a science because it is based on systematical and logical framework of understanding the body, disease and its relationship with the environment. The correlations between body, disease and its treatment are well explained under the theory of *Jung-wa-lna*. It is an art because the diagnostic techniques and composition of medicines are based on the creativity, immensity, delicacy and compassion of the medical practitioner. *Sowa-Rigpa* is a philosophy because it follows the key Buddhist principles of self-sacrifice, karma and ethics etc. The holistic approach of body and mind to lead a healthy way of life with complete harmony with nature has made *Sowa-Rigpa* a comprehensive guide to lead a healthy way of life with balanced body and mind. But like any art or science, *Sowa-Rigpa* also needs to be revalidated and redefined with time and change of social systems, therefore, a proper scientific research and use of modern tools can provide a new dimension for public health and concept of preventive and curative health through this time tested ancient science of healing.

References

Anonymous, (1970). Gyud-bzh. Tashigang Publishers, BPO Nimo, Leh.

Dash, Year Vaidya Bhagwan. Tibetan medicine with special reference to Yoga Shataka, Library of Tibetan works & Archives Mechliod Gang Daramsala (H.P).

Dorjai, Gawai (1995). Thungs-Pai Drimed Shail Gi Melong. Lhasa Tibet: Meregs Pai Tun Khang.

Phuntsog, Dilmar Dgeshes bStanzin (1983). Dri Med Sel Gong & Sel-Phring. Tashigang Publishers, BPO Nimo, Leh.

Chaurasia, O.P. and Gurmet, P. (2003). A checklist on medicinal and aromatic plants of trans-Himalayan Cold Desert (Ladakh and Lahoul-Spiti), Field Research Laboratory (DRDO) and Amchi Medicine Research Unit (CCRAS), Leh.

Gurmet, Padma (2002). Ayurveda in Tibet, Compendium of invited lectures World Ayurveda Congress 2002, pp. 117-122. Kochi-kerala: Swadeshi Science Movement.

Gurmet, Padma (2004). "Sowa-Rigpa": Himalayan art of healing Indian Journal of Traditional Knowledge 3(2):212-218.

Gurmet, Padma (2007). The Amchi System of Medicine; The Art and Science of Healthy Way of Life, Recent Research on Ladakh 2007, J&K Academy of Art, Culture & Language, Leh, pp. 125-133.

Gurmet, Padma. Amchi Chikitsa Ekk Parechaya, Sindu Darshan. Vol-2 (Yearly Hindi magazine) Field Research Laboratory (DRDO) C/O 56 APO.

Gyatso, Desrid Sangyas (1969). Gyud-bzi Gril pa Vaidurya Snonpo. Tashigang publisher, BPO Nimo Leh.

Men-Tsee-Khang. Fundamental of Tibetan Medicine, Macleod Gang, Daramsala, HP.

□□□

24

Traditional Food and Beverages

DORJEY ANGCHOK AND SANJAI K. DWIVEDI*
Defence Institutre of High Altitude Research (DRDO) C/o 56 APO
* CEPTAM (DRDO), Metcalfe House, Delhi - 110054

William Moorcroft, a veterinary surgeon traveled in the 1820s, was the first Englishman to give a detailed account of Ladakh. Moorcroft describes three meals a day: first: tea, which would be salty not sweet; second: porridge, made with *tsampa* and water and probably allowed to ferment slightly, or tea; third: meat, rice, vegetables and bread for the upper classes or soup, porridge and bread for the lower classes. The poor people drank *chhung* (local barley beer), often in place of the more expensive tea; the wealthier, in addition, drank grape juice and water or *sherbets*. These were either from the dried and powdered fruits of apricot or apple. A drink called *tsigu chhu* is still made today, from ground apricot kernels and water, at least in Dhomkar area of lower Ladakh. He also described a wild fruit he called *chasta ruru*, now usually known as *tsestallu* (Seabuckthorn - *Hippophae rhamnoides*), but it appears this was not used for juice and jam, as is being tried out recently although it was used medicinally.

G.T. Vigne (Vinge, 1842), traveled in the 1830s, told us about items that passed through Ladakh on the trade routes from Yarkand to Hindustan and vice versa. Tea came in three kinds from Yarkand, two black, one green; there were also dried cinnamom flowers. These traders colony from Central Asia also had influenced the diet especially that of the Muslims, in respect of bread such as *kulcha* and *bakir kham*, still made by Muslims baker and meat dishes such as *Yarkendi Polau*. David Fraser, in his amusing account of travels around the turn of nineteenth century, found eggs, fowl and vegetables as part of the diet in *Lamayuru* (west of Ladakh). The Moravian missionaries came to Ladakh in the last quarter of the nineteenth century and brought with them vegetables such as potatoes, spinach, cauliflower, radish, green bean, Brussels sprout and tomato. These were probably the first significant additions to the Ladakhi diet for many years - turnip, Swedes and peas were predominated until then. It was probably the Moravians who demonstrated that root vegetables could be stored in pits in the ground, thus improving the variety of foods available in the winter.

Recent initiatives : Regarding the interventions made by various institutions like Field Research Laboratory (DRDO), State Department of Agriculture and Horticulture, there has been a tremendous improvement in the diversity, quantity and quality of cultivated crops (especially vegetables). It has been demonstrated (on field conditions) that around 64 different kinds of vegetables (both European and Asiatic type) can be successfully grown in Ladakh. Presently, local farmers in Ladakh are successfully growing newly introduced crops like kale, parsely, celery, summer squash, okra, brinjal and various cucurbits. The increase in crop diversity has helped them to increase their income and improve the nutritional uptake. At present 50% of Army's (a major market for local produce) fresh vegetables requirement is fulfilled by the local farmers (local farmer's cooperative is supplying 20 different kinds of vegetables to the Army). It not only provides income to the local farmers but also saves large defense expenditure (Rs. 27/kg fright charge - when road closed; and Rs. 05/kg - when road open during the year 2008) on part of the Government.

Traditional Foods and Beverages

Most of the foods and beverages described in this chapter are made throughout the region. However, the elevation of the cultivable land affects both the length of the growing season and the choice of crops produced. The staple diet of region is huskless barley or grim, which is roasted and ground for use as flour, *tsampa* (Tibetan) or *namphey* (Ladakhi), or for making the local 'beer' brew, *chhang*. For a quick, nutritious and warming breakfast or lunch it is hard to beat butter tea with *khokak* (described below). These foods are easy to make, simple, least fuel consuming; locally available ingredients are preferred. While in the early days the food habits might have been different but over the centuries interaction between different culture and races have resulted in the development of a system, which is truly representative of the region. The majority of the dishes, that are as follows, are sustaining food for the people of a remote region. As such they are outstanding, and a first class lesson on how to stay alive on the produce of a six month growing season.

1. Bakeries

Breads are made from wheat flour (*paqphey*) as well as barley flour (*narjen* - meaning uncooked barley) or a mixture of the two. At times breads are also made from pea or lentil flour. Sometimes pea and wheat are mixed and ground into flour which makes it more nourishing and palatable.

1.1 Tagi Khambir or Skyurchuk (Browned Sour Dough Bread)

A bread (*tagi* is the general name for bread) eaten throughout Ladakh. *Skyur* means sour, *chuk* means mixed with. *Skyurchuk* is the name by which the bread is generally known in the villages; *thagi khambir* is its Leh name. Before baking powder was available, Ladakhis used local soda (*pul*) from the Nubra valley. In a traditional kitchen the dough would first be cooked on a cast iron plate placed over one of the cooking holes in the stove, under which is the fire of wood or cow dung. Then it will be finished off on the ambers inside.

1.2 Tagi Buskhuruk (Puffed Unleavened Bread)

Buskhuruk means unleavened. The dough is the same as for *tagi shrabmo* (similar to chappati), but the rounds are made thicker and smaller in size. They are cooked briefly as above, but are allowed to puff up, and are then, if possible, put in the ashes of stove or fire. Traditionally, it is backed on a plate placed on a cow dung fire, covered with another plate and more dung.

1.3 Tagi Thalkhuruk (Bread Uncovered and Baked in Ashes)

This is lighter than *tagi thalshrak* (covered and backed in ashes) and is therefore good for people who are not well. The dough is made in the same way but the bread is cooked uncovered on a plate, on the ashes–*thalkhuruk*.

1.4 Tagi Mer-Khour

It is thick round bread, dough is made off flour mixed with molten butter and well beaten egg white. To make the bread more attractive and shinny, egg yolk is spread on the upper surface of the bread. They are baked by placing them over flat sheets of iron covered with hot ash from all sides. After it is fully backed it gives a mouth watering flavour and is removed from the iron sheet.

1.5 Tagi Thal-Khour

It is similar to *tagi mer-khour* except the butter and egg. It is backed directly in hot ash, charcoal or dung fire.

1.6 Tagi Tain-Tain

It is soft to eat and resembles a *dosa* except that it is thicker and is served with milk or tea. It is made by putting a thick paste of wheat flour (fermented overnight) over moderately heated iron plate (tava) spread with a thin layer of fat or oil.

1.7 Tagi Kiseer/Giziri

It is made as *tagi tain-tain* where wheat flour is substituted by buckwheat flour.

1.8 Skien/Mayaro

It is made by making dough of flour and sheep fat, given the shape of an ibex horn and then backed on charcoal. While consuming it is broken, and the bread along with the fat is heated in a bowl and mixed together. It is presented by close relatives on the birth of a boy. It is more prevalent in the Dard area of Da-Hanu (West Ladakh).

1.9 Kaptsey/Makhori

Thick round bread made of wheat flour. The edge is turned, twisted and pressed to give a woven appearance, cooked in oil medium, and sugar may be added as per the taste. It is presented when a girl child is born. In Baltis and Shin tribes (in Kargil district), it is also presented by the groom's family to the bride's family during the betrothal and marriage ceremonies. At *Losar* (Ladakhi New Year - generally falls in December-January) the Dard or Shin tribals of Drass, stuff the *makhori* with animal fat and send it as present to the close female relatives like daughters, sisters, and close cousins.

1.10 Tagi Tsabkhur (Ground Sprouted Wheat Bread)

It is very easy to cook and the taste is very sweet, but making the flour from the sprouted wheat or barley is a little complicated and interesting. *Tsabkhuruk* means sprouted grain and cooked covered (as opposed to *thalkhuruk,* which means not sprouted and uncovered). It is said to be good for pregnant women.

1.11 Sephe Tagi (Freshly Sprouted Wheat Bread)

This bread is made in springtime in the Zanskar valley. The grain is sprouted as above, but is used fresh when the sprouts are about one centimeter long. The sprouts are crushed in a pestle and mortar and then mixed with water. This mixture is added to flour to make moist/ loose dough. The dough is cooked with a little fat, or dry, on a heated stone or griddle. It swells, like *tagi khambir,* and is eaten split with a little fat or butter added inside.

1.12 Khura (Sweet Deep Fried Biscuits)

It is made especially for *Losar.* When a married girl visits her parents home during *Losar,* she will carry with her a plate of *khura* for the family members. In the old days, local flour was used, and may be still in use in some villages. Now-a-days few people use rice flour to achieve an even finer consistency. The technique for giving shape to this biscuit is quite tricky. They can be salty, in that case salt syrup is used instead of the sugar.

2. Ready to Serve

2.1 *Khokak* (*Tsampa/Namphey* Mixed in Butter Tea)

It is the most commonly used food and involves no cooking. This dish is a quick and easy breakfast, lunch, or anytime food. The flour is made by grinding the roasted grains of barley (generally known as *tsampa* in Tibetan and *namphey* in Ladakhi), foxtail millet (*tse-tse phey*) or wheat (*chhoptsos*) to a fine powder which is generally referred to as *phey. Khokak* is made (Fig. 1) by adding the powder to any liquid or semi-liquid and bought to a consistency where it does not stick to the hand through proper mixing and accordingly named:

2.2 Tsiri Khokak (Diluted Chhang Khokak)

In some villages especially in the sham area they boil peas, then dry and grind them into flour, *shranphey,* to be used with *namphey* or *khokak.* In the Zanskar valley, in place of tea, *khokak* may be made with *tsiri* (or *singri*), which is diluted *chhang.*

2.3 Sbangphe (Chhang Residue Khokak)

The *chhang* (fermented barley drink) residue is dried and ground together with roasted barley. It is then mixed with butter tea, or sometimes water, and left overnight to be eaten the next morning, with *snamthuk.*

2.4 Chuu Khokak

Chuu means water. *Phey* is added to water with a pinch of salt or sugar. Mostly used by hunters or travelers when nothing else is available to eat.

2.5 Cha Khokak

Cha means tea. *Khokak* made by adding *phey* to tea (made from green tea leaves and salt in place of sugar).

2.6 Der Khokak

Derba means butter milk. This is made by adding *phey* to butter milk.

2.7 Chhang Khokak

Chhang means local barley made fermented drink. *Phey* is added to *chhang.* It has a sour taste and mostly preferred by travelers.

2.8 Phemar (Khokak for the Sweet-Tooth)

Phemar is usually served only to guests or on special occasions such as weddings. *Phey* is added to salt tea with a lump of butter. Sugar can be added depending upon individual's choice.

2.9 Chubtsos (One of the Major Ingredients of Phemar)

For *chubtsos,* wheat grain is boiled and made soft. Then it is strained, dried in the sun and roasted in a big pan. It is then mixed half and half with *namphey* and kept specifically for this dish (*Khokak*). To make *khokak, chubtsos* is added to tea with a large lump of butter and sugar to taste. It taste better than the rest of the *khokaks,* and is mostly served on special occasions only (as the ingredients are quite a luxury and it is considered a very valuable food).

2.10 Kushi Phey Khokak (Crushed Dried Apple with Namphey)

This is made in villages for wedding and the parties to celebrate a baby's first year.

2.11 Chuli Phe Khokak (Powdered Dried Apricot with Namphey)

This is a refreshing food mostly used by travelers. It is a mild laxative.

Some of the dishes which are often served with *khokak* are given below:

- Baril (Walnut and apricot kernel dip): This is from Sham area, or anywhere where there are walnuts. It can be eaten for breakfast or lunch, with bread or *khokak.*
- Thud (Butter and dried cheese brick): Common in Chang-thnag and Zanskar valley, it is made by mixing together local butter, ground dried cheese and sugar. This mixture is then shaped into a brick. It can be keept for a long time, especially in winter.
- Ruskhu (Soup made from bones): It can be eaten with *khokak* or *paba,* or on its own.
- Other: Yogurt, curd and vegetables (especially roots like radish and swede). *Khokak* can be taken as such with tea, butter, vegetable/meat preparations, *tsamik, chhang* etc.

2.12 Paba (Mixed Grain and Legume Flour Pudding)

Another most commonly used food in Ladakh is *paba* and is popular in all the classes all over the region. It is another healthy dish often taken to the fields for lunch at sowing or harvesting time, or made in the evening when the housewife has been busy with other tasks and doesn't have time for more elaborate cooking. *Paba* flour is called *yotches* (Fig. 2) and can have different ingredients, depending on what is grown in the area. It can include barley or wheat or both with any of the legumes like peas, broad beans or lathyrus to make it more nutritious and palatable which is half roasted and ground. In some place they grind caraway (locally known as *kosnyot*, grows wild in some of the villages) seed with the flour to give a good flavour. Traditionally, *paba* is cooked in a special stone pot called *doltok*. Some of the dishes or soups that are oftenly taken with *paba* are given below:

2.13 Tsamik

It is a sauce made from coriander and mint leaves, radish and onion. All the items are pounded in a mortar. To this, chilly powder and salt is added. Sometimes curd or butter milk is added to obtain a thick paste. The simplest *tsamik* is a mix of dried and crushed chilies with salt. According to Crook, in the Zanskar valley radish leaves as well as dried leaves of wild plants are used. It is a yogurt dip for eating with *paba*.

2.14 Tangthur (Green-Leafed Vegetables and Buttermilk or Yogurt)

Quite a variety of green-leafed vegetables are grown. As well as spinach, there is *ldums* (cos lettuce), mongol, salad (Chinese cabbage), sugar beet (leaves only), radish leaves, celery (leaves only) and cabbage. Some people also use chrysanthemum leaves (*pato*). In villages, *tangthur* is served, made with wild green vegetables the proportion of vegetables to butter milk (or yugurt) depends on choice/taste.

2.15 Zathuk (Nettle Soup)

Nettle (*Urtica hyperboria*) grows high up in Ladakh. In winter, especially in villages, people take it with *paba*. People in Tibet and Ladakh believe that the great Tibetan yogi Milarepa, while meditating in the mountains survived by eating nettle.

2.16 Tsong Chhu (Onion Soup)

This is very simple and quick to make.

2.17 Markhu (Tsampa with Melted Butter)

A little *tsampa* is added to melted butter.

2.18 Sharjen (Thick Dried or Frozen Meat Soup)

In former times this soup was common in most part of Ladakh, but now-a-days it is found mostly in Chang-thang and Zanskar areas.

2.19 Tsha Chhu or Shespa (Meat Broath)

This is another soup from the Zanskar valley and it has two versions. In one, meat, salt and spices are boiled in water with a small amount of *paba*, and eaten with it. In the other it is simply a soup, made without *paba* and eaten with it.

2.20 Thukpa

Thukpa is the generic term for a variety of soups (using wheat or barley flour in many cases) and what we might call stews with wheat flour noodles. Literally, it means over cooked. It is a thick soup or noodle cooked to taste. It is made from *namphey*, wheat flour, or rice and is accordingly named. In the villages, the flour used is of the local variety. *Chhurphe*, the dried cottage cheese is what gives those dishes a special zip. The procedure varies with the main ingredients and is supposed to provide warmth in winter, cure cold and as an elixir for curing constipation in the elderly people.

2.21 Tsap-Thuk

Tsap is flour made from pre-germinated barley. The barley grains are washed and kept in a sunny place till they germinate. The germinated barley seeds are dried and then ground into flour. The flour is cooked in plain water. No salt or sugar is added. It has a sweet taste. Its use is restricted to the winter months and is believed to be good for stomach and effective in couch control. It is a food used during extreme cold months.

2.22 Chhan-Thuk

Half-cooked barley grains are beaten in a pestle and mortar. They are rolled into small balls. These balls are frozen and stored for use during long winter months. Small quantities of this *chhan* are added to *thukpa* and mostly cooked with the head and leg parts of goat/ sheep.

2.23 Ngamthuk (Tsampa soup)

This soup can be eaten at any meal, generally at breakfast, when it makes a cheering start, especially in winter. It is given whenever somebody catch a cold (quite a frequent occurrence in a Ladakhi winter even if otherwise healthy). To make it more nutritious small pieces of meat, local pea and *chhurphe* is also added.

2.24 Jamthuk (Unroasted Barley Flour Soup)

This soup is the same but is made with unroasted barley flour.

2.25 Pakthuk (Wheat Flour Noodle Soup)

Pakphe means wheat flour or dough. This dish can be eaten at any time of the year and is very cheering in the winter. In this season the *labuk* (local radish-large pink-skinned), dug up from an underground storage pit and the spinach reconstituted from what was dried on the roof of the house during the growing season; those who have a green house, have fresh green leafy vegetables, *Pakthuk* in the Zanskar valley is made using butter instead of the vegetable oil. For this dish Ladakhis do not use potatoes, carrots or swedes, as these are sweet. Before onions were widely available, villagers would have used *skotse* (*Allium przewalskianum*) - wild garlic, as flavouring, and in some remote area they still use it. Now-a-days some people add tomatoes to the onion. The simplest version omits cheese, peas, and eggs.

2.26 Guthuk (Nine Ingredients Soup)

Made like *pakthu*, this is special soup for one of the days of the *Losar* celebration. These ingredients, which are some of the nine for the soup, would be specially obtained for *Losar*. It is traditional to use some of the dough to make little balls containing different things, one at least for each person present. The one you get, which you are not obliged to eat, says something about your character: charcoal (*sola*), your mind is dark; the sun (*nyima*) or moon (*ldawa*) - written on pieces of paper, your heart is open, light and beautiful; salt (*tsa*), you never speak the truth; chilli (*tshante*), you always speak sharply about people and you are restless, or you might be told to sing a song.

2.27 Thenthuk (Soup with Hand-made Noodles)

Thenthuk, the Tibetan name for this dish, meaning something torn. The dish is like *pakthuk* but with short noodles.

2.28 Gyathuk (Chinese Thukpa)

Chinese cuisine has influenced this sophisticated modern version of *pakthuk*, often made for parties. It is sometimes served with meat only.

2.29 Lama Pakthuk (Monks' Wheat Flour Soup)

This dish is often made for monks when they came to the house to say special prayers.

2.30 Trimthuk (Soup with Hand Rolled Noodles)

A dish which is made mainly for people who are sick.

2.31 Trapthuk

To make this, add salt to the liquid in which *drapu* dumplings have been boiled (*drapu* is dumplings with ground apricot kernel sauce). It is more common in apricot growing areas, such as Sham. *Drapu* is a very nutritious dish, eaten for dinner on its own, particularly after threshing, or on fast days. The liquid in which the dumplings have been cooked is known as *trapthuk*. The sauce can be kept for two to three days.

3. Stew

3.1 Chhu Tagi (Bow-Tie Noodle Stew)

The literal meaning is 'water bread' - *chhu* means water and *tagi* means bread. This dish would originally have been very simple with just a little salt and perhaps onion to flavour the *tagi*. These days vegetables are generally added to it. The dough can be made into many different shapes and the vegetables used should be ones that take about the same time to cook. Potatoes and carrots are a common combination, although one can have it with green leafy vegetables and potatoes. Adding meat to it is also common (for which the meat has to be cooked earlier before putting the shaped dough, so that both are cooked simultaneously). In the Zanskar valley this dish is known as *chubtse* (*chubtsos*).

3.2 Skyu (Cap-Shaped Noodles Stew)

This dish is similar to *chhu tagi* and, like it, originally very simple. The ingredients are as for *chhu tagi*. Generally, less water is added, compared to *chhu tagi*, but it should certainly not be as dry as *khokak* or *paba*. If meat is added, then it has to be pre-cooked with ginger or garlic, or both as in *chhu tagi*. In the Sham area apricot kernels are sometimes used if there is no meat available. There is another version of *skyu* known as *o-skyu* in which milk (*oma*) is added with water for the sauce. It is made mainly in villages where there is a ready supply of fresh milk. It is a very tasty variant.

3.3 Paktsa Marku (Noodle Balls in Water)

This is a version of *skyu* often made at the beginning of the first Tibetan month for fasting, during which many Buddhist devotees do not eat 'hot' food like onions or garlic. It is also made in *gompas* (monastery), when it is known as lama *paktsa*. There the sugar, and sometimes butter, will be omitted.

4. Steam Cooked

4.1 Timok (Steamed Bread Twist)

This steamed bread dish was originate in Tibet and now is made mostly in the eastern (Chang-thang) area of Ladakh. *Timok* is good with cooked vegetables or dal.

4.2 Mok Mok (Steamed Dumplings filled with Meat or Vegetables)

A dish that was originated from Tibet, probably with Chinese influence. The meat version is often served as a starter at parties or on special occasions when it can be eaten with pickle or chutney. As a super dish for the family it may probably be accompanied by a simple soup made from either mutton bones or vegetable stock, also served with salad of tomatoes and onions (a modern introduction). *Mom mok* now-a-days is available in many of the metropolitan cities across the world.

4.3 Loko Mok Mok (Cup-Shaped Dumplings)

This was originated in Tibet and now-a-days is mostly confined to the Zanskar valley. The dough is made as for *chhu tagi*, moderately dry/stiff. Break off small pieces to make into balls, and then push in with thumb to make a shape like a handless cup. The dumplings are boiled, or steamed like *mom mok*, and eaten with vegetables or meat, whichever is available.

5. Beverages

5.1 Gur Gur Cha or Shrusma Cha (Butter Tea)

This is the tea no household is without. Usually a supply will be made in the morning, put in a thermos or in the villages a clay samovar heated with cow dung, and is served at frequent intervals throughout the day. Often extra butter be added in the tea to be soaked by bread; *tsampa* may also be added at times, anything from just one pinch to quite a lot.

5.2 Khunak (Salted Black Tea)

This is the basic tea *chhathang* (boil the tea leaves in an open pot and make a concentrate, is then poured off and stored in a jug or jar as the basis for the drink. It can be kept for a few days) to which water is added and boiled with salt but not churned. People in the Kargil area tend to drink this and milk tea rather than *gur gur cha*.

5.3 Cha Shrul (Butter Tea with Tsampa)

This is butter tea with enough *tsampa* mixed in to make it soup-like.

5.4 Chuli Chhu (Apricot Juice)

Chuli is one of the seedling types of apricot and is usually dried with the stone removed. In early spring when nothing much is available, the dried fruit is soaked and the juice is drunk. A drink made out of apricot kernel is *tsigu chhu*. This drink is made with apricot kernels (*tsigu*) ground and mixed with water. It comes from the Sham area, where apricot grows abundantly.

5.5 Chhang (Fermented Barley Drink)

In Ladakh, *chhang* is the general word for alcohol, but here it refers to the barley drink. *Skyems* is the polite word for *chhang*. No good party is without it, although now-a-days there seems to be a tendency, among the men at least, to favour hard liquor such as rum or whisky. The Ladakhi word for yeast is *phabs*. It mostly comes from the Nubra valley and can be found in the market in nuggets form. There are various ways of making this brew, resulting in different strengths. This drink would be submitted to carefully tasting by one of the older ladies and pronounced good or not so good. Mostly, liberal dollops of *tsampa* will be put on top of the poured drink and this is called *chhang shrul*. As a sigh of honour, guests are served *chhang* with a small piece of butter on the beautiful brass pot from which *chhang* is usually served.

5.6 Arak (Distilled Chhang)

It is made by collecting the boiled-off vapour from alcoholic drinks. *Chhang* and *arak* is made in small quantities throughout Ladakh (only by the Buddhist households).

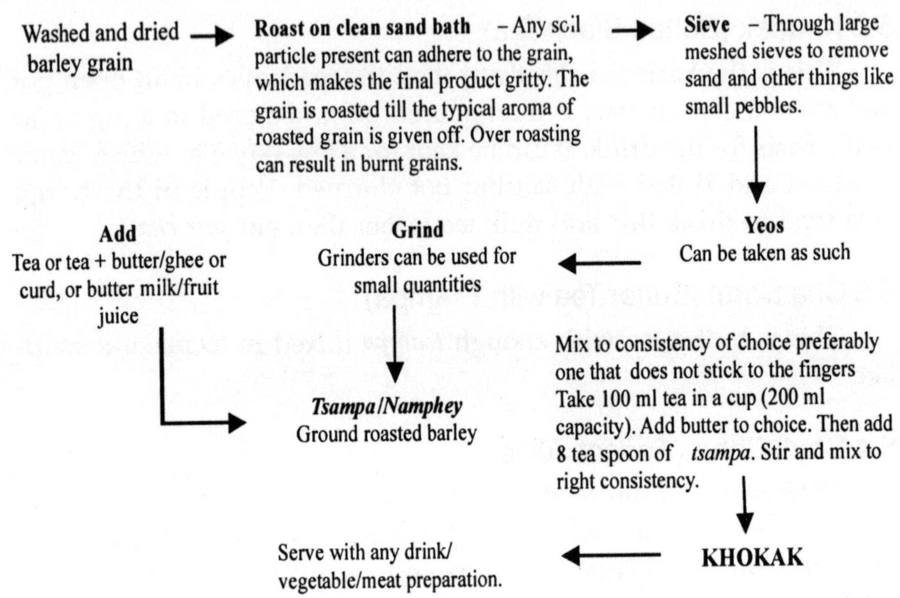

Fig. 1: Flow chart of *Khokak*

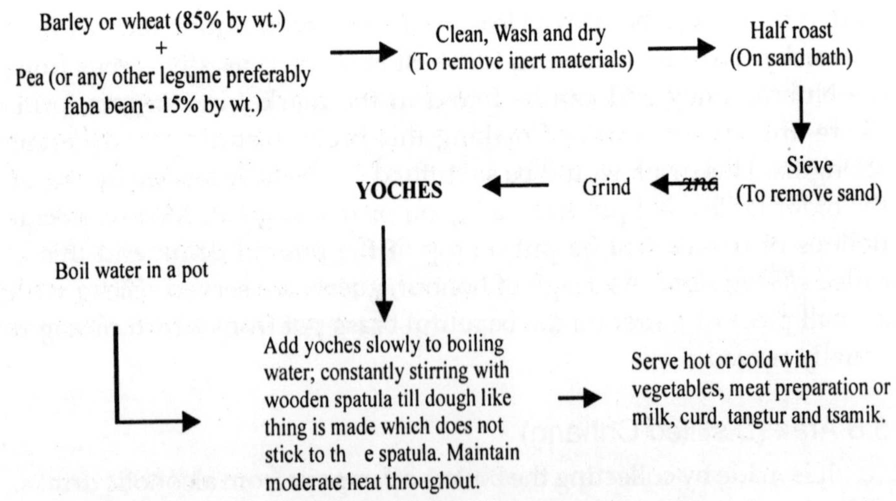

Fig. 2: Flow chart of *Paba*

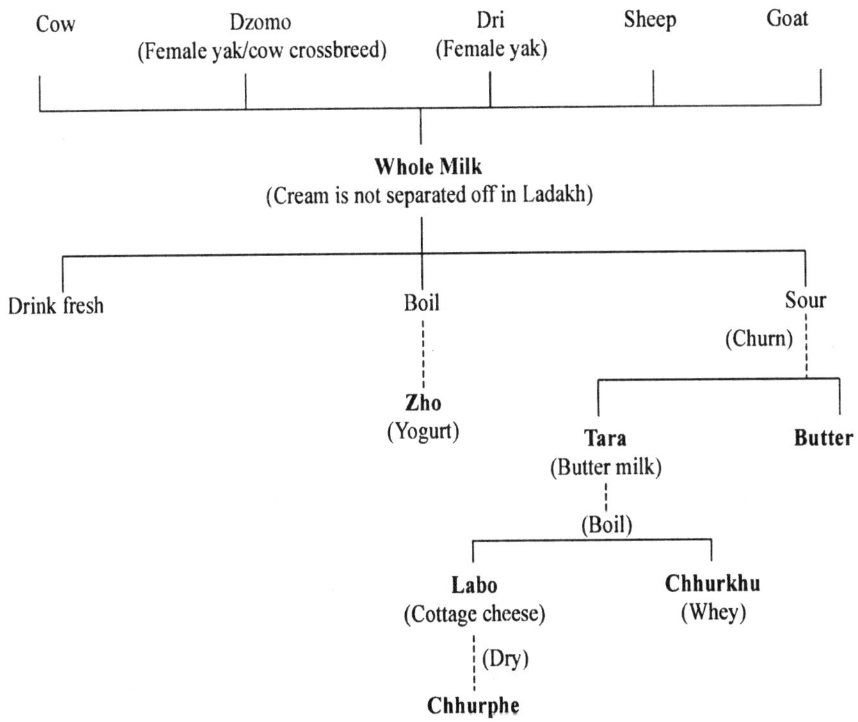

Fig. 3: Milk products in Ladakh, adapted from, Reifenberg (1998)

5.7 Rguntshang (Fermented Grape Drink)

Grapes are grown in the lower valleys of Ladakh, in Da-Hanu areas, where the *Brokpa,* or Dards, live. This drink is mildly alcoholic.

6. Milk Products

Milk and its products are major and important ingredients in many of the food and beverages (e.g. *o-skyu* and *gur gur cha,* respectively) in Ladakh. Fig. 3 shows the various products of milk, and how they are made. Some of the products are described below:

6.1 Sri (Post Natal Milk)

It is the milk obtained from the cow, *dri* or *dzomo* for a day or two after she has given birth. The milk is heated and is then left to stand, after which it is eaten with *tagi shrabmo* and little sugar or salt. The mother of the house (in whose house the calf has been born) tries to makes it available to every member of the family and next door neighbours, especially the children.

6.2 Labo (Cottage Cheese)

It is eaten fresh with chappati or *khokak;* sometimes sugar is added.

6.3 Chhurkhu (Whey)

It is very nutritious and has the property to give instant energy. Generally, farmers take it just after they have finished their work (like ploughing).

6.4 Chhurphe (Dried Cottage Cheese)

It is made by forming *labo* into nuggets and then dried. It is mostly used in *thukpa*.

7. Imposing Change

Ladakh is an interesting example of how policy can be deliberately set up to replace a local economy. First, the combination of subsidised food (provision of the Government to develop this area) and the new infrastructure accelerated a mass migration of men-folk from the farms into Leh (town), the capital, to serve the burgeoning tourist industry and the new urban economy. Second, subsidised food grain distribution programme brought vast quantities of subsidised rice, sugar (leading to the drinking of more sweet tea than previously), refined wheat flour, and kerosene. These subsidized food items are trucked across long distances into the Himalayas to be sold at very low prices – in many cases at lower prices than goods produced locally. Consequently, Ladakh's local economy which has provided enough food for its people for two thousand years is now being invaded by produce from industrial farms located on the other side of the Himalayas. Due to which for many Ladakhis, it is no longer worthwhile to continue farming. For instance rice, which had previously been a luxury in the Ladakhi diet, but now it is so cheap, it quickly became a staple. And as the desire to appear modern grows, people are rejecting their own culture. Even the traditional foods are no longer a source of pride. In this context Norberg Helena rightly quotes- "Now when I'm a guest in a village, people apologize if they serve the traditional roasted barley, *namphey,* instead of rice or instant noodles *etc*."

All these forces have between them introduced 'convenience' foods and new change in diet has let to increasing incidence of tooth and gum decay, high blood-pressure and heart disease. In addition to these, these changes carry with them the danger inherent in any cash crop scheme the world over: that such crop will replace those that are needed for the survival of the local population. Despite all these, there have also been positive changes to the quality of life and in attitudes to traditions and the culture. The government has helped with irrigation schemes to bring more land under cultivation. The establishments of various research institutes related to agriculture, with expertise in sustainable agriculture and the use of renewable energy has let to the wide spread and beneficial introduction of green house technology, one instance where government subsidies are well spent. This, where they have been constructed, have already made it possible to extend the growing season, providing, for most of the winter, fresh green leafy vegetables such as spinach, celery and lettuce, together with herbs such as mint and coriander. The inevitable monotony of dried vegetables has thereby been much relieved, and the diet, has improved in respect of fresh vegetables especially during winter season - the bleakest time of year as far as fresh produce is concerned, when ground-stored and dried vegetables have mostly been used up and the roads are not yet open.

In spite of the influx of a cosmopolitan culture as a result of boost in tourism, and better connectivity with the outside world with improving communication infrastructure (virtual and real), people in Ladakh are still fond of their traditional foods. The traditional food system is followed especially in marriages and other cultural ceremonies. One perfect example related to it is the 'Ladakhi Kitchen', a recently opened restaurant (owned and run by a group of educated unemployed youths of Ladakh) which is the only restaurant in Leh town which serves traditional Ladakhi foods. These traditional foods of Ladakh are prepared by making use of the indigenous agricultural products and are simple and easy to make with rare use of spices. The foods are indicative of the simplicity of their ways of life and also an indicator of the impact of surrounding environment of the people living in this area. Among the foods, *khokak, paba, skyu* and *khambir* are the most common foods. While the consumption of meat is more among the nomads of Changthang, whose main occupation is pastoral farming. Meat consumption is comparatively more in Muslims, especially during winter season.

References

Statistical Digest (1993). (Government of Jammu and Kashmir).

Angchok, D. and Singh, P. (2006). Traditional irrigation and water distribution system in Ladakh. Indian J. Traditional Knowledge 5(3): 397.

Reifenberg, G. (1998). Ladakhi Kitchen. Melong Publication of Ladakh, Leh.

Vigne, G.T. (1842). Travels in Kashmir, Ladakh, Iskardo. II: 344-45.

Fraser, D. (1986). Transhimalaya Unveiled (originally The Marches of Hindustan, ca. 1906), I (reprint 1986) 106.

Mir, M.S. and Mir, A.A. (2000). Ethnic foods of Ladakh (J&K) India. In: Dynamics of Cold Arid Agriculture, Sharma, J.P. and Mir, A. eds. Kalyani Publication, New-Delhi, 297-306.

Crook, John H. and Osmaston, Henry A. (1994). Himalayan Buddhist Villages: Environment, Resources, Society and Religious Life in Zangskar, Ladakh. Motilal Banarsidass Publication, Delhi.

Norberg, H. (1992). Ancient Futures: Learning from Ladakh. Oxford University Press, New-Delhi.

□□□

25

Local Knowledge System for Sustainable Development

DORJEY ANGCHOK
Defence Institute of High Altitude Research (DIHAR), Formerly Field Research Laboratory (DRDO) C/o 56 APO

These are days of perpetual talk and prolonged admiration on the rapid growth of science and technology taking place in the West. But what is our position as a remote high altitude cold arid desert in this context? Too many questions are raised and too many contradictory remarks are made in this connection. Do we know that Ladakh had a glorious scientific and technological tradition in the past? If yes, then why did Ladakh not develop her own science and technology? Why should Ladakh depend on or borrow everything from the West? When we have a look at the past, we look highly sophisticated and really civilized. How did this come about? Where from? These are another set of questions heard today in public forums. The direction in which the western world is developing is not one which all must inevitably emulate. The unique socio-economic and environmental condition of Ladakh do not favour the "copy and paste" model of technology intervention. The fact that western science can not find an explanation for many of the traditional and/or ancient practices people of Ladakh, more specifically farmers use, does not in any way imply that the practices or technologies are not "scientific". It is just that western science cannot take all the factors into account that occur on a farmer's field or in the environment at a particular time (Angchok, 2000). Only in the recent past, a growing number of scientists and academicians have started to analyze the consequences of their development efforts and started giving due importance to the traditional wisdom which had been prevalent since time immemorial. In the words of Warren (1993), "developing countries have a valuable, but largely untapped reservoir of indigenous agricultural knowledge." Taking into consideration these facts, through this chapter an effort has been made to present the concept of local knowledge system (LKS), how to incorporate it in research and extension network, and finally how to preserve it, so that a viable strategy can be designed to sustain, spawn, and stimulate the sustainable LKS in context of Ladakh.

1. What is Local Knowledge System (LKS)?

The term "Local Knowledge" (LK) is used synonymously with "Traditional" and, "Indigenous Knowledge" to differentiate the knowledge developed by a given community from the International Knowledge System, sometimes called as "Western System", generated through universities, government research centers and private industries. LK refers to the knowledge of local people as well as any other defined community. "Indigenous" means native born, originating or born naturally in a country, and is opposite to exotic. It also means innate and inherent. More specifically LKS refers to knowledge that is unique to a given culture and society. This knowledge is the information base of that society, codified in the language of the society and it facilitates communication and decision-making. While defining Indigenous Knowledge, Atteh (1989) was of the opinion that it is the untapped and unwritten body of language which is held in different brains, languages and skills in as many groups, languages, cultures and environments.

1.1 Local Knowledge (LK) Vs Western Knowledge

Johnson (1993) in her book LORE has made an attempt to compare and contrast the Local Knowledge (LK) with Western Knowledge:

i. LK is recorded and transmitted through oral tradition, while western science is taught and learned in a situation usually abstracted from the applied context, ii.L K is learned through observation and hands-on experiences, while western science is taught and learned in a situation usually abstracted from the applied context,

iii. LK is based on the understanding that the elements of the environment (earth, air, fire and water etc.), which are classified as inanimate by western science, also have a life force,

iv. LK does not view human life as superior to other animate and inanimate elements; all life forms have kinship and are independent, unlike western science; humans are not given the inherent right to control and exploit the nature for their own interests at the expense of other life forms,

v.LK is holistic, while western science is reductionist, i.e., western science deliberately breaks down data into small elements to understand whole and complex phenomena. For LKS, all elements are viewed as interconnected and cannot be understood in isolation,

vi. LK is 'intuitive' in its mode of thinking, while western science is 'analytical',

vii. LK is mainly qualitative and sometimes quantitative, but western science is mainly quantitative,

viii. LK is based on data generated by resource users. As such, it is more inclusive than western science, which is collected by a specialised group of researchers who tend to be more selective and deliberate in the accumulation of facts,

ix. LK is based on diachromic data (long time series of information on one locality), while western science is largely based on synchronic data (short time series over a large area),

x.L K is rooted in a social context that sees the world in terms of social and spiritual relations between all life forms, while western science is hierarchically organised and vertically compartmentalized,

xi. LK explanations of environment phenomena are often spiritual and based on cumulative and collective experiences, while western science employs the methods of generating, testing and verifying hypotheses and establishes theories and general laws as its explanatory basis.

It seems that still there lies confusion in the minds of many about the use of two words "Indigenous Knowledge" and "Traditional Knowledge". However, the viewpoints put forth by Parasar (1994) were remarkable. He stated, "All indigenous knowledge are traditional by nature, but not the *vice-versa* is true". The logic of the above statement, according to him, was that a knowledge developed locally may have scope to be transmitted from generation to generation which after some years may be given status of tradition, or considered as "traditional knowledge". While, in this way, some traditional knowledge an exclusive indigenous base, yet some have both the indigenous and exotic components in them. Although the difference between these two terms is too thin and indiscernible to sustain in all conceivable cases, there may not be any semantic confusion in using the expressions "traditional knowledge" and "indigenous knowledge" somewhat interchangeably also.

1.2 Constraints and Limitations of Local Knowledge

There are a number of constraints, in particular the transfer and use of information is likely to be constrained and error-prone, since it has to be passed on orally or by direct experience and is held in the heads of practitioners.

i. The scope for improvements via pure LKS is limited to what can be done with the local pool of techniques, materials and genetic resources;

ii. Many genetic possibilities are not explored within the informal system, such as the crossing of self-pollinating crops where specific plant breeding techniques are required;

iii. The capacity of individuals to generate, implement and transfer LKS varies; and

iv. Difference in the function of social groups will influence the type and extent of LKS developed by each.

Again, each individual of a community possesses only a part of the community's LK. Specialized knowledge is often kept secret or known only to a selected few, such as elders, midwives, or healers; In many cases, peasants do not document their knowledge, so that it can be made available to strangers. Their knowledge may be implicit within their practices, actions and reactions, rather than a conscious resource (Reintjes *et al.*, 1992). Kalyansundaram (1993) stated that the practices, their intricacies and rationale are known to uneducated and aged people in the villages and scientists by and large have scant respect for the utility of such practices; and scientific elite methods such as publication, seminar etc. are needed to create awareness. Moreover the State does not officially recognise the ancient/indigenous knowledge in the same way as it does modern scientific knowledge (Parasar, 1994).

1.3 Incorporation of LKS in Research & Extension Network (REN)

There is a need for developing a framework for incorporating local knowledge into the Research and Extension system because:

i. The transfer of technology (TOT) model fits badly with the needs and priorities of resource-poor farmers;

ii. Agricultural extension programmes are still biased towards techniques and strategies, which are capital-intensive;

iii. Resource-poor farmers (RPF) are scattered and are not able to make their needs and priorities readily known and felt; and

iv. The TOT model cannot easily handle the complex interactions of RPF farming; links between crops, especially with intercropping and multiple tiers; agro-forestry and livestock-crop-tree complementaries; and the progressive adjustments required in the field in the face of seasonal and inter-annual fluctuations.

A framework for incorporating LKS into agricultural research and extension should have following features:

i. Strengthening the capacities of regional research and extension organizations,

ii. Identifying the need for extension scientists/social scientists in an interdisciplinary regional research team,

iii. Formation of a sustainable technology development consortium to bring farmers, researchers, NGOs and extension workers together well ahead of the process of technology development; and

iv. Understanding that it is impractical to depend entirely on research stations for innovations, considering the inadequacy of human resource capacity of the regional research system.

In this connection Parasar (1994) developed a comprehensive model for incorporating local knowledge in REN. The model fits well for Ladakh, with little or no modification. Major features of the suggested model are:

i. To set up a Center for Indigenous Knowledge (CIK), assigning broadly at least three-fold work, *viz.*, (a) collection, (b) documentation and (c) preservation of LK in a kind of Knowledge Bank;

ii.O ther activities of CIK should include coordination and monitoring, management of research outputs and related publications for developing literature on 'Indegenous Knowledge System', and dissemination of center's activities;

iii. Curricular sanction for local knowledge in the school and undergraduate courses and integration in the pre-service training courses of village level agricultural extension workers to remove attitudinal barrier against IKS;

iv. Support of general academia in addition to agricultural disciplines;

v.Involvement of NGOs in the process;

vi. In order to popularise LKS among the scientists, extension functionaries as well as the farming communities, adequate scope for publicity through electronic and print media should be provided; and

vii. The collected local agricultural knowledge should be subjected to preliminary screening and ready assessment in different professional forums before final recommendation.

1.4 How to Conserve LKS?

"When a knowledgeable person dies, a whole library disappears"
_ *African Proverb*

The irrationality of western science is exposed due to its inherent nature of un-sustainability. The final proof of its irrationality can only came, when it destroys all of us. Thus there is an urgent need to safeguard and reaffirm our rich local knowledge system.

(i) Consumers Can Conserve

The economists say, "There is no free lunch". They imply that someone or the other pays for anything and everything that we use.

But when we do not pay for it, the resource concerned tends to get eroded. The resource-poor farmers who are engaged in conserving LKS are poor economically because they are not able to use chemical inputs, and the chemical intensive technological alternatives are not suitable for their heterogeneous conditions (Gupta, 1997a). There is an urgent need to provide incentives to such people to improve their income; use innovative, non-chemical technologies; and adopt land-use practices that maintain the soil health, so that they could continue their age-old practices of conserving the biodiversity.

But the question is how do we generate markets for their produce? Will the consumers pay for conserving diversity, the health of the ecosystem and most importantly their own health and well-being? In this context, there are four kinds of consumers (Gupta, 1997a):

- *Pioneer mobilizers*: those who practice environmental friendly consumption behaviour (EFCB) and persuade others to do the same.
- *Populist mobilizers*: those who don't practice EFCB but persuade others to practice it.
- *Aloof practitioners*: those who practice EFCB but don't persuade others.
- *Indifferent & indolent*: those who neither practice nor persuade others.

The strategy to mobilize each category of consumers obviously will have to be different. We must begin by asking ourselves, what proportion of our purchasing power we spent on buying those products? The answer may be very less or nil. Whatever may be the reasons (*e.g.* not aware of their availability or not satisfied with the quality *etc.*) if we don't buy those products, then the farmers will not have that surplus to continue with the practices of conserving the ecosystem. These farmers are poor because of their superior ethics and our inferior individualistic ethics makes us richer. Why can't society rewards good ethics, excellence and efficiency with good economic opportunity as well? So, here lies the opportunity to pay our dues towards the pioneer of conservation.

(ii) Recognition

Those farmers who are practicing traditional agricultural practices should be given due recognition. This will act as feedback or they will function as motivators for them. Both, material and non-material types of recognition should be given.

(iii) Viable Network

According to Gupta (1997b), a network should be established to link together the innovator, institution, innovator and enterprise. For example the Grassroots Innovation Augmentation Network (GIAN) was established in 1997. It is an instrument that will help those innovators whose knowledge, practices and innovation can be commercialised or disseminated by the State or NGOs among potential users. It tries to achieve market and technological linkage.

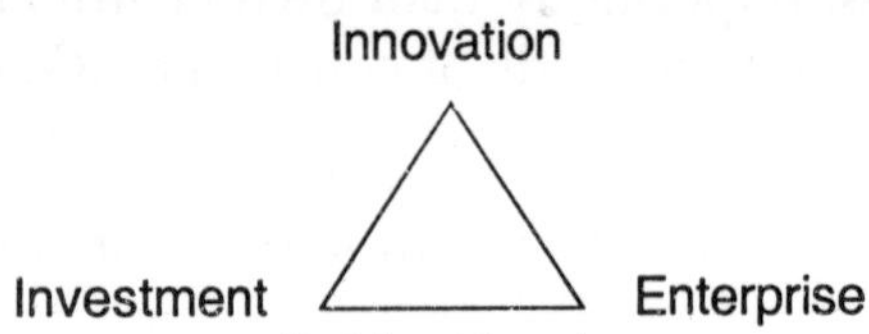

Golden triangle

This model of golden triangle enables an entrepreneur sitting in New York to invest in an innovation from India to set up an enterprise in China. Or, regionally, an entrepreneur in Delhi can invest in an innovation from Ladakh to set up an enterprise in Mumbai. Or locally, an entrepreneur in Nubra can invest in an innovation from Zanskar to set up an enterprise in Leh.

(iv) Academic Treatment of LKS

There is a tendency to scrutinize local knowledge with scientifically dominant parameters for their legitimacy and credibility and moreover some of the so called modern scientific personnel have developed a negative attitude towards LKS. Instead, effort should be made to provide an environment where those age-old practices are put on to a rigorous trial. Whole of this process should be participative in nature, because farmers sometimes differ in their criteria in judging the rationality behind their practices, e.g. in cold-arid desert of Ladakh a weed locally called as *dambu*, is preferred by some farmers, because, for them this plant (relative weed) improves the nutritive value of the straw (cattle feed). Even, if it is not found useful today, it may prove useful to solve future problems. Another example is the local breeds of cattle. Though their milk yield is quite low, as compared to those high yielding breeds, but many a times local farmers prefer their own local breeds because the high yielding breeds do not fall into their selection parameters (like cold tolerance, open grazing, sure footedness *etc.*).

(v) Databank of LKS

In today's world, after the outset of globalization and liberalization, it is necessary to maintain a databank of available knowledge and practices, so that any unwanted exploitation from outsiders can be prevented. In this regard, panchayats (a form of democratic decentralized village institution) can be effectively used, especially after the 93rd amendment, which gave them constitutional status.

(vi) Redraft the Textbooks

Especially at the school level and in the curriculum of in-service training to field level extension worker, LK should be included. It will sensitize students and extension workers towards the richness of tradition of those who conserve biodiversity and the precious germplasm of animal as well as crops.

(vii) Change in Nomenclature

In many institutions it has been seen that merely changing the name of the positions, being held by the personnel contributes to increase in excellence and efficiency and moreover it is found that their commitment towards the organization increases, *e.g.* Hygiene inspector instead of sweeper; Homemaker instead of housewife. Likewise we can designate the so-called "laggards of green revolution" by some other name like "conservator of biodiversity" or "ecosystem protector".

(viii) Dedication, Honesty and Perseverance

It has been observed that a researcher while pursuing his/her research in the field of LKS, especially documentation, takes the help of questionnaire or a schedule. These instruments have certain limitations, but apart from this the instrument by itself would not be able to get a clear picture of the whole scenario. It demands something more, extra. The researcher has to identify and assimilate himself with the people, their culture, ethics, values etc. Some LK especially the non-material part are not always at the conscious level. They are embedded in their practices and daily activities. Merely visiting the farmers with a schedule, conducting interviews and discussions would not do justice to this wonderful endeavour.

(ix) Equality of Status

For parties in dialogue, equality of status is a pre-requisite for effective interaction. It will enable the two-way flow of information between the scientist and the farmers and both the communities will know each other's situation much better.

(x) Platform to Share Experiences

Provide an environment, where the local knowledge generators and potential users, belonging to different geographic areas can freely exchange their experiences and thereby they can learn from each other and apply those experiences to solve their own problems.

(xi) Emphasis on Non-Material Part of LKS

Interest in LKS has been growing in the past few decades. In the 90's it was in full flow and everywhere people talked of LKS, organic food *etc.* Rather it was fashionable to be a part of LKS. Many things have been done in this direction, but much more is still left to be done. Particularly in agriculture, whether it was documentation or validation, much of the emphasis was given on the tangible aspects. On the other hand intangible parts of the LKS were ignored or given less importance e.g. belief and value system. The intangible part of LKS though, is more difficult to document and validate, but in some cases it is the major factor behind a system, *e.g.* in Phey village (15 km. from Leh), there is an apricot tree, locally called as *baba chulli,* which has the late fruit maturing characteristic. The grandmother of the household believes that more the number of people who eat its fruit, more will the tree bear fruits the next season. Further, the owner believes that the moment this tree's fruits are commercialised it will loose its efficacy, so the tree is for everybody to relish its fruits. Another example is the practice of *tsetar.* Among the nomads people of Changthang it is a common practice to offer their best sheep and goat to a deity (the offered animal is not killed by them). This way they believe that lots of merits are collected and unwanted calamity does not fall on them. We may not realise a rational reason behind those belief system (the intangible part of LKS), but this practice has certainly helped in the conservation of the desirable traits in their stock.

(xii) Create Micro Ventures Capital Fund, to Finance

The *incubation process i.e.* converting the innovation into product and *Business development i.e.,* converting the product into enterprise.

(xiii) Reforms in Patent Laws

During an international conference on creativity and innovation at grassroots (ICCIG, IIMA, 1997), Professor Thomas Cottier asked Kareembhai (a natural healer), how he would feel if some multinational drug company developed a drug based on his knowledge and sold it around the world. Kareembhai replied that, would it not be so nice that many people would benefit from his knowledge. Like Kareembhai there are so many other practitioners in Ladakh, who are so generous that they do not charge for their advice or medicines. But today, the global policies are such that people like them need to be given incentives, so that the knowledge system survives. In this regard, the present patent laws should be given a new look in favour of our LKS conservators. Thus, the patent laws should have the following features (Gupta, 1997c):

- Every applicant for a patent must declare that the material or information used in the application has been obtained lawfully and rightfully. Here lawfully means, in compliance with the law and rightfully means, a fair agreement for benefit sharing between the knowledge provider and applicant.
- The applicant must declare the use of LK, if any, on the subject and the improvements made therein.
- The patent office must consult various database on LK, medicinal plants, and contemporary innovations, before giving any sort of patent. The databank should contain the identity of the local community or the individual.
- Minimize legal processes and expense on registration.
- A public fund to facilitate filing of patents application by small innovators, small farmers, artisans etc. must be set up immediately on the pattern of legal aid committees of the courts. Patent lawyers may be encouraged to provide voluntary help with the understanding that if these innovations get commercialized, they could recover their fees at that stage.

Patents on LK are not just filed by Indians or Westerns in America, but also by Indians in India, *e.g.* Dasaprakash Pvt. Ltd. has been granted patent for process of making idli, dosa, pongal (Gupta, 1997c). It implies that unless we put our own house in order, we will not be able to make any dent in the global policies. Many great scholars of the past and present are/or were followers of all sorts of strange doctrines. Newton was a follower of alchemy; Alfred Russel and Canon Doyle were advocates of spiritualism; Kepler followed astrology and so on. These days the formally educated people have developed a "dual mentality" of consulting astrologers and criticizing astrology (Angchok, 2000).

Ladakh has a valuable, but largely untapped reservoir of Local Knowledge, particularly in context of agriculture. This rich pool of LKS accompanied by strong social networking (in a high diversified and low population density) through local institutions helps farmers in efficient management of their scarce resources in a sustainable manner. It is argued that in today's global scenario it is necessary and immensely imperative for Ladakh that we develop a viable strategy to protect ourselves from the exploitative global policies, and harness future opportunities. Such a strategy is inevitable for the sustenance of LKS in Ladakh to conserve its rich local knowledge heritage and to integrate it into the mainstream development process.

References

Angchok, D. (2000). Traditional Method of Weather Forecasting in Tibetan Astrology and its Relevance in Today's Agriculture. Unpublished M.Sc. Thesis, Department of Extension, Education, Banaras Hindu University, Varanasi, India.

Atteh, David O. (1989). Indigenous Local Knowledge as a Key to Local Level Development: Possibilities, Constraints and Planning. In: The Seminar on 'Reviving Local Self Reliance: Challenges for Rural/ Regional Development in Eastern and Southern Africa', 21-24 Feb. 1989, Arush, Tanzania.

Gupta, A.K. (1997a). Consumers can conserve: will you? Honey Bee 8 (4): 1.

Gupta, A.K. (1997b). From grassroots to global: linking innovations, institutions, investors and entrepreneurs. Honey Bee 8 (2): 1-2.

Gupta, A.K. (1997c). Neem, turmeric, Pongal and patents: Traditional Knowledge and contemporary institutions. Honey Bee 8 (3): 1-2, 14.

Johnson, M. (1993). LORE : Capturing Traditional Environment Knowledge. Dence Cultural Institute, IDRC, Canada.

Kalyanasundaram, N. (1993). Preceiving Farmer's Experiences in Agricultural Research for Promoting Sustainability. In: The National Seminar on 'Indignenous Technologies for Sustainable Agriculture'. March 23-25, 1993, IARI, New Delhi.

Parasar, B. (1994). Indigenous Agricultural Knowledge of Tribals in Eastern Ghat High Land Zone of Orissa. Unpublished Ph.D. Thesis, Department of Extension, Education, Banaras Hindu University, Varanasi, India.

Reijnties, C.; Haverkort, B. and Bayer, A.W. (1992). Farming for Future: An Introduction to Low External Input and Sustainable Agriculture (LEISA), ILEIA, Leusden, The Netherlands, pp. 52 & 210-218.

Warren, D. Michael (1993). Indigenous Knowledge, Biodiversity Conservation and Development. In: International Conference on Conservation of biodiversity in Africa: Local Initiatives and Institutional Roles, Nairo.

□□□

Parasar, R (1994). Indigenous Agricultural Knowledge of Tribals in Eastern Ghat High Land Zone of Orissa. Unpublished Ph.D. Thesis, Department of Extension Education, Banaras Hindu University, Varanasi, India.

Reijntjes, C., Haverkort, B. and Bayer, A.W. (1992). Farming for Future: An Introduction to Low External Input and Sustainable Agriculture (LEISA). ILEIA, Leusden, The Netherlands, pp. 52 & 210-218.

Warren, D. Michael (1993). Indigenous Knowledge, Biodiversity Conservation and Development. In: International Conference on Conservation of Biodiversity in Africa: Local Initiatives and Institutional Roles, Nairobi,

ABOUT THE EDITORS

Dr. Deepa H. Dwivedi (Email-deepahansraj@rediffmail.com), presently Reader at Department of Applied Plant Sciences, BBA University, Lucknow (UP), is an alumni of GB Pant University of Agriculture & Technology, Pantnagar. She started her carrier as Assistant Professor (Horticulture) from Sher-e-Kashmir University of Agricultural Sciences and Technology, Srinagar (J&K) and worked for over four years on fruit and flower crops in the cold arid Ladakh. She also handled research projects on characterization of apricot oil. She has guided number of Ph.D. & M.Sc. thesis research besides having 23 research publications and 5 book chapters to her credit. She is life member in 5 professional societies and is a member in executive council/editorial board of two journals. Her current research focus is on the underutilized crops and plant bioregulators.

Dr. Sanjai K.Dwivedi (Email-sanjaifrldrdo@rediffmail.com) is a Scientist in Defence Research and Development Organisation (DRDO), Government of India. He has studied the Higher Himalayas for propagation, conservation of genetic diversity and processing of horticultural crops, especially seabuckthorn and apricot. He has 5 patents and 2 books, 50 research publications to his credit besides 10 chapters in books, 5 monographs/bulletins and 75 popular articles. Dr. Dwivedi is life member in 6 professional societies, member of editorial board, and executive council in 3 societies. He is recipient of 5 gold medals, HSI Golden Jubilee Students' Award and two prestigious National Awards, namely, Fakharuddin Ali Ahmed Award (ICAR) and DRDO Young Scientist Award.

Dr. Sandhya Gupta (Email-sandhya_gupta87@ yahoo.com) is a Senior Scientist at Tissue Culture and Cryopreservation Unit, National Bureau of Plant Genetic Resources (NBPGR), New Delhi. She did her M.Phil. from University of Delhi and Ph.D. from IIT Delhi. She has been involved in the *in vitro* conservation of temperate fruit crops since 1995. She has been awarded Overseas Biotechnology Associateship from DBT, Govt. of India for research on cryopreservation of *Rubus* species at US Department of Agriculture-ARS-NCGR, Corvallis, Oregon, USA. She has many publications to her credit. She is editor of Journal of Indian Plant Genetic Resources. Presently, her area of research work is cryopreservation of *in vitro* germplasm of temperate & minor fruits. She is member of International Society of Horticultural Society (ISHS) and Indian Society of Plant Genetic Resources and The Horticultural Society of India.